1994

# BIOLOGICAL POLLUTION:

## The Control and Impact of Invasive Exotic Species

# BIOLOGICAL POLLUTION:

## THE CONTROL AND IMPACT OF INVASIVE EXOTIC SPECIES

Proceedings of a Symposium held at the University Place Conference Center, Indiana University-Purdue University at Indianapolis on October 25 & 26, 1991

Bill N. McKnight
Editor



Indiana Academy of Science
Indianapolis

*Manufactured in the United States*

*Library of Congress Cataloging-in-Publication Data*

*Biological pollution : the control and impact of invasive exotic species : proceedings of a symposium held at the University Place Conference Center, Indiana University-Purdue University at Indianapolis on October 25 & 26, 1991 / Bill N. McKnight, editor.*
*p. cm.*

*Includes bibliographical references and index.*
*1. Pest information—North America—Congresses.*
*2. Biological Invasions—North America—Congresses.*
*3. Pests—Control—North America—Congresses.*
*I. McKnight, Bill N.*

*SB990.5.N7B56 1993 632′.097—dc20* *93-3889*
*ISBN 1-883362-00-8 (hard cover)* *CIP*

*1 2 3 4 5 97 96 95 94 93*

# Table of Contents

# Acknowledgments

I extend appreciation to the members of the Indiana Academy of Science (IAS) Biological Survey and Executive committees for supporting the concept of an exotic species symposium and allowing it to happen. The IAS has a track record for producing symposia but this event was a magnitude greater than anything previously attempted. An event of this scale requires the talents and time of many. The other members of the Steering Committee were Cornelius Pettinga, Edward Squiers, and Robert Waltz. Their promptness and wisdom made it possible to do the planning and production from a dead start to finish in less than ten months. Edward Frazier, IAS Treasurer, played a major role. He was always available, extremely punctual in dispersing funds, and offered valuable advice. Rose-Mary Gibson assisted with all aspects. She provided companionship, served as a sounding board, did mindless clerical tasks, et cetera. Without her help I am convinced the meeting would not have happened. I thank my wife (Katherine) who had to live with this frustrated perfectionist during the planning and production of this meeting. She also made numerous substantial contributions with respect to the planning and production. Among the other people and organizations who made key contributions are: Pat Butcher, Roberta Donahue (Indiana State Museum Society), Illinois Department of Conservation, Ellen Jacquart, Peter Jenkins, Gary Johnston, Duvall Jones, Ulrika Peterson, Aaron Rosenfield, John Schwegman, Sea Grants Program (Illinois, Ohio, Michigan, Minnesota, and Wisconsin), Ann Sweeden, Bill Weeks, and J. David Yount. I apologize for any omissions. Numerous other individuals provided assistance by reviewing draft manuscripts or providing assistance with graphics. They include Robert Anderson, John Bacone, Greg Croy, Culver J. DeLoach, Rebecca Dolan, Robert Doren, John Ebinger, James Gammon, Bill Glass, Thomas Green, Joe Hennen, Sharon Hilgenberg, George Hudler, Louis Iverson, Barbara Knuckles, Christine Kohler, Wallace LaBerge, Douglas Ladd, John Lapp, Lloyd Loope, Bill Merrill, Paul MacMillan, Imants Millers, Jack Munsee, Roger Purcell, Jay Rendall, Stanley Rice, Victor Riemenschneider, Paul Rothrock, Thomas Simon, Michael Sinsko, Ronald Stuckey, Mark Wetzel, John Whitaker, and Robert J. Wolff. Finally, I would express thanks to the disparate and august group of presenters for their willingness to participate and for excellent presentations.

The management and staff of the Conference Center and Hotel, University Place, Indiana University–Purdue University at Indianapolis, allowed access to their outstanding facilities and provided world-class service. For this we are grateful.

This symposium was made possible with financial support from the following organizations: Ball Corporation, Eli Lilly & Company, The Nature Conservancy, USDA National Park Service, US Environmental Protection Agency–Environmental Research Lab (Duluth), and Wayne–Hoosier National Forest. The Natural Areas Association (NAA), through their President (Francis Harty) and Executive Director (David Paddock), provided funding that facilitated the production of this publication.

Bill N. McKnight
Chair, IAS Biological Survey Committee
& Symposium Steering Committee

# Introduction

On October 25 & 26, 1991, about 300 people from 35 states and abroad gathered in Indianapolis for a conference entitled *Biological Pollution: The Control and Impact of Invasive Exotic Species.* This symposium celebrated the 100th anniversary of the Indiana Academy of Science Biological Survey Committee, originally entitled "Committee for the Restriction of Weeds." It also coincided with the approaching quincentennial of Columbus' "discovery" of the Americas. This discovery was recently celebrated as a federal holiday and many other commemorative activities will occur around the world during the next year. But to some there is little reason for rejoicing. In fact, it has been suggested that the anniversary of this eventful first voyage and its aftermath should be observed, but not celebrated. As the Europeans and their slaves attempted to subjugate this new land, making it resemble the Old World as they attempted to survive, native American cultures suffered egregious damage and many, like the Arawaks, were obliterated. The native fauna and flora also have been victimized.

The delicate balance of the North American ecosystems has been permanently disrupted by the intercontinental post-Columbus exchange of nonnative life forms. Interestingly, not all groups of organisms have participated in this invasion; apparently there are few exotic bryophytes or lichens, and exotic reptiles and amphibians are almost totally absent from the middle portion of the United States. However, about 25 percent of the flowering plant species in North America are not native. Likewise, one does not have to look far to find numerous problematic exotics among the birds, fishes, and insects. Actually we should be amazed that five-hundred years of contact has not resulted in more homogeneity along with a decline in the native biodiversity or more out-and-out extinctions or extirpations.

Because these exotics (invasive and noninvasive) are so common, it was difficult to select organisms and regions to highlight at this meeting. But one species could not be overlooked; the ultimate invasive weedy species (*Homo ignorans* Wagner). Not only is man often the vector, he is also the primary cause, since invasions are typically successful only if the original ecology of an area has been degraded. Overpopulation, greed, ignorance, and ill-advised land use decisions are undeniably at the heart of the problem. In thousands of square miles of North America indigenous plants have been almost eliminated or restricted to uncultivated strips along thoroughfares. Moreover, in some instances invaders are capable of overtaking healthy systems, causing environmental chaos as well as considerable economic loss and occasionally public health threats. It is this last group that causes special concerns.

The last century has seen the largest introduction of nonnative species since the 1492-1600 period. Some of these are natural occurrences, some have been unintended, but many have been intentional introductions; and no region has escaped the invasion. Tom Crisman has brought our attention to a recent, localized exotic in Florida, namely *Peromyscus mickeyensis.* This illustrates the fact that invasive exotics may occasionally be profitable, but this utility should not blind us or prevent us from making wise decisions regarding our natural resources.

Will this biotic homogenization continue? What should we do? What role can science play in eliminating or slowing this alien invasion? We hope to answer some of these questions during this meeting by offering a diverse group of speakers that includes researchers, land-managers, policy-makers, and educators. The presentations will feature a litany of horror stories with case studies for both animals and plants from aquatic as well as terrestrial systems. We feel a better understanding of the issue will be gained by this mix and approach. As the presenters discourse on their various topics we hope you will be enlightened and motivated to appreciate what some, including some geographers and biologists, deny is a problem and others would proclaim as the jeopardy of our greatest treasure. After all, the Americas were christened the New World because of the unique animals and plants found there. We also want to find out what is being done to stymie the invasion and to learn what is not being done, and why. Certainly one of the primary goals is to determine a way to improve communication between state and federal agencies, between states, and within states, realizing up front that there must be a framework at the federal level for any hope of regulation. Equally important is the need to emphasize education and the appropriation of funds to combat the invasives (e.g.; several public agencies and private organizations were unwilling to support this meeting, suggesting that it was too applied and not a clearly understood environmental issue).

Hopefully this exchange of information and ideas will help us deal with the problems caused by invasive species. Clearly there is much to do. Biological pollution is one of the least publicized environmental issues facing us. And, unless we redouble our efforts here and abroad, the ecological chaos being caused by exotics does not seem likely to abate.

Bill N. McKnight, Park Tudor School
Editor & Chair, IAS Publications Committee

# Problems with Biotic Invasives: A Biologist's Viewpoint

Warren Herb Wagner, Jr.[1]

For one individual to express a viewpoint representing all biologists regarding invasive species is quite impossible; especially from a botanist who's main familiarity with this subject is limited to Hawaii and North America. There is as yet no composite viewpoint but one thing is certain, the scientific study of biotic invasives is quite complex, involving the interactive intricacies of population biology, evolution, and biogeography. The simplest definition of a biotic invasive that I can think of is "a foreign taxon that enters an established ecosystem and contaminates it." The word "contaminate," however, has various connotations: it may mean that the introduction simply changes the "purity" of the ecosystem and merely adds a new alien species to the fauna or flora, which then becomes naturalized and settles in as a regular and well-behaved component. The word contaminate can also mean that the invasive organism seriously upsets the system. The alien becomes strongly undesirable in this case in that it competes with, and even smothers, its associates. In agroecosystems, the invasive may be a damaging herbivore of serious economic significance. Each invasive has its own species-specific characteristics. The invaders may come from the same general geographical area, but typically they arrive from some other continent. In North America a high percentage of them come from Eurasia.

From the biologist's standpoint, invasive plants may seem easier to study than invasive animals and microorganisms because of their simpler biology and their overall visibility. However, there are many "cryptic" features involving the physiological processes of a particular plant species. The habitats of invasive plants are usually disturbed and harbor rather few native species; the latter (native invaders) are more or less successional and normally follow disturbance. The exotics compete with natives, occupying their habitats and forcing them out. There is no uniformity, however, and special complications may occur; contingency creates different problems for each situation. Soil requirements, light and shade, associated species, and many other facets of biology are involved. For example, many different reproductive systems are used. Some invasive plants are dispersed by berry-like, fleshy fruits, others by achene- or nut-like dry fruits or seeds, the major dispersal agents being birds or small mammals. Others are carried by the wind on glider-like or plumose propagules. Some are asexual, such as seed apomicts, or species that reproduce underground by stem fragmentation. In introduced animals, volant species may tend to spread more rapidly than nonvolant, but some terrestrial or aquatic types frequently invade with great speed, often assisted by farm equipment or vehicles. It is surprising how many of our invasive species, plants and animals, are aquatic or semi-aquatic. Thus the manner of dispersal of newly arrived organisms is highly diverse and illustrates one facet of the overall problem of reducing the biology of invasives to a few simple factors. There is so much *contingency* involved among organisms that we regard as invasive, that their study has to be essentially a case by case analysis. No two situations are alike.

1. Department of Biology, University of Michigan, Ann Arbor, MI 48109

## NATURAL BIOTIC UPHEAVALS IN THE PAST

Perhaps one reason that more biologists do not study invasive species is the tradition that biotic upheavals and readjustment have always taken place routinely in the history of life on earth. Scientists and even non-scientists sometimes argue that the phenomenon of invasives is not at all new, but simply a human-induced variation on an ancient theme. Migrations of newly evolved or already existing species by long-distance dispersion, and major physical upsets, such as mountain building, volcanism, island formation, and new riverine pathways, have produced ecosystematic and biotic perturbations through all biological time. There are even individuals today who would deny the validity of such concepts as "equilibrium" or "balance of nature." To abandon these concepts altogether however leads to oversimplication and denies that the extent of stasis that does exist has any significance.

Paleobiologists are constantly trying to estimate what the world of life was like before civilization. Disciplines like stratigraphy, continental drift analysis, palynology, and paleontology, have all demonstrated many aspects of history of life on the earth. We have also learned much by examining the pristine vegetation and ecosystems that exist today, areas still essentially undisturbed by people, steep mountain gorges, remaining wild forests, and newly explored primitive areas like the mountains of South America and New Guinea. From such observations made today we can extrapolate back into the past and correlate them with the fossil strata.

What may be called "normal ecosystems" are, of course, no longer regarded as strictly fixed categories. Their fluidity and variability have been recognized at least since the times of Gleason (New York Botanical Garden) and Curtis (University of Wisconsin). No longer do biologists conceive of a wholly fixed checkerboard of communities, each type rigorously and invariably defined by an absolutely fixed ensemble of species. An oak-hickory community or a distinctive sphagnum bog community—neither is a fixed entity. Today we consider them as parts of "continua" and accept a state of dynamic stability, now fixed and now shifting. Changes may be abrupt or gradual. Nevertheless, such communities are extremely valuable "standards" or "norms" that practically all students of North American vegetation readily accept, albeit with a full recognition of their variability. In many ways definition of communities is more loose than definitions of taxonomic species or genera. Even if the species or generic definitions in a given group are often gross and have frayed boundaries, we nevertheless do not abandon the taxonomic system. The conclusion is that there is a reality to community structure and there is a dynamic stability that pervades the forms of vegetation in North America and in the world. Like a Van Gogh painting of a flower, it is by no means perfect. However, there is a highly significant underlying element of organization and constancy, and this has existed with overall and local changes here and there, and now and then, over hundreds of thousands or millions of years. I, therefore, disagree with those who insist that we should not worry about human-induced species invasions because change and disruption have occurred throughout the past history of life. The invasive situation we witness now as a result of human commerce is truly different, in respect to rate of introduction, mode of origin of introductions, occasionally the type of organism, and, in general, the abruptness of change. Using the vernacular, it is a "new ball game."

The communities and the successional sequences that we have today resulted from an enormous amount of evolution in the past, and we believe that they would in the absence of civilization continue to change slowly as old species disappear and new ones replace them by evolution and immigration. There is always availability of new sites after destruction; the physical nature of the new habitats and the arrival of propagules decide what will develop. Natural communities and successional sequences are today more or less stable for a given time and place. Likewise, predictable geographical clines in biotas and communities occur in accordance with particular latitudes, altitudes, and longitudes. Climate, especially the relative extent of precipitation, always plays a major role—for example, in western North America, where the Great Plains are differentiated from the mountains. More locally, topography and edaphic factors always play a strong role. Soil chemistry alone, including acidic, neutral, basic pH reactions, to say nothing of underlying rock types like limestone or serpentine, help to determine what species will become established in a given site.

The natural successions involve natural fugitives, species that move about to wherever new disturbances have occurred. Deer and bison pastures allow certain patterns of transformation to occur when they are abandoned. Major floods may devastate the vegetation and generate a distinctive sequence of successional species. Some more or less persistent stages in succession may exist for thousands of years in certain venues like the snowfields of the Rockies that melt and become inhabited by plants and animals year after year or river banks that are exposed and re-exposed continuously. Devastations of fire, lava flow, and landslides lead to repeatable sequences of species and communities.

Even in successional situations, there are more or less constant types, repeated again and again, and there are more or less sustainable "mature" vegetation types—bogs, fens, hardwood swamps, conifer forests, beech-sugar maple communities, and so on. Today, even though biologists will teach that all of these are subject to variation, there is still a major element of stasis and continuity, and one of our biological goals is to understand and explain them. The natural ecosphere, lacking the human influence, is infinitely diverse, but its multiformity can be described in detail and can engender a variety of predictions. Of course, major natural upheavals have always occurred in the past, some local, some widespread in their influence. Such natural perturbations as glaciation, earthquakes, landslides, and mountain formation have always happened and always will happen. The results of the various glaciations on the vegetation of North America are examples of such events. There are many, still unsolved mysteries, such as why there are so many floristic affinities between eastern North America and eastern Asia, mysteries that will continue to attract the attention of biologists for a long time to come.

## CIVILIZATION AND THE WORLD'S SUPER INVADER

The species *Homo sapiens* itself is without question the super invader of all time. In spite of numerous local genotypes gathered into the ancient races, the human line of evolution has not speciated and does not promise to, at least in the near future, in part because of constant hybridization and introgression. The origin of people was a gradual process, and the origination of modern global civilization was much later. Civilization itself developed slowly, probably in Asia and Asia Minor. The changes must have been subtle, and no one can say when the transition from non-civilization to civilization actually took place because it was accumulative. Sporadic at first, civilization became more or less recognizable perhaps as early as 5,000 years ago in eastern China and 2,500 years ago in western Europe. The most explosive development, particularly of technology and science, in the western world of the past five centuries has been extraordinary and at a rate that is still constantly zooming. This one primate species has not only increased in numbers to many billions but has taken control of the environment far beyond any other species in the history of life. This single dominant taxon has remodeled the patterns of biodiversity to an extent almost inconceivable in the relatively stable biosphere of the past. Human intervention into the natural patterns of the world's biota is quite recent in the large picture of the earth's biodiversity. In a short time whole ecosystems are being destroyed. The human population itself is hurtling into an uncertain and dubious future, the details of which we cannot, or will not, preconceive. The study of wholesale contamination by invasives is only one of many dozens of the elements of an anticipated holocaust of the entire globe.

A major responsibility of biologists is now and will be in the future to create understanding of what the broad implications of civilization are ahead and to see the big picture. Present problems like AIDS, environmental contamination by invasives, loss of individual species or destruction of particular habitats—all of these problems and many more need to be subsumed under a broad conceptual scheme from which we can make reliable predictions. The biologists must acquire the methods and the ability to convincingly influence the other participants in civilization by educating our young people, informing our administrators and politicians, and restraining our business structure.

Until now we have experienced three main effects of the dramatic growth of *Homo ignorans*. The first is the destruction of natural ecosystems by such interruptions as farming, grazing, logging, removal of forest, production of billions of house lots, vacant lots, shopping malls, parks, playing fields, golf courses, trails, railroads, and highways. Compare Marion Co., Indiana, 500 years ago with Marion Co. today, and we will have a microcosm of the influences of humans.

A second potential effect of our expansion has been to produce genetic changes in species. We know too little about this: has there really been widespread selection of natural fugitives or weed species for optimal association with our destructive actions? Do weeds become better weeds? Almost certainly this is so. In some cases indeed, weeds evolved (or were artificially selected) to become useful to us. Surely Queen Anne's lace = carrot (*Daucus*) and wheat (*Triticum*) evolved to their present state, inadvertently or deliberately, by the activities of people. Some became our pets and our food like pigeons (*Columba*), cats (*Felis*), dogs (*Canis*) and pigs (*Sus*). Others became "civilized pests" like house sparrows (*Passer*) and starlings (*Sturnus*). Today some of the best known and most publicized genetic changes in species are among insect pests, in which we have not only transported and established them, providing them with new habitats and ranges but, in the attempt to control them with chemical pesticides, we have created novel forms that are resistant.

A third effect of our spread and commerce has been to alter the normal potentiality of natural dispersal mechanisms, the ancient sources of long-distance dispersal. Our commerce now enables major long-distance migrations of species—if animals, their gravid adults; if plants, their propagules; or simple wholesale transmissions of one or more individual animals, plants, fungi or microbes. Much has surely been unintentional through such agents as marketing, wagons, seed sales, ships, trains, dirty towels, autos, and adulterants. The airplane is probably going to become a leading source of major accidental introductions in the future; even the wheel wells of jets can harbor mammals or snakes.

Intentional introductions are made with some purpose in mind, but when the introduction becomes a serious invasive, it often comes as a surprise. Who would have thought that the beautiful garden plant *Lantana* would become a serious pest? We have no idea how many rare native species of dry gulches and lowland forests were destroyed after the introduction of this species into Hawaii and its escape into the native vegetation around 1860. Because its fruits are readily dispersed by birds, *Lantana* spread throughout the islands and was one of the most serious contaminants of the lowland vegetation until a century later when it was brought under biological pest control. Not only garden plants but vegetables introduced for food, or animals for hunting, or trees for lumber—many of these have escaped and become troublesome invaders. What is even worse is that counter-control organisms have themselves in some cases become invasive and have switched from their original pest prey to prized native endemics. Of course many contrary examples exist. An excellent example of successful biological control involves the "molesting salvinia," an aquatic floating fern which created serious problems in Australia, India, New Guinea, southeastern Asia, and southern Africa (Moran 1992; Thomas and Room 1986). This is an example of which I am especially fond because (1) it involves a fern and (2) it demonstrates the crucial value of taxonomy. A misidentified *Salvinia,* capable of doubling its size in a little more than two days, was completely clogging waterways. It turns out the villian was *Salvinia molesta,* an endemic undescribed species from Brazil, and not *S. auriculata.* Investigators observed that *S. molesta* was kept in control in its native areas by herbivorous insects. Field work resulted in the discovery of an endemic previously undescribed species of curculionid weevil (*Cyrtobagous salviniae*) that has now been successfully used as a biocontrol agent in the infested areas.

## THE NATURE OF HUMAN-FOSTERED INVASIVES

General characteristics of the invaders resulting from civilization are as follows: they are usually exotics far from their original homes. Some, like the low ragweed (*Ambrosia artemisiifolia*), are native to North America, but by far the majority are from Asia and Europe. Usually there are but one or a few species from a given taxonomic genus that are invasives even though the genus itself may be species-rich and taxonomically complex. Among the estimated 60 species of dandelions, only one, *Taraxacum officinale*, is a truly serious weed in North America; among over 100 species of buckthorns, only two (*Rhamnus cathartica* and *R. frangula*) are considered pests; and from over 200 species of honeysuckles, three (*Lonicera japonica*, *L. maackii* and *L. tatarica*) have become problematic. Many of our invasive species have more or less weed-like tendencies in their original homes, and this is a subject that biologists could well investigate in detail statistically, asking the question "To what extent are exotic invasives in North America derived from taxa that are already adapted to successional habitats in their original home ranges?" So far there is little solid evidence that invasive populations will ultimately come into equilibrium with their new associates and then function as non-aggressive members of their communities. From a theoretical standpoint it should be determined whether alien invaders will sooner or later become less aggressive and more compatible with their associates, and this is another question that biologists should be prepared to study. Probably the most general problem of invaders is that some have become completely out of hand locally and require drastic and expensive controls if they are to be held in check at all. In the present symposium we will hear of a number of examples of such aggressive adventives and of the daunting challenges of containing them.

One more focus of future research for biologists trying to paint in the broad picture of aggressive aliens is the extent to which different families contribute to the global situation. In plant invasives, we in North America think right away of the enormous contributions of the pink family, Caryophyllaceae (e.g., species of *Cerastrium*, *Lychnis*, *Saponaria*, *Silene*), the mustard family, Brassicaceae (*Alliaria*, *Barbarea*, *Brassica*, *Capsella*), and the knotweed family, Polygonaceae (*Polygonum*, *Rumex*). These families of course contain numerous second-growth and ruderal taxa in their original homes, so it is no surprise that they contribute so many weeds. Among birds, we think of passerines especially; among mammals, various rodents. But some groups have few or no invaders, such as butterflies (only a few major ones in North America); orchids (very few), ferns (limited mainly to a dozen or so taxa in southern Florida), and bryophytes.

Most of modern biological expertise which is consulted concerning problems of the environment comes from the category we can loosely refer to as systematic biologists, people concerned with biodiversity, evolution, comparative life cycles and, above all, the specificity that is involved in the dynamics and interactions of species of given taxonomic groups and their particular habitats. These are the biologists who most often deal with matters of conservation. For this reason there is presently a strong movement to increase the number of systematic biologists and group-oriented specialists. Systematists have declined in recent years in number and in specialties, but the scientific public has recognized this fact, and many leaders in biology are urging that more systematists be trained and more jobs created to accommodate them. This is a plea that has been gaining in strength over the past decade. The matter comes down to this: we simply must have more biologists who know and understand particular groups of organisms.

## BIOLOGY AND MANAGEMENT

Most botanists and zoologists are more interested in the overall phenomenon of invasives as such than in their management and control. The entire subject of pest control is an unsatisfactory one, biologically speaking, because of the variety of factors, known and unknown, that are involved. It is not a subject that can be readily reduced to a series of basic and predictive principles. There have been without doubt a number of major successes in control here and there around the world. Some of the leading examples of successful arrestment of naturalized exotics come from the continent of Australia and the islands of Hawaii, but in those areas, as in the rest of the world, failures are more numerous. Each taxon has its own special control demands, and the methods available have to be applied with awareness of its life-cycle requirements. Without such awareness of the particular biology of a given target species, time and money may be spent uselessly. Manual control by pulling up plants may serve in some cases, but two life-cycle factors are often overlooked. Many plants have extensive underground root or stem proliferation systems, enabling them to form vigorous clones. The underground organs often have abundant stored food and are actually stimulated to form numerous new upright branches when the old ones are pulled up. Other invasive plants create an accumulated seed bank that may lie dormant for years until the soil is overturned. Walking through a field or lawn and pulling up shoots may stimulate the banked seeds to germinate and start up a vigorous new population, larger in extent than the original.

Counterpests are always chancy, and in many instances the control organism designed to eliminate the invasive does not succeed. The worst case scenario with biological controls is that they may not only not succeed in eliminating the target pest, but they may spread into the local biota and kill off or severely damage local treasured species. Chemicals today tend to be highly controversial even among the general public because they are often unpleasant and even produce dangerous effects upon people as well as on native species and communities. As indicated above, certain invaders especially in the insect world may evolve resistance and become immune to the chemical factors. Chemical pesticide companies have come under scrutiny by conservation biologists, to say nothing of journalists, consumer advocates, public health experts, agriculturists, and exporters, to ensure that they exercise care in promoting the use and application of their products. Genetic control of invasive species is the latest contribution of biology to pest control, but it has yet to prove itself in any major or broadly applicable way, and many of the hoped-for benefits have not materialized. Genetic controls include introduction of resistant genes in valuable crops to keep pests from destroying them and inserting sterility factors in pest breeding to reduce or stop their reproduction. Public alarm over manipulation of the genetics of species is an important sociological problem, and there is still much work to do in devising appropriate genetic controls and in determining a priori whether they may prove to be dangerous. Many biologists have come to regard genetic controls as a sophisticated method of management that bears little inherent hazard. We expect to learn much more about potential dangers as the field of biotechnology advances.

## SOME FINAL GENERALIZATIONS

The biologist is looking for overall principles and generalizations as well as predictability. The practical manager seeks solutions to particular problems and requires that actions be taken. So far there seem to be relatively few generalizations that can be made and certainly too few to make reliable predications in any given case of invasive contamination. A close interaction between the "ivory tower" biologist and the "real world" problem-solver is called for. In particular, biological specialists should help supply information about the life-cycle features of any trouble-making invasive and work closely with the individuals responsible for controlling that invasive. On their side, management authorities should feel free to seek any expert insights that the specialist might provide. In my opinion, all too often the appropriate specialists are *not* called in, and the agencies responsible for biotic contaminants are not taking advantage of the specialist knowledge from their biological colleagues. Much of modern conservation effort has been initiated by these systematic specialists, the biologists who focus on the comparative natural history of particular groups. An entomologist can often give useful facts about the life cycle of a particular insect. A herpetologist can provide relevant information about an island-hopping, highly destructive snake. A dendrologist can be expected to provide us, for example, with data on vegetative reproduction of a given species of tree or shrub. Looking at the broad picture, we can come up with a statement of known or possible generalizations about invasive biocontaminants, as follows:

**Geography**; the most serious invaders usually originate long distances from the area of invasion. Those in North America come mainly from Eurasia.

**Ecology**; it appears that functionally healthy and established communities are able to resist aliens. The more complex the community, the more resistant it seems to become. Successional phases will tend to be especially prone to contaminants from afar. Disruptions of the ecosystem that open them up and upset the natural processes make them especially vulnerable.

**Convergence of diversity** may result, at least in some parts of the world, from increasing omnipresence of "world travelers," species that invade many continents. The same set of contaminant species may ultimately become standard for any given climate (and may even reduce world biodiversity by eliminating endemics).

**Evolution**; we can perhaps anticipate that invasive contacts will ultimately stimulate new genetic and evolutionary adjustments, both in the invader and in the host ecosystem. A sort of co-evolution may thus take place between the newly associated species aggregations.

**Taxonomy**; we can assume that in any given genus, all or most species are unable to invade a new part of the world and that at best they will become only briefly established and then die out. Other related species, for whatever reasons, can become powerful invaders and spread widely. Certain whole genera and families may contain unusually high numbers of invasive species, and all the component species must be watched closely when bringing in new taxa.

**Harmonization** may result from invaders coming into equilibrium with their new surroundings through adjustments with local parasites and predators. (Over-use of pesticides may actually delay such accord between exotics and natives.)

**Pest Control** is still primarily an art. Each invader has its own most effective treatment(s), but all treatments tend to have potential disadvantages. Manual methods may promote regrowth of plants, for example, because of clonal properties. Chemical controls may destroy native biota or lead to acquired resistance on the part of the pest. Counterpests, the classical "biological controls," may run amok and destroy valuable crops and/or native species. There is no sure way yet of predicting whether this will happen.

**Genetics** gives us various tools that we can potentially utilize. Biotechnology will bring knowledge of genetic phenomena to problems of control, but there is still a possibility that unanticipated damage may result.

Over the past three decades, academic biologists have become increasingly interested in the problems of biotic invasives. The book by C. S. Elton in 1958 was a pioneering effort and has become a classic. The more recent work is summarized for North America and Hawaii by the compendium of 16 excellent reviews and bibliographies edited by H. A. Mooney and J. A. Drake of Stanford University, published in 1986. There is no question that biologists have much to contribute to the resolution of the problems of conservation, agriculture, and management created by invasives. Biogeographers, ecologists, evolutionists, systematists, physiologists, biochemists, and geneticists all have a role to play in the big picture. For any special problem involving a particular invasive taxon, the expertise of appropriate systematic specialists—microbiologists, phycologists, mycologists, vascular plant systematists, protozoologists, malacologists, ornithologists, and so on—should be employed. Biologists are well aware that we are facing a serious problem, and they are willing to contribute whatever knowledge they possess to present or alleviate the destructive effects of invasive alien taxa.

## REFERENCES

Elton, C. G. 1958. The ecology of invasions by animals and plants. Methuen, London.

Mooney, H. A. and J. A. Drake (eds.). 1986. Ecology of biological invasions of North America and Hawaii. Springer-Verlag, New York, N.Y.

Moran, R. C. 1992. The story of the molesting salvinia. Fiddlehead Forum 19:26–28.

Thomas, P. A. and P. M. Room. 1986. Taxonomy and control of *Salvinia molesta*. Nature 320: 581–84.

# Biocontrol of Multiflora Rose

James W. Amrine, Jr. and Terry A. Stasny[1]

Multiflora rose (*Rosa multiflora* Thunb.) is native to Japan, Korea, and Eastern China (Rehder 1936). It was introduced to North America in the early 1800s for breeding roses, for root stock, and as a garden plant. Its hardy growth, relative resistance to disease and insects, abundant fragrant bloom, and persistent, colorful, nutritious fruit made it a desirable plant for many growers, hobbyists, and wildlife enthusiasts (Fig. 1). In the 1930s and 1940s it was promoted by the USDA, Soil Conservation Service, and many state departments of natural resources as an ideal plant for living fences, wildlife food, and erosion control. As a consequence, millions of multiflora roses were given to farmers and conservation groups and intentionally planted throughout the eastern United States. In West Virginia over 14 million were planted from the 1940s to 1960 (Dugan 1960), and in North Carolina some 14 to 20 million plants were set out (Nalepa 1989). Only a few states, like Kentucky, opted to restrict use of the plant and thus largely avoided widescale introduction of this noxious weed.

In the 1940s and 1950s, the weedy characteristics of multiflora rose became apparent. A few voices warned of the impending problem, like pioneer Hoosier botanist Charlie Deam, who carefully commented in a letter in 1948: "I understand they are strongly recommending now that all the old cemeteries be planted with multiflora rose. When Gabriel sounds his horn, I am afraid some will be stranded and not be able to get thru the roses. Please do not recommend the multiflora rose except for the bonfire" (Kriebel 1987, p. 135). Klimstra (1956) stated: "Emphasis now being placed on multiflora rose . . . might well result in the establishment of another nuisance plant . . . ."

A medium-sized multiflora rose is capable of producing 500,000 to 1,000,000 seeds in a good year (Amrine unpubl.), and the seeds are widely dispersed by songbirds such as robins, mockingbirds, and red-wing blackbirds. Seeds pass through songbirds undamaged while gallinaceous birds (chickens, turkeys, pheasants, etc.) use stones in their powerful gizzards to grind seeds to a meal. In Massachusetts, Lincoln (1962) showed that seed can pass through robins unharmed and that the rate of germination was actually doubled. Other animals such as deer can also spread the seed. By the early 1960s the potential spread of multiflora rose became obvious, especially in "marginal land" (i.e., hilly pastures, roadsides, fence rows, rights-of-way, and other sites where tractors cannot pass).

Farmers and others soon tried to eliminate the plant by cutting, burning, pulling out with tractors, bulldozing, applying herbicides, and grazing with goats (Hindal and Wong 1988; Bryan and Mills 1988). They also began campaigning for legislation to declare multiflora rose a noxious weed and to prohibit its sale and propagation. In West Virginia, multiflora rose was declared a noxious weed by the State Legislature in the Plant Pest Act of 1967, which was amended in 1972, and in the Noxious Weed Act of 1976. Several neighboring states soon followed with similar legislation, and multiflora rose is now declared a noxious weed in Illinois, Indiana, Iowa, Kansas, Maryland, Missouri, Ohio, Pennsylva-

1. Division of Plant and Soil Sciences, P.O. Box 6108, West Virginia University, Morgantown, WV 26506-6108

FIGURE 1.—*Rosa multiflora* in bloom. (photo by Dale Hindal)

nia, and Virginia. In 1975 the West Virginia Department of Agriculture (WVDA) began testing herbicides to control the weed. In 1980 an aerial survey of West Virginia indicated that 36,500 ha (149 $mi^2$) of land was heavily infested with multiflora rose (Fig. 2). State-wide experimental control programs in 1980 and 1981 indicated that a ten-year multiflora eradication program using herbicides would cost over $40 million (Williams and Hacker 1982). Chalamira and Lawrence (1984) reported that multiflora rose was the highest priority agricultural problem in West Virginia.

In the early 1980s Hindal and Wong (1988) began a survey for plant pathogens and/or insects in West Virginia that could be used as potential biocontrol agents of multiflora rose. Their search of the literature revealed two agents showing potential for control of multiflora rose: (1) the rose seed chalcid, *Megastigmus aculeatus* var. *nigroflavus* Hoffmeyer, a small torymid wasp that lays eggs in and kills developing rose seeds, and (2) a "virus," Rose Rosette Disease (RRD), transmitted by the eriophyid mite, *Phyllocoptes fructiphilus* Keifer. RRD was reported killing roses in the Midwest and appeared to hold great potential as a biocontrol agent (Crowe 1983; Doudrick and Milligan 1983). Amrine and Stasny (unpubl.) found a third agent, *Agrilus aurichalceus* subsp. *aurichalceus*

FIGURE 2.—Aerial photo of West Virginia field infested with multiflora rose. (photo by Dale Hindal)

Redt., a buprestid beetle whose larva girdles multiflora rose stems and thus kills the canes and seeds and reduces spread of the weed.

## ROSE SEED CHALCID

*Megastigmus aculeatus* var. *nigroflavus* was reported from New Jersey in 1917, causing heavy infestation of multiflora rose seed imported from Japan for producing rootstock for ornamental roses (Weiss 1917). Later, Milliron (1949) reported that the rose seed chalcid "now appears to be well-established in parts of the Atlantic Seaboard." Scott (1965) found large numbers of the rose seed chalcid at the Patuxent National Wildlife Refuge near Washington D.C., and infestation rates were around 95 percent. Mays and Kok (1988) reported on a survey of multiflora rose seed for the chalcid in 1985 and 1986, finding average infestation rates of 26.5 percent (range: 2–59 percent) and 23.9 percent (range: 2–52 percent), respectively, in 50 of 51 counties surveyed. Hindal and Wong (1988) found the chalcid in West Virginia and mentioned its potential for biocontrol.

Amrine (unpubl.) surveyed multiflora rose seed (dissecting all seed from each of 20 rose hips per sample) from 67 sites (in 21/55 counties) in West Virginia in 1984–1985 and found an average of 49.7 percent (range: 0–100 percent) of viable seed infested with the torymid. A similar survey from 16 sites in Maryland, Missouri, Oklahoma, Pennsylvania, Tennessee, Texas, and Virginia had an average infestation of viable seed of 46.7 percent (range: 0–95 percent). In the samples dissected, the hips had an average of 5.75 (range: 1–21) full-sized seeds. Amrine found that the chalcid oviposits in the developing receptacle just after petal-fall in June. The wasp takes a position along the shriveled stamens and inserts her ovipositor from a single point and guides it into each of the developing ovules (Fig. 3). A typical oviposition required 45 minutes to complete.

Larvae develop in the ovules during July and August, consuming the con-

FIGURE 3.—Adult rose seed chalcid on multifora rose hip. (photo by Dale Hindal)

tents of the seeds and killing them. They reach full size by late September and then enter diapause. During the winter, larvae die when exposed to temperatures below −20°C (−5°F), suffering 20–80 percent mortality from extended periods when temperatures fall below −26°C (−16°F). Amrine (unpubl.) found that chalcids in rose hips close to the ground and in other protected sites survived low temperatures better than those hips on upper canes exposed to ambient low temperatures. By late May the larvae transform to pupae. At about petal-fall (this occurs about the third week of June in West Virginia) adult wasps eclose within the seed and begin to chew their way out of the seed and the hip, leaving neat round exit holes.

Dispersal studies by Shaffer and Amrine (Shaffer 1987) indicated that the insect has poor abilities to fly to newly established rose plantings. Apparently most dispersal is by movement of infested seed by birds which would explain the apparently low colonization of rose seed in West Virginia (49 percent). The millions of multiflora roses planted in the state were set out as rooted cuttings—they were not planted from seeds, thus no chalcids were disseminated. Therefore the chalcid has had a monumental "catchup" job to find and colonize all the potential rose sites in the state. We believe that the chalcid will eventually find and utilize 90 percent or more of the multiflora rose seed in West Virginia and surrounding states and act as a powerful biological control for the weed, helping to reduce its rate of spread to a low level. But we believe this process will be slow and may require 20 or more years.

## ROSE STEM GIRDLER

Amrine and Stasny (unpubl.) found several sites in Indiana, Ohio and West Virginia where the rose stem girdler, *Agrilus aurichalceus* (Coleoptera: Buprestidae) was fairly abundant and affected some degree of control of multiflora rose. At sites along Interstate 71 in Ohio (Fayette Co.), near Woodsfield

(Monroe Co., Ohio), and in a pasture near Morgantown, West Virginia, upwards of 12 percent of canes were found attacked by the girdler in 1988–1989, with often two or three larvae girdling each cane. All tissue distal to the girdle was killed including developing rose hips and seeds (Fig. 4).

The overwintering borers, still in the previous season's girdled canes, pupate in April and emerge as adults in May. The adults are small, bronze-colored metallic beetles that can be found on sunlit foliage in the mornings. They find new shoots in May and oviposit just under the bark. The larvae hatch and begin burrowing under the bark, spiraling outward from the oviposition site. The initial burrowing does not kill the cane, but by late July the infested stems begin to die and appear as brown "flags" on rose bushes. Since no infested site has been found in which more than 12 percent of canes were affected, we believe that this insect will only have minor importance as a biocontrol agent of multiflora rose.

## RRD AND *PHYLLOCOPTES FRUCTIPHILUS*

The eriophyid mite, *Phyllocoptes fructiphilus,* was first described in California in 1940 and was found on the fruits of *Rosa californica;* the mite was not then

FIGURE 4.—Damage to multiflora rose cane by spiral-boring larvae of the rose stem girdler. Cane will die back over winter. (photo by Dale Hindal)

associated with RRD. Rose rosette disease (RRD) was first found in California, Wyoming, and Manitoba, Canada in 1941 on ornamental roses and the wild rose, *Rosa woodsi,* which is common throughout the western states (Thomas and Scott 1953). Their tests showed that it could be graft-transmitted to other species of roses but not to other plants in the family Rosaceae or to members of other plant families. Thomas and Scott did not observe the eriophyid vector.

Symptoms in multiflora rose include a red or purplish vein mosaic (diagnostic), production of bright red lateral shoots, dwarfed foliage, and premature development of lateral buds, producing many compact lateral branches forming "witches' brooms" (Amrine and Hindal 1988) (Fig. 5). Symptoms on ornamental roses include a yellow mosaic pattern on leaves, greatly increased thorniness of stems, wrinkled foliage, and witches' brooms; however, the bright red lateral shoots and vein mosaic seen in multiflora rose do not usually occur (Fig. 6).

Allington et al. (1968) conducted research on RRD of roses in Nebraska in the 1960s and showed that *Phyllocoptes fructiphilus* transmits RRD to several species and varieties of roses, including multiflora rose. Doudrick et al. (1986) in Missouri erroneously concluded that *P. fructiphilus* could not transmit RRD to multiflora rose. In 1985 Hindal and Amrine traveled to Missouri to obtain RRD-symptomatic plants for transmission tests and found RRD and *P. fructiphilus* in several counties in western Kentucky, southern Illinois, and Missouri. In 1987 and 1988 RRD and *P. fructiphilus* were found in Indiana and southern Ohio (Hindal et al. 1988). Extensive surveys were conducted of multiflora roses in West Virginia for both RRD and eriophyid mites from 1985 to 1988; populations of *P. fructiphilus* and a closely related species, *P. rosarum,* were found in several counties each year, **but not RRD**. In 1989 RRD was finally found in West Virginia in 10 western counties extending from Huntington, north to Wheeling. RRD was found in 8 additional counties in 1990, 11 more during 1991 and, thus far, 3 more in 1992 (Fig. 7).

Amrine et al. (1988) proved transmission of RRD by *P. fructiphilus* to multiflora rose. In several tests, transmission was 100 percent, with symptoms appearing as soon as 17 days after mites were placed on new shoots of dug and pruned transplants. Transmission tests with *Phyllocoptes rosarum,* spider mites, and several insects (aphids, leafhoppers, planthoppers, and thrips) proved negative. In field tests, mite transmission took as long as 90+ days. Transmission by "patch" grafting involves placing small pieces of diseased tissue under the bark of target plants (6 to 12 grafts per plant) and requires 30 to 90+ days for symptoms to appear.

Research at West Virginia University shows that development of *Phyllocoptes fructiphilus* is similar to that of other eriophyids. Females are extremely small, averaging 0.2 mm (200 microns) in length and 45 microns in width. They overwinter on rose canes, either under bud scales (if buds are opened slightly, which varies from year to year), within residual clumps of foliage, or in cracks and crevices; the mites must stay on green living tissue to feed during mild periods. In early spring the mites move onto developing shoots to lay eggs; females live about 30 days and can lay about one egg per day. Eggs hatch in 3–4 days and each immature stage (protonymph and deutonymph) requires about 2 days. Thus, in warm weather one generation may be produced per week. Development is continuous throughout the season until weather turns cold in the fall. The mites disperse aerially on warm, sunny days by standing on their tail ends and actively waving their legs and being blown off the foliage in winds of $\geq$ 4 miles per hour.

In May 1987 Amrine et al. (1990) began a long-term study at Clifty Falls State Park (Jefferson Co.) in southeastern Indiana. The site was heavily infested with both healthy and RRD-symptomatic multiflora roses, and the topography

FIGURE 5A.—Blotch mosaic and vein mosaic on multiflora rose leaves contrasted with a healthy green leaf underneath. Mosaic is a diagnostic symptom of rose rosette disease (RRD). (photo by Dale Hindal)

FIGURE 5B.—Proliferation of lateral shoots forming "witches' brooming" caused by rose rosette disease on multiflora rose. (photo by Dale Hindal)

FIGURE 6.—Increased thorniness, wrinkled and dwarfed foliage, misformed buds and flowers on an ornamental greenhouse rose (Cara Mia); all symptoms of rose rosette disease on ornamental roses. (photo by Dale Hindal)

and environment are similar to that of West Virginia. This was the nearest known site of RRD in 1986–87. The study was initiated to determine the long-term effects of RRD on multiflora rose; especially to learn how rapidly the disease spreads, what happens to infected plants, the approximate number of mites present on diseased compared to healthy plants, and how mite populations change during the year and from year to year.

A total of 180 multiflora rose plants were marked and visited monthly during the growing season for the next five years. The initial average density was 1,200 plants per acre (3,000 per ha), and at the beginning of the study 30 percent of plants were symptomatic and 1 percent had been killed by RRD. The infection increased each year and leveled off at 94 percent in September 1991 with a mortality of 88 percent. The average longevity of infected plants was 22.4 months (range: 3–48) (Fig. 8).

*Phyllocoptes fructiphilus* were found in petiole bases, especially on tender growing shoots, and inside and between unfolding leaflets. Mite populations

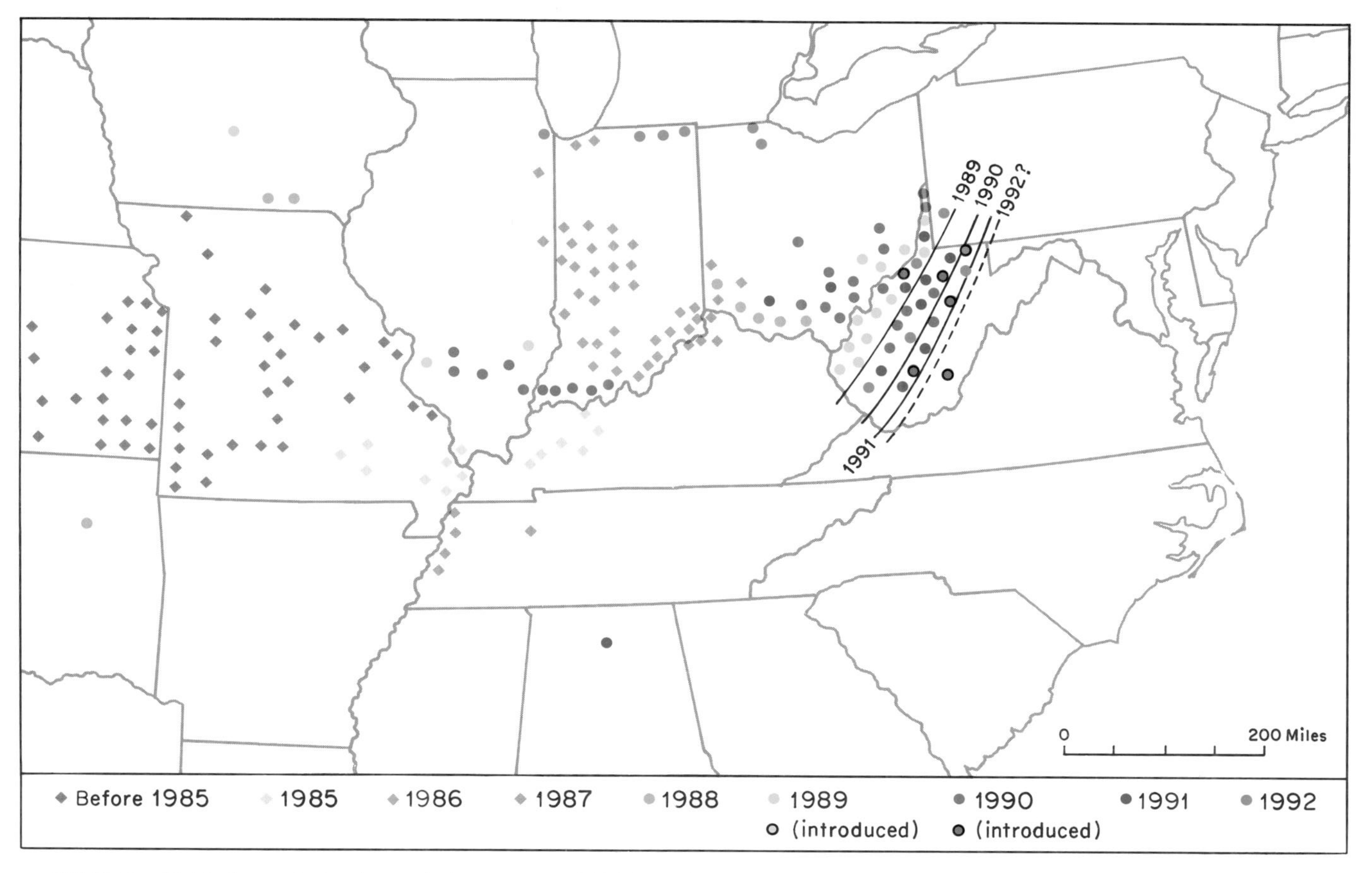


FIGURE 7.—Distribution and rate of spread of rose rosette (RRD) in the Eastern U.S., 1985 through August 1992.

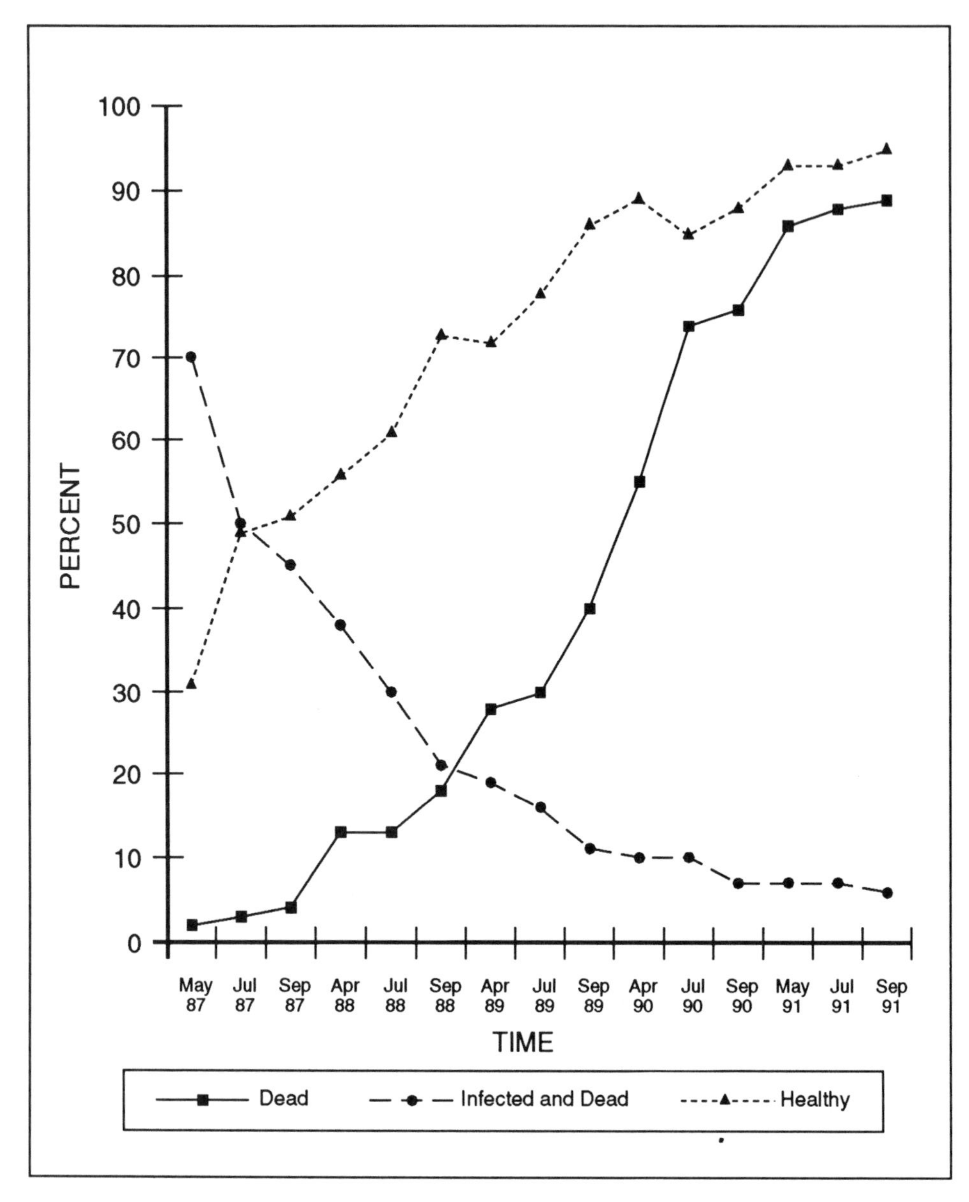


FIGURE 8.—Morbidity of multiflora rose infected with rose rosette disease (Madison, Indiana) May 1987 to September 1991.

were 14 times larger on symptomatic compared to healthy plants in 1987 and 1988. Mite populations began low and sporadic in April and gradually increased to peak populations by September in most years. At peak population nearly all RRD-symptomatic plants were infested with mites. The average number of mites per symptomatic shoot in September of each year (1987–90) was 112, 30, 112 and 6.6, respectively. The low average number in 1988 (30) resulted from severe drought which caused death of mites on desiccated foliage; the mites rebounded to an average of 72 per shoot by October 1988. The low average numbers in September 1990 (6.6) and during the entire year resulted from unusually cold weather in December 1989 (−31°C or −24°F) that killed nearly all aboveground symptomatic canes and thus killed most of the overwintering mites. Few nonsymptomatic multiflora canes were killed by the cold temperatures and mites from these plants repopulated RRD-symptomatic plants which had produced new growth from their crowns.

It was also observed that approximately 15 percent of infected multiflora rose became asymptomatic during the course of the disease. That is, plants showing unmistakable symptoms at one time, later became totally free of observable symptoms. This period of "reversion" to normal foliage varied from one to several months, but in all cases the plant eventually became symptomatic again and eventually died. The same phenomenon but to a greater degree was observed in ornamental roses. In a greenhouse study, several 'Cara Mia' had two or three alternating periods of symptomatic and then asymptomatic foliage; the plants lived about four years before succumbing to the disease.

The most difficult part of this research has been attempting to identify and characterize the disease agent. Investigation of the "virus" has been conducted both at West Virginia University (R. Frist, Biology Department) and by researchers at Iowa State University (A. Epstein, R. Di et al., Plant Pathology Department). The group at Iowa State demonstrated a double-stranded RNA virus as the apparent disease agent (Di et al. 1990) and this was corroborated by Frist (pers. comm.). Attempts to clone the virus and to produce a DNA probe have been negative at both labs. The agent is apparently atypical for a plant virus and will need careful and extensive future research in order to be characterized.

To date we have no simple and precise test to prove that plant material is infected with or free of RRD. The potential of shipping infected but asymptomatic stock is very real. This can have major consequences for export of rose tissue from the United States to Europe, Asia, and other areas of the world where RRD is not present. Arrival of the mite vector and RRD in East Asia could be disastrous for multiflora rose which is a valuable plant in those countries; it brings to mind the forest devastation caused by chestnut blight, Dutch elm disease, and gypsy moth when they arrived in this country. Graft transmission of suspected RRD-infected plants into healthy multiflora roses is the most reliable technique for detecting the presence of RRD. However, this test often requires more than 90 days to produce diagnostic symptoms. The USDA should be funding research to characterize and assay for the agent of RRD and to better understand the mode, time, and extent of dispersal of the vector.

Many persons ask the question, "Will RRD spread to other important plants?" So far, apple, apricot, blackberry, black cherry, mountain ash, peach, pear, plum, raspberry, and strawberry have been tested in our laboratory repeatedly for several years, and *none* of the challenged plants have developed symptoms. Doudrick (1984) tested several rosaceous plants and attempted transmission to potential index plants including *Vinca major,* all unsuccessfully. Thomas and Scott (1953) also attempted to transmit RRD to other rosaceous and non-rosaceous plants with negative results. We have made backgrafts from challenged plants to multiflora rose, and these too remain uninfected and have shown no decline. We therefore believe that only plants in the genus *Rosa* can be infected with RRD. Our transmission tests have also shown that some species of roses, especially *R. carolina, R. palustris,* and *R. setigera,* native to eastern states cannot be infected with RRD. Thomas and Scott (1953) showed that *R. californica* and *R. spinossissima* could not be infected. Thus, sources of roses are available for breeding resistance to RRD into ornamental roses. Of all rose varieties tested, *Rosa bracteata,* the weedy McCartney rose of Louisiana and Texas, is the only species on which *Phyllocoptes fructiphilus* cannot develop; however it is susceptible to RRD (by grafting). The McCartney rose may serve as a source for resistance to the vector mite.

To learn how to protect valuable ornamental roses, tests were conducted to evaluate miticides for the control of *P. fructiphilus*. Six chemicals were evaluated, and the three most effective materials were carbaryl, amitraz, and diazinon, in that order. We recommend treatment of ornamental roses with one of the three materials every week, from mid-May until early September.

## SUMMARY

The eastern United States is heavily infested with the noxious, introduced weed, multiflora rose, as a direct consequence of USDA, SCS, and various state departments of natural resources "conservation programs." This weed is currently under attack by several biocontrol agents, especially a native mite and virus: the vector *Phyllocoptes fructiphilus* and rose rosette disease (RRD). The mite and RRD are native to the western states and began to spread eastward after millions of multiflora roses were planted across the Great Plains as wind breaks. At present RRD has spread eastward up the Ohio River Valley to central West Virginia and is expected to spread to the east and north on the prevailing winds to the Atlantic coast and into New England (Fig. 7). Results of field experiments indicate that RRD has the potential to eliminate over 90 percent of the multiflora roses in areas of dense stands.

A major problem that remains is identifying and characterizing the agent of RRD. This task will require major funding and considerable virological and DNA-probe investigations to find the answer to "just what is the graft-transmissible, mite-transmitted agent" that causes RRD? Understanding the complex biology of the disease agent and its relationships to the host plant and the mite vector may help us understand similar disease-mite complexes such as Currant Reversion Disease of black and red currant (*Ribes nigrum* and *R. rubrum*), its eriophyid mite vector, *Cecidophyopsis ribis*, and the yet unidentified but graft-transmissible "viral" agent (Adams and Thresh 1987).

Important questions remain unanswered about the time, mode, and rate of dispersal of the mite vectors. Answers to these investigations would allow farmers to better understand expected rates and times of spread of RRD among noxious multiflora roses. Conversely, the same knowledge would better enable rosarians to time applications of control materials and/or to better space their plantings relative to multiflora roses growing in their neighborhoods.

The excellent work in breeding resistance to the mite vector and causative agent of Currant Reversion Disease into black currants in Great Britain and elsewhere (Adams and Thresh 1987) leads us to believe that resistance to *Phyllocoptes fructiphilus* and to RRD could be bred into ornamental roses; however, this would be a long-term endeavor.

Finally, the rose seed chalcid, *Megastigmus aculeatus* var. *nigroflavus*, is increasing in numbers in the eastern United States and is expected to eventually infest nearly 90 percent of the multiflora rose seeds of the plants that survive RRD. The combined impact of these two biocontrol agents is expected to greatly reduce both the numbers and rate of spread of multiflora rose in the eastern United States. But this expected level of biocontrol will take place over an extended time scale: 20–30 years? Sources of funding of research on biocontrol must realize that both the research and the expected results occur over an extended period of time. The "business school mentality" of a quick return for short duration input does not, and never will, apply to biocontrol.

## REFERENCES

ADAMS, A. N. and J. M. THRESH. 1987. Virus and virus-like diseases of black currant. Mite-borne diseases. Reversion of black currant. Pages 133–136 *in* R. H. Converse (ed.). Virus diseases of small fruits. USDA, ARS, Agric. Handbook No. 631.

ALLINGTON, W. B., R. STAPLES and G. VIEHMEYER. 1968. Transmission of rose rosette virus by the eriophyid mite *Phyllocoptes fructiphilus*. J. Econ. Entomol. 61:1132–40.

AMRINE, J. W., JR. and D. F. HINDAL. 1988. Rose rosette: a fatal disease of multiflora rose. Circular 147, Aug. 1988. West Virginia Univ. Agric. and For. Exp. Sta., Morgantown.

AMRINE, J. W., JR., D. F. HINDAL, T. A. STASNY, et al. 1988. Transmission of the rose rosette disease agent to *Rosa multiflora* by *Phyllocoptes fructiphilus* (Acari: Eriophyidae). Entomol. News 99 (5):239–52.

AMRINE, J. W., JR., D. F. HINDAL, R. WILLIAMS, et al. 1990. Rose rosette as a biocontrol of multiflora rose. Proc. Southern Weed Sci. Soc. 43:316–19.

BRYAN, W. B. and T. A. MILLS. 1988. Effect of frequency and method of defoliation and plant size on the survival of multiflora rose. Biol. Agric. and Hort. 5:209–14.

CHALAMIRA, L. R. and L. D. LAWRENCE. 1984. Agricultural research needs and priorities as perceived by West Virginia Vocational Agriculture Teachers and County Agents. West Virginia Univ. Agric. Exp. Sta. Misc. Publ. 11.

CROWE, F. J. 1983. Witches' broom of rose: a new outbreak in several central states. Plant Dis. 67: 544–46.

DI, R., J. H. HILL and A. H. EPSTEIN. 1990. Double-stranded RNA associated with the rose rosette disease of multiflora rose. Plant Dis. 74(1):56–58.

DOUDRICK, R. L. 1984. Etiological studies of rose rosette. M.S. Thesis, Univ. of Missouri, Columbia.

DOUDRICK, R. L., W. R. ENNS, M. F. BROWN, et al. 1986. Characteristics and role of the mite, *Phyllocoptes fructiphilus* (Acari: Eriophyidae) in the etiology of rose rosette. Entomol. News. 97:163–68.

DOUDRICK, R. L. and D. F. MILLIKAN. 1983. Some etiology and symptomological aspects of rose rosette. Phytopathology. 73:840. (Abstract).

DUGAN, R. F. 1960. Multiflora rose in West Virginia. West Virginia Agric. Expt. Sta. Bull. 447. p. 1–32.

FRIST, R. 1992. Biology Department, Brooks Hall, West Virginia Univ., Morgantown, 26506. Personal communication.

HINDAL, D. F., J. W. AMRINE, R. L. WILLIAMS, et al. 1988. Rose rosette disease on multiflora rose (*Rosa multiflora*) in Indiana and Kentucky. Weed Technol. 2:442–44.

HINDAL, D. F. and S. M. WONG. 1988. Potential biocontrol of multiflora rose, *Rosa multiflora*. Weed Technol. 2:122–31.

KLIMSTRA, W. D. 1956. Problems in the use of multiflora rose. Trans. Illinois State Acad. Sci. 48:66–72.

KRIEBEL, R. C. 1987. Plain ol' Charlie Deam: pioneer Hoosier botanist. Purdue Univ. Press., West Lafayette, Ind.

LINCOLN, W. C., JR. 1962. The effect of the digestive tract on the germination of multiflora rose seed. Newsletter, Assoc. Off. Seed Analysts 52:23.

MAYS, W. T. and L.-K. KOK. 1988. Seed wasp on multiflora rose, *Rosa multiflora*, in Virginia. Weed Technol. 2:265–68.

MILLIRON, M. E. 1949. Taxonomic and biological investigations in the genus *Megastigmus*. Amer. Midl. Nat. 41:257–420.

NALEPA, C. A. 1989. Distribution of the rose seed chalcid *Megastigmus aculeatus* var. *nigroflavus* Hoffmeyer (Hymenoptera: Torymidae) in North Carolina. J. Entomol. Sci. 24(4):413–16.

REHDER, A. 1936. On the history of the introduction of woody plants into North America. Nat. Hort. Mag. 15:245–57.

SCOTT, R.F. 1965. Problems of multiflora rose spread and control. Trans. 30th North American Wildlife and Nat. Reserv. Conf. 30:360–78.

SHAFFER, D. F. 1987. A study of the biocontrol of *Rosa multiflora* Thunb. utilizing the rose-seed chalcid wasp *Megastigmus aculeatus* var. *nigroflavus* Hoffmeyer (Hymenoptera: Torymidae) in West Virginia. M.S. Thesis, West Virginia Univ., Morgantown.

THOMAS, E. A. and C. E. SCOTT. 1953. Rosette of rose. Phytopathology. 43:218–19.

WEISS, H. B. 1917. *Megastigmus aculeatus* Swed., introduced into New Jersey from Japan. J. Econ. Entomol. 10:448.

WILLIAMS, R. L. and J. D. HACKER. 1982. Control of multiflora rose in West Virginia. Proc. Northeast Weed Sci. Soc. 36:237.

# Dogwood Anthracnose Disease: Native Fungus or Exotic Invader?

Margery Daughtrey[1]

The flowering dogwood, *Cornus florida* L., is one of our best-loved native ornamental trees. Its showy bract display in the spring and red fruits in the fall provide a feast for the eyes and sustenance for wildlife (Fig. 1). It is appreciated by hunters as an important source of food for game species such as turkey, bobwhite quail, squirrels, rabbits, and white-tailed deer (Krasny 1991). Birders appreciate the dogwood's value as a preferred food source for many non-game species (Table 1). The natural range of *C. florida* is the eastern third of the United States, from southwestern Maine to northern Florida, and northeastern Mexico (Gleason and Cronquist 1963) (Fig. 2). In the West the Pacific dogwood, *Cornus nuttallii* Aud., which is also prized as an ornamental, occupies a range from southern British Columbia to northern California (Sudworth 1967), including a pocket in Idaho (Roper 1970).

In the late 1970s homeowners in the New York city area noticed signs that dogwoods were not healthy. Dieback of the lower branches was occurring on landscape specimens, limiting the showy bract display to only the upper part of the tree (Fig. 3). Dead leaves from the previous season remained clinging to the branches in the spring. Gardeners also reported excessive watersprouting (epicormic branches) to their local Cooperative Extension horticulturists (Fig. 3).

A disease outbreak was also observed on *C. nuttallii* in Washington State beginning in 1976 (Byther and Davidson 1979). At first it was thought that the dogwood problems in both the Northeast and the Northwest were due to previously described fungus diseases. The infection was attributed to *Colletotrichum gloeosporioides* (Penz.) Penz. & Sacc. in Penz (Pirone 1980) in the Northeast and to *Gloeosporium corni* Green (Byther and Davidson 1979) in the Northwest. Further study of the disease affecting dogwoods in both locations led to the discovery that a fungus in the genus *Discula* was associated with the disease on both coasts (Salogga 1982; Salogga and Ammirati 1983; Hibben and Daughtrey 1988). This fungus was later described as a new species of *Discula*: *Discula destructiva* Redlin (Redlin 1991).

This new dogwood disease belongs to the grouping of 54 tree diseases called anthracnoses. These diseases are generally host-specific, cause leaf blighting and sometimes twig infection, and are most severe in cool, wet springs. Rain facilitates spore dispersal, and prolonged leaf wetness is essential for spore germination and penetration.

## THE EFFECT OF THE NEW ANTHRACNOSE DISEASE ON FLOWERING DOGWOOD

Whole-tree effects of dogwood anthracnose first caught the public eye. Close examination of diseased dogwoods disclosed a number of associated symptoms

1. Cornell University, Long Island Horticultural Research Lab, 39 Sound Avenue, Riverhead, NY 11901

FIGURE 1—Closeup of flowers and bracts of flowering dogwood. (photo by Richard Fields)

on leaves, twigs, and trunks (Daughtrey and Hibben 1983; Daughtrey et al. 1988). Brown, purple-rimmed leaf spots with yellow haloes were apparent on trees with lower branch dieback and also on seedling dogwoods on the forest floor (Fig. 3). In some cases, rather than discrete spots, larger blotchy areas of dead tissue developed, sometimes encompassing the entire leaf. At times, spotted bracts were also evident.

Twig dieback was also apparent on affected trees (Fig. 4). Previous season's shoots were killed back one or more nodes during the winter, and these dead areas turned tan. Shoot infection began at a node where a diseased leaf had been

TABLE 1. Birds for which dogwood fruits are a preferred food.[1]

| | |
|---|---|
| TURKEY | BOBWHITE |
| COMMON FLICKER | PILEATED WOODPECKER |
| RED-BELLIED WOODPECKER | YELLOW-BELLIED SAPSUCKER |
| HAIRY WOODPECKER | MOCKINGBIRD |
| BROWN THRASHER | AMERICAN ROBIN |
| WOOD THRUSH | HERMIT THRUSH |
| SWAINSON'S THRUSH | GRAY-CHEEKED THRUSH |
| EASTERN BLUEBIRD | CEDAR WAXWING |
| STARLING | YELLOW-RUMPED WARBLER |
| SUMMER TANAGER | CARDINAL |
| EVENING GROSBEAK | PINE GROSBREAK |

[1]DeGraaf and Witman, 1979

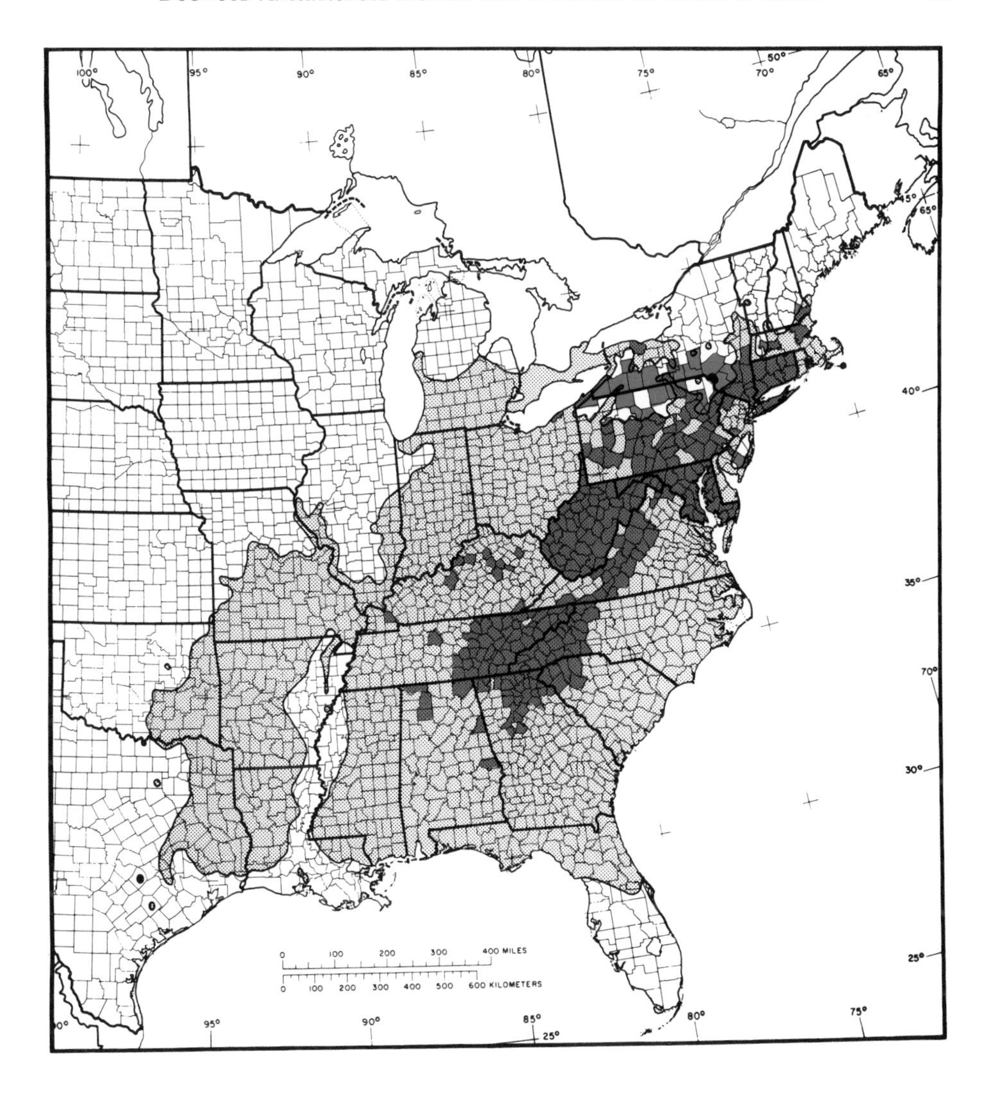


FIGURE 2.—The natural range of *Cornus florida:* counties from which dogwood anthracnose has been reported are indicated in red. The red area represents the extent of the disease in the eastern United States 12 years after it was first reported in the New York City area. (Disease distribution information courtesy U.S. Forest Service; *C. florida* range map from Silvics of North America, Vol. 11. Hardwoods. USDA Agricultural Handbook 654.)

attached, followed by death of the distal portion after the cambium of the shoot was girdled.

Disease symptoms on leaves and twigs were more pronounced in the lower canopy, a pattern typical of anthracnose diseases. The leaves on the epicormic branches, shaded and low in the canopy, were exposed to rain-carried spore inoculum from above. This made them particularly vulnerable to the blighting that was the precursor to twig cankers (Figs. 5 & 6).

Cutting away the bark at the juncture of a killed epicormic branch and a trunk or branch would often reveal discoloration and death of the cambium in the shape of a dark brown ellipse. These were "annual" cankers, able to expand only during a single dormant season. If enough of these cankers occurred to-

FIGURE 3.—Landscape specimen of *Cornus florida* showing dieback of lower branches caused by dogwood anthracnose infection. (photo compliments of C. R. Hibben)

gether on the same portion of the trunk, they girdled the tree. Complete girdling was most often observed on lower branches or on the trunks of small trees. Water and nutrient uptake were presumably disrupted on larger trees by the additive effect of many trunk cankers.

Tree mortality within one to three years was sometimes the end result of the leaf blighting, twig cankers, and trunk or limb cankers that were caused by *Discula destructiva.* Although the public outcry was most often associated with lower branch dieback in ornamental dogwoods, the greatest mortality occurred in natural woodland populations.

The most thorough documentation of the impact of anthracnose in a native population of *Cornus florida* is a study conducted at Catoctin Mountain Park in Maryland by Mielke and Langdon in 1984 (Mielke and Langdon 1986). This was followed by re-examination of these same plots by Schneeberger and Jackson in 1988 (Schneeberger and Jackson 1989). The dogwood population was much reduced over the four-year period of the study. Mortality advanced from nearly 33 percent in 1984 to 79 percent in 1988, with the least mortality seen in the largest trees. There were an average of 276 living dogwoods per acre in (682 per ha) 1984 and only 32 per acre (89 per ha) in the 1986 survey. None of the dogwoods within the study area in 1988 were in the category referred to as "apparently healthy"; all were showing some effect of the disease. Mortality from dogwood anthracnose has also been reported in a natural stand of *C. nuttallii* in British Columbia (Van Sickle and Wood 1989).

Diseased flowering dogwoods in open to lightly-shaded landscape plantings in the Northeast have fared relatively well. Although some valuable specimens were killed in years in which environmental conditions strongly favored the disease, in many cases only the lower branches have been lost to the disease, and well-maintained ornamental trees have survived. Dogwoods growing in more open locations along the roadside have been more likely to escape mortality than

FIGURE 4.—Epicormic branches (watersprouts) often appear on the trunks and main branches of *Cornus florida* with extensive dieback from dogwood anthracnose.

trees growing in more shaded understory locations, but even in the woodland understory there are a few surviving dogwoods apparent in the Northeast.

## THE EXPANDING RANGE OF THE DISEASE

By 1984 dogwood anthracnose symptoms on flowering dogwood were reported from southern Connecticut to eastern Pennsylvania (Daughtrey and Hibben 1983). The disease continued to spread south along the Appalachian Mountains. By 1986 it had progressed as far south as Greenbrier Co., West Virginia (Hibben and Daughtrey 1988).

Then in October 1987 dogwood anthracnose took a great leap southward. It was discovered in the Cohutta Wilderness area of the Chattahoochee National Forest in the mountains of northern Georgia (Anderson 1991). Soon thereafter, the disease was detected in North Carolina, South Carolina, and Tennessee. This was a major extension in the number of states concerned but not in geographic range, since the portion of the states affected was primarily within the area of

FIGURE 5.—Leaf infection on *Cornus florida,* showing spotting and blotching caused by infection with the fungus, *Discula destructiva.*

high elevation where the states adjoin. Public concern over the disease increased as it moved into the south. *Cornus florida* is the focus of spring celebrations in the south, the *raison d'etre* of many dogwood festivals.

As of 1992 the disease is known to occur in Alabama, British Columbia, Connecticut, Delaware, Georgia, Idaho, Kentucky, Maryland, Massachusetts, New Jersey, New York, North Carolina, Ohio, Oregon, Pennsylvania, South Carolina, Tennessee, Virginia, Washington, and West Virginia in the eastern United States. In the southeastern part of its range, dogwood anthracnose is still largely limited to higher elevations (Fig. 2).

## POSSIBLE ORIGINS OF THE DISEASE

There has been a great deal of speculation regarding the origin of the fungus pathogen, as there is no record of its occurrence in North America prior to the 1970s. The coincidental appearance of the disease on both coasts suggested that *Discula destructiva* was introduced from another country, since the disease was first noticed in the vicinity of Seattle and New York, two major ports of entry. In addition, both cities have arboreta which periodically import ornamental plants. Dogwood anthracnose has spread rapidly from its original area of discovery in southeastern New York over 10 years ago. Introduced pathogens may spread rapidly when a large, continuous population of a susceptible host is available.

The alternative possibility is that *D. destructiva* is not new to native *Cornus* species but has previously had an obscure ecological position, causing no remarkable effects on its host plant. Changes in climate, including man-engendered effects such as acid rain and fog, can be postulated to have altered the ecological balance in such a way that a weak pathogen assumed a new, more aggressive role. As in all of nature, mutations and natural selection are operative in fungus populations. Pathogens with greatly enhanced virulence have periodically arisen to plague agricultural crops in the past (Schumann 1991). Either environmental or genetic changes might suddenly have favored *D. destructiva* or eliminated populations of other microorganisms that previously were in competition with this fungus, keeping its population in check.

## COMPARISON TO OTHER EXOTIC TREE DISEASES

As newspaper reporters created scary headlines for stories on the new dogwood disease, unfortunate comparisons were made to previous horror stories of introduced tree diseases that had severe ecological and societal impact (Table 2). In particular the public asked whether this dogwood disease was "another chestnut blight."

TABLE 2. Introduced disease-causing fungi of North American forest trees.

| ALIEN FUNGAL PEST | NATIVE HOST SPECIES | ORIGINAL RANGE | FIRST REPORTED |
|---|---|---|---|
| *Ascocalyx abientina* SCLERODERRIS CANKER | *Pinus banksiana* (Jack pine) plus other pines, firs, larch, and spruces | Europe | Great Lakes Region 1950s |
| *Cronartium ribicola* WHITE PINE BLISTER RUST | *Pinus strobus* (Eastern white pine) | Europe from Asia | New England 1906 |
| *Cryophonectria parasitica* CHESTNUT BLIGHT | *Castanea dentata* (American chestnut) and some *Quercus* spp. (oaks) | Asia | New York City 1904 |
| *Discula destructiva* DOGWOOD ANTHRACNOSE | *Cornus florida* (flowering dogwood) and *C. nuttallii* (Western flowering dogwood) | Asia, possibly | Washington (state) 1976; SE New York (state) early 1970s |
| *Glomerella miyabeana* BLACK CANKER OF WILLOW | *Salix* spp. (willows) | Europe | Canadian Maritime Provinces 1927 |
| *Gymnosporangium fuscum* PEAR TRELLIS RUST | *Juniperus virginiana* (red cedar) | Europe, N. Africa & Asia | Victoria, B.C. 1960 |
| *Lachnellula willkommii* LARCH CANKER | *Larix laricina* (Eastern tamarack) | Europe | Massachusetts 1927 |
| *Nectria coccinea* var. *faginata* BEECH BARK DISEASE | *Fagus grandifolia* (American beech) | Europe | Nova Scotia 1890 |
| *Ophiostoma ulmi* DUTCH ELM DISEASE | *Ulmus americana* (American elm) | Holland | Ohio & east coast early 1930s |
| *Phytophthora lateralis* PHYTOPHTHORA ROOT ROT | *Chamaecyparis lawsoniana* (Port Orford cedar) | Asia, possibly | Washington (state) 1923 |
| *Sirococcus clavigignenti-juglandacearum* BUTTERNUT CANKER | *Juglans cinerea* (butternut) | unknown, on most *Juglans* of Asia and Europe | Midwest early 1900s (cause properly identified 1979) |
| *Venturia saliciperda* WILLOW SCAB | *Salix* spp. (willows) | Europe | Canadian Maritime Provinces 1927 |

The chestnut blight disease, caused by the exotic fungus *Cryphonectria parasitica* (Murrill) Barr was discovered in 1904 at the New York Zoological Park in the Bronx (Merkel 1906). It is believed that the disease was imported on chestnut seedlings brought to the adjacent New York Botanical Garden by a plant collector (Schumann 1991). The new disease spread rapidly, helped by birds and insects. Areas in Long Island, Connecticut, Massachusetts, New Jersey, and Pennsylvania were seriously affected by 1908 (Roane et al. 1986). There were sightings of the disease in Delaware, Maryland, and Virginia as well. By 1914 the disease had progressed as far south as northern Georgia. Within 50 years of its first observation, it had extended throughout the range of chestnut (Sinclair et al. 1987). The resulting widespread chestnut mortality profoundly altered the structure of the eastern forests.

On close examination the dogwood anthracnose fungus interacts with its host in a different manner than the chestnut blight fungus, so that our experience with chestnut blight cannot be used to predict the long-term effects of dogwood anthracnose on dogwood. Chestnut blight has no foliar phase, and the pathogen requires the assistance of wounds (made by insects, primarily) to gain entry to the cambium of the tree. Leaf infections are not the precursor to cankers as they are in the case of dogwood anthracnose. *Cryphonectria parasitica* causes cankers which kill the bark and vascular cambium of the chestnut, although it may also be an occasional parasite or common saprophyte on shagbark hickory, red maple, staghorn sumac, and certain oaks (Sinclair et al. 1987). *Discula destructiva* is at this time known to survive only on three species of dogwood. Another important difference is that chestnut blight cankers continue to enlarge year after year once they have been initiated, so that the branch or trunk is eventually girdled and killed (Sinclair et al. 1987). Rather than causing a girdling canker, dogwood anthracnose produces an annual canker that enlarges during one season only.

Another tree disease with a high public profile that is caused by an exotic pathogen is Dutch elm disease, which has all but eliminated ornamental plantings of American elms lining our city streets and college campuses. Dutch elm disease is believed to have originated in Asia, since Japanese and Chinese elms are resistant to the disease. It is known to have been introduced at least three times before 1930, to Ohio, New York City, and Quebec, on elm veneer logs imported from Europe (Sinclair et al. 1987). In only 50 years, the disease spread to 42 states and 6 Canadian provinces (Sinclair et al. 1987). An estimated 40 million elms have been lost to Dutch elm disease in the U.S. (Schumann 1991).

The Dutch elm disease, which is caused by the fungus *Ophiostoma ulmi* (Buisman) Nannf., has little biological similarity to dogwood anthracnose. Dutch elm disease requires elm bark beetle vectors to convey it from elm to elm. To our knowledge, there is no specialized vector for dogwood anthracnose. Dutch elm disease is also distinct in that the pathogen causes a systemic infection of the xylem, while dogwood anthracnose infections are more localized.

Dogwood anthracnose impacts trees' leaves, thus interfering with photosynthesis and plant energy reserves. Trees have the potential to withstand some leaf loss, but successive years of leaf attack may lead, directly or indirectly, to mortality. Dogwood anthracnose is more sensitive to environmental conditions than either chestnut blight or Dutch elm disease. Protracted rainy weather in the spring increases disease severity on the foliage, which can lead to extensive twig blighting and cankering during the following dormant season.

## A CANDIDATE FOR INTRODUCTION OF THE DISEASE

If dogwood anthracnose is an exotic invader, then one likely conduit for its introduction is the Oriental large-bracted dogwood, *Cornus kousa.* Since

FIGURE 6.—Epicormic branches are frequently killed back to their base by dogwood anthracnose.

1875, Kousa dogwood has been imported to this country, where it is prized for its ornamental qualities (Dirr 1983). *Cornus kousa* blooms approximately one month after *C. florida* and provides a showy display of pointed bracts against a green background of fully-expanded leaves. *Cornus kousa* may show leaf and twig symptoms of infection by *Discula destructiva* but only under environmental conditions which are unusually favorable for disease development. For example, gardeners employing overhead irrigation that directly impacts *C. kousa* foliage have reported blighting and dieback (Holmes and Hibben 1989).

Under normal landscape conditions with less intensive irrigation practices, *C. kousa* growing beside severely diseased *C. florida* has shown only minor leaf and bract spotting, if any. Although the introduction of a disease-causing organism on a less-susceptible host is a plausible mechanism for the introduction of a new disease, there is no hard evidence that *C. kousa* is responsible in this case. The time delay is also hard to explain: the fungus has not been reported from *C. kousa* or from adjacent flowering dogwood in the century-plus since the exotic tree's introduction.

## THE CURRENT STATUS OF DOGWOOD ANTHRACNOSE DISEASE

Dogwood anthracnose caused by *Discula destructiva* is now fully established on *C. nuttallii* in the northwestern U.S. and southwestern Canada and on *C. florida* in much of the northern and high-elevation southern portions of its range. Efforts at eradication would be futile at this point. Skeletal standing dead dogwoods are still evident in the woodlands, reminders of the original wave of dogwood anthracnose disease in the late 1970s. Some of these dead stems have toppled and are beginning to decay. Many thousands of dogwood trees have disappeared from the eastern deciduous forest. The impact of this loss on wildlife populations has not yet been assessed, but it is expected to be serious.

In the northeastern U.S., where dogwood anthracnose has been present for more than 12 years, we see signs of dogwood survival in the forest. On some surviving trees, probably those situated on the most favorable sites, cankers are closing in and disappearing as successive seasons of annual ring deposition wall in the area cankered by *D. destructiva.* A few seedlings are appearing in sunny clearings. Dogwoods along roadsides, where they receive maximum sun exposure and better air circulation, are still providing scenic beauty along some of the parkways in the Northeast. Well-maintained trees in home landscape situations have also survived, particularly when they are in less-shaded locations and have been protected from drought stress. Mortality does not invariably follow dogwood anthracnose infection, but the disease is capable of killing trees in shaded areas and/or those weakened by drought stress, given favorable environmental conditions.

The current feeling of those who have studied this disease since its first appearance is that dogwood anthracnose in the future will fluctuate in its impact according to rainfall patterns in early spring. Extended periods of fog with drizzling rain are especially conducive to infection during leaf expansion. We still have much to learn about this new disease. Fortunately, there are now a number of scientists who are attempting to learn more about dogwood anthracnose, and who are undertaking research with the practical objective of preserving dogwoods in our forests and managed landscapes. University, state, Park Service, and Forest Service employees are working together and exchanging information towards the common goal of saving our dogwoods.

Exciting hybrids of *Cornus kousa* X *C. florida* recently introduced by Dr. Elwin R. Orton at Rutgers University also offer a hope of ornamental dogwoods with reduced disease susceptibility (Orton 1990). In one study that tested *C. florida* grown from seed collected in different geographic areas (Santamour et al. 1989), there were no indications of strong geographic variation in dogwood susceptibility, but genetic resistance within the population remains a possibility for future selection or breeding efforts. Seed is now being gathered from all over the range of the dogwood for resistance screening trials, and surviving, possibly resistant, trees are being sought in the forests where dogwood anthracnose has been present for over a decade.

It is not yet clear whether dogwood anthracnose represents an exotic invader or an altered native host or parasite. Additional research is necessary to allow us to learn more about the disease, so that we can discover how to minimize its impact in both natural stands and ornamental plantings of our ecologically and socially important native dogwoods.

## REFERENCES


Anderson, R. L. 1991. Pages 1–5 *in* Results of the 1990 dogwood anthracnose impact assessment and pilot test in the southeastern United States. Compiler, R. L. Anderson. USDA Forest Service, Southern Region, Protection Report R8–PR 20.

Byther, R. S. and R. M. Davidson, Jr. 1979. Dogwood anthracnose. Ornamentals Northwest Newsletter 3:20–21.

Daughtrey, M. L. and C. R. Hibben. 1983. Lower branch dieback, a new disease of northeastern dogwoods. Phytopathology 73:365. (Abstract).

Daughtrey, M. L., C. R. Hibben and G. W. Hudler. 1988. Cause and control of dogwood anthracnose in northeastern United States. J. Arboric. 14:159–64.

DeGraaf, R. M. and G. M. Witman (eds.). 1979. Trees, shrubs and vines for attracting birds, a manual for the Northeast. Univ. of Mass. Press, Amherst.

Dirr, M. A. 1983. Manual of woody landscape plants. Stipes Publishing Co. Champaign, Ill.

Gleason, H. A. and A. Cronquist. 1963. Manual of vascular plants of northeastern United States and adjacent Canada. Van Nostrand, New York, N.Y.

Hibben, C. R. and M. L. Daughtrey. 1988. Dogwood anthracnose in northeastern United States. Pl. Dis. 72:199–203.

Holmes, F. W. and C. R. Hibben. 1989. Field evidence confirms *Cornus kousa* dogwood's resistance to anthracnose. J. Arboric. 15:28–29.

Krasny, M. E. 1991. Wildlife in today's landscape. Cornell Coop. Exten. Publ. L-5–20.

Merkel, H. W. 1906. A deadly fungus on the American chestnut. N.Y. Zool. Soc. Amer. Rep. 10:97–103.

Mielke, M. E. and K. Langdon. 1986. Dogwood anthracnose threatens Catoctin Mt. Park. Park Sci. Winter: 6–8.

Orton, E. R., Jr. 1990. *In* Eastern region new plant forum. Proc. Intl. Plant Prop. Soc. 40:635–36.

Pirone, P. P. 1980. Parasitic fungus affects region's dogwood. The New York Times, New York. Feb. 24, p. 34, 37.

Redlin, S. C. 1991. *Discula destructiva sp. nov.* cause of dogwood anthracnose. Mycologia 83(5):633–42.

Roane, M. K., G. J. Griffin and J. R. Elkins. 1986. Chestnut blight, other Endothia diseases, and the genus *Endothia.* APS Monograph, APS Press, St. Paul, Minn.

Roper, L. A. 1970. Some aspects of the synecology of *Cornus nuttallii* in northern Idaho. M.S. Thesis, Univ. Idaho, Moscow.

Salogga, D. S. 1982. Occurrence, symptoms and probable cause, *Discula* species, of *Cornus* leaf anthracnose. M.S. Thesis. Univ. Washington, Seattle.

Salogga, D. and J. F. Ammirati. 1983. *Discula* species associated with anthracnose of dogwood in the Pacific Northwest. Pl. Dis. 67:1290.

Santamour, F., Jr., A.J. McArdle and P. V. Strider. 1989. Susceptibility of flowering dogwood of various provenances to dogwood anthracnose. Pl. Dis. 73:590–91.

Schneeberger, N. F. and W. Jackson. 1989. Impact of dogwood anthracnose on flowering dogwood at Catoctin Mountain Park. Pl. Diagnostician's Quart. 10:30–43.

Schumann, G. L. 1991. Plant diseases: Their biology and social impact. APS Press, St. Paul, Minn.

Sinclair, W. A., H. H. Lyon and W. T. Johnson. 1987. Diseases of trees and shrubs. Cornell Univ. Press, Ithaca, N.Y.

Sudworth, B. 1967. Forest trees of the Pacific slope. Dover Publications, Inc. New York, N.Y.

Van Sickle, G. A. and C.S. Wood. 1989. Pacific and Yukon Region. Page 86 *in* Forest insect and disease conditions in Canada, 1988. Compiler, B. H. Moody. Forest Insect and Disease Survey, Forestry Canada, Ottawa, Ont.

# Biological Pollution Through Fish Introductions

Walter R. Courtenay, Jr.[1]

## INTRODUCTION

> "When a species is introduced into an area where it has not lived before, it is almost impossible to foretell the consequences, although it is quite probable that it will either succeed gloriously or eventually fail entirely." (Graham 1944).

This quotation tells only part of a complex story, one that all too often assumes success means beneficial to human interests. Its author was correct about "consequences" in more ways than he may have known.

The human propensity to move plants and animals beyond native ranges goes far back into our prehistoric past. This activity probably began about the same time hunter-gatherer groups found that certain plants could be cultivated and some animals domesticated. It was development of trade with neighboring groups, as well as migrations by early humans and their agricultural descendants, that initiated transfers of biotas.

Cultivars and many domesticated animals have rarely become invasive beyond the extent intended. Nevertheless, humans have not generally distinguished between moving such organisms and introducing stocks of feral species. Many of the latter have become invasive, some becoming serious pests to our interests and threats to natural habitats and native species (Elton 1958; Laycock 1966; Courtenay 1979; Courtenay and Robins 1989). The economic and natural resource costs of such mistaken introductions continue to grow, although most were not contemplated prior to introductions. Similarly, world lists of introduced organisms continue to swell, demonstrating little if any forethought by introducers as to the consequences that impact humans and the web of nature we depend upon.

Much of the information on international transfers of invasive fishes was summarized by Welcomme (1981, 1988). There have been numerous listings produced in recent decades, too numerous to list here, of fishes and other aquatic species introduced into and within various nations. The purpose of this contribution is to review why invasive fishes were introduced, in many instances how those introductions occurred, and to indicate some consequences. Because what has happened in North America regarding invasive fishes has been or is being repeated in many other parts of the world, I concentrate on that continent.

## PERIODS OF INTERNATIONAL INTRODUCTIONS

Aquaculture dates back to prehistoric times in eastern Asia and perhaps to the period of the late Roman Empire in Europe, when common carp (*Cyprinus carpio*) were introduced into central and western Europe for culture in ponds as a food fish (Misik 1958; Balon 1974). Goldfish (*Carassius auratus*) were probably brought to Europe in the same historical period, as an ornamental species for

1. Department of Biological Sciences, Florida Atlantic University, Boca Raton FL 33431–0991

pond and indoor culture. From the beginning, fish culture has been accompanied by escapes, and sometimes intentional introductions, into natural waters; this occurred with both carp and goldfish. Welcomme (1981) cited the period of the European Middle Ages as the first phase, or period, of international transfers of invasive fishes in both Europe and Asia. The number of species introduced during this time was miniscule in comparison with later periods of transfers. The extent of early introductions, however, may have been much greater than we suspect today, particularly in Asia.

A second period of transfers began with colonization of several parts of the world, mostly by European immigrants, and with the early development of international trade. Welcomme (1981) suggested this period of fish transfers began around the middle of the 19th century (although it probably originated in the late 17th century) and ended as World War II began. Numerous introductions were made, particularly during the latter half of the 19th and early 20th centuries, from Europe to the Western Hemisphere, to Africa (Bruton and van As 1986) and Australia (McKay 1984, 1989; Kailola 1990), and to parts of Asia (Di Sylva 1989). Much of this activity appears to have stemmed from a desire by immigrants to have fishes from waters with which they were familiar in their homelands, or what Welcomme (1981) termed introductions for reasons of "sentiment." With exploration of the interior of new areas, especially North America, native fishes were discovered that had excellent sporting qualities; these were shipped eastward for introduction, and many, including species from inland waters in the east, were exported to other parts of the world.

During the late 1800s and early 1900s, "acclimatization societies" were created in many places of the world, particularly in the Western Hemisphere, for the sole purpose of introducing intercontinental species. It was a time of "introduction madness," which occurred before the advent of modern, rapid, international transportation. It was a period, however, when invasive biota was desired and not recognized as potentially damaging to native biota, the latter often being considered inferior and, therefore, expendable or at least alterable. Unfortunately, some of that same thinking persists today in public, political, and managerial circles.

In 1872 the world's first national park, Yellowstone National Park, was established in the western United States of America (U.S.). This was a beginning of U.S. conservation efforts. Ironically, one of the first activities to be conducted in Yellowstone was exploration to assess its native fish resources and to recommend what nonnative fishes should be introduced for purposes of sport fishing (Jordan 1891).

In the years immediately preceding and during World War II, several invasive fish species were moved into ecosystems new to them. The Japanese introduced tilapia, primarily *Oreochromis mossambicus,* into some Pacific Rim countries and island nations. The U.S. Army Air Corps (predecessor of the U.S. Air Force) aided other U.S. and allied military units during the war in spreading mosquitofishes (*Gambusia* spp.) into many foreign waters for the purpose of reducing or eliminating a disease vector, mosquitos. That mosquitofishes failed as a biological control for mosquitos in most areas where they were introduced and became serious pests, to the point of extirpating native fishes and threatening others, is now evident (Courtenay and Meffe 1989). Introductions of mosquitofishes, accomplished by many governments and agencies, did not end after World War II but actually increased over the following two decades.

The third major period of introductions began after World War II, mostly during the 1950s and into the early 1970s, following the advent of commercial, transcontinental jet cargo aircraft, and continues today. This period was initially characterized by explosive growth of the aquarium, or ornamental, fish industry and hobby (Courtenay and Stauffer 1991). Species previously known to serious hobbyists only through drawings or photographs in scientific and some popular literature could now be purchased at many pet shops around the world. Culture of aquarium fishes grew dramatically to the point that fishes native to parts of Africa and Central and South America, for example, were being bred and reared in large quantities in widely separated places such as Singapore and southern Florida. Many fishes escaped or were released from culture; others were released by hobbyists, thinking this was the humane thing to do with unwanted aquarium fishes. These factors greatly and rapidly added to the comparatively low number of aquarium fish species previously established in the wild.

This third period has been complicated with both sanctioned and illegal introductions of fishes of foreign origin for purposes of sport fishing and biological control, mostly of aquatic pest plants, many of which are themselves of foreign origin, released from the aquarium plant industry and by aquarium fish hobbyists. Adding to this complexity is a developing interest by modern-day promoters, government and private, in aquaculture, where there is great interest in "something different," namely foreign species; this is almost a repeat of the origin of aquaculture in Europe before the Middle Ages but on a much larger scale and usually for different reasons.

As noted previously, where there is culture there will be releases, and many aquaculture species will become invasive aquatic species. Recently there has been, at least in the U.S., a renewed interest by some state government agencies responsible for managing sport fishing and by private bait dealers in utilizing intercontinental introductions, further proof that history tends to repeat itself.

## OVERVIEW OF INTERNATIONAL TRANSFERS OF INVASIVE FISHES

Welcomme (1981) listed about 170 species of fishes that had been transferred on an international basis. He reported on sources and targets of introductions and, where known, dates of these transfers, the status of these fishes (whether or not they had become established), and reasons for releases. In some instances he included comments from the United Nations Food and Agriculture Organization (FAO) regional bodies or governments providing the records as to their opinion on the introductions. He noted that his list was incomplete and did not include intranational transfers or numerous introductions of aquarium species not released into the wild. His updated listing (Welcomme 1988) cited 237 species of inland fishes as having been introduced beyond their historical ranges into 140 countries, the result of 1,354 known introductions. His new listing lacks information on the extent of invasion, natural or otherwise, within nations and does not include marine species or translocations within nations.

Welcomme's (1988) listing included species that became established and those that failed. Fishes have been most frequently introduced internationally for aquaculture, secondly for sport, and thirdly for improvement of wild stocks. Accidental introductions and intentional stockings for biological control placed fourth and fifth, respectively. In recent decades, introductions for improvement of wild stocks have overtaken releases made for sport purposes. Species most widely introduced were common carp, Mozambique tilapia, and rainbow trout (*Oncorhynchus mykiss*).

Translocations of fishes within many nations have been significant. For example, Courtenay and Taylor (1984) indicated 168 species of fishes as having been transferred beyond their native ranges within the contiguous United States. Courtenay (1991) noted 158 species known to have been moved and another 70 or more species that have been introduced (Table 1).

TABLE 1. Exotic fishes established in the United States of America.

| FAMILY & SPECIES | COMMON NAME[1] | STATE[2] | YEAR OF RELEASE |
|---|---|---|---|
| CLUPEIDEA: | | | |
| *Herklotsichthys quadrimaculatus* | GOLDSPOT HERRING | HI | 1958 |
| *Sardinella marquesensis* | MARQUESAN SARDINE | HI | 1955 |
| SALMONIDAE: | | | |
| *Salmo trutta* | BROWN TROUT | all except AL, FL, HI, KS, LA, MS, OK, TX | 1884 |
| OSMERIDAE: | | | |
| *Hypomesus nipponensis* | WAKASAGI | CA | 1959 |
| CYPRINIDAE: | | | |
| *Brachydanio rerius* | ZEBRA DANIO | WY | >1984 |
| *Carassius auratus* | GOLDFISH | most except AK and FL | 1680s |
| *Ctenopharyngodon idella* | GRASS CARP | AR, KY, IL, LA, MO, MS, TN, TX | 1960s |
| *Cyprinus carpio* | COMMON CARP | all except AK | 1831 |
| *Hypophthalmichthys nobilis* | BIGHEAD CARP | MO | >1986 |
| *Leuciscus idus* | IDE | ME | >1877 |
| *Puntius semifasciolatus* | GREEN BARB | HI | 1940 |
| *Rhodeus sericeus* | BITTERLING | NY | 1920s |
| *Scardineus erythropthalmus* | RUDD | KS, ME, NB, NY | 1890s |
| *Tinca tinca* | TENCH | CA, CO, CT, ID, NM? | >1877 |
| COBITIDAE: | | | |
| *Misgurnus anguillicaudatus* | ORIENTAL WEATHERFISH | CA, FL, HI, ID, MI, OR | 1930s |
| CLARIIDAE: | | | |
| *Clarias batrachus* | WALKING CATFISH | FL | 1960s |
| *Clarias fuscus* | WHITESPOTTED CLARIAS | HI | <1900 |
| LORICARIIDAE: | | | |
| *Ancistrus* sp. | | HI | 1980s |
| *Hypostomus* sp.[3] | SUCKERMOUTH CATFISH | FL | <1958 |
| *Hypostomus* sp.[3] | SUCKERMOUTH CATFISH | NV | 1960s |
| *Hypostomus* sp.[3] | SUCKERMOUTH CATFISH | TX | 1960s |
| *Pterygoplichthys multiradiatus* | SAILFIN CATFISH | FL, HI | 1960s |
| BELONIDAE: | | | |
| *Strongylura kreffti* | LONG TOM | HI | 1988 |
| CYPRINODONTIDAE: | | | |
| *Rivulus harti* | GIANT RIVULUS | CA | 1960s |
| POECILIIDAE: | | | |
| *Belonesox belizanus* | PIKE KILLIFISH | FL | 1957 |
| *Poecilia mexicana* | SHORTFIN MOLLY | CA, ID, MT, NV, TX; HI | 1960s; 1922 |
| *Poecilia reticulata* | GUPPY | AZ, CA?, FL, HI, ID, NV, TX, WY | |
| *Poecilia sphenops* | | HI | <1950s |
| *Poecilia vittata* | CUBAN LIMIA | HI | <1950s |

TABLE 1. (cont.)

| FAMILY & SPECIES | COMMON NAME[1] | STATE[2] | YEAR OF RELEASE |
|---|---|---|---|
| *Poeciliopsis gracilis* | PORTHOLE LIVERBEARER | CA | <1965 |
| *Xiphophorus helleri* | GREEN SWORDTAIL | FL, ID, MT, NV[4], WY; HI | <1962; 1922 |
| *Xiphophorus maculatus* | SOUTHERN PLATYFISH | FL, NV[4]; HI | 1960s; 1922 |
| *Xiphophorus variatus* | VARIABLE PLATYFISH | FL, HI, MT | 1960s |
| Synbranchidae: | | | |
| *Monopterus albus* | SWAMP EEL | HI | <1900 |
| Serranidae: | | | |
| *Epinephelus argus* | BLUESPOTTED GROUPER | HI | 1956 |
| Percidae: | | | |
| *Gymnocephalus cernuus* | RUFFE | MN, WI | 1986? |
| Lutjanidae: | | | |
| *Lutjanus fulvus* | BLACKTAIL SNAPPER | HI | 1955? |
| *Lutjanus kasmira* | BLUESTRIPED SNAPPER | HI | 1955 |
| Sciaenidae: | | | |
| *Bairdiella icistia* | BAIRDIELLA | CA | 1950 |
| *Cynoscion xanthulus* | ORANGEMOUTH CONVINA | CA | 1950 |
| Cichlidae: | | | |
| *Astronotus ocellatus* | OSCAR | FL; HI | 1958; 1952 |
| *Cichla ocellaria* | PEACOCK CICHLID | FL; HI | 1986; 1961 |
| *Cichlasoma bimaculatum* | BLACK ACARA | FL | 1950s |
| *Cichlasoma citrinellum* | MIDAS CICHLID | FL | <1981 |
| *Cichlasoma managuense* | GREEN GUAPOTE | UT | 1980s |
| *Cichlasoma meeki* | FIREMOUTH CICHLID | FL; HI | 1970s; 1940 |
| *Cichlasoma nigrofasciatum* | CONVICT CICHLID | ID, NV, WY | 1960s |
| *Cichlasoma octofasciatum* | JACK DEMPSEY | FL | 1960s |
| *Cichlasoma spilurum* | BLUE-EYED CICHLID | HI | 1984 |
| *Cichlasoma urophthalmus* | MAYAN CICHLID | FL | 1980s |
| *Geophagus surinamensis* | REDSTRIPED EARTHEATER | FL | <1982 |
| *Hemichromis letourneauxi* | AFRICAN JEWELFISH | FL | 1960s |
| *Oreochromis aureus* | BLUE TILAPIA | AZ, CA, FL, GA?, NC, OK, TX | 1960s |
| *Oreochromis macrochir* | LONGFIN TILAPIA | HI | 1958 |
| *Oreochromis mossambicus* | MOZAMBIQUE TILAPIA | AZ, CA, FL, TX; HI | 1960s; 1951 |
| *Oreochromis urolepis hornorum* | WAMI TILAPIA | CA | 1970s |
| *Pelviachromis pulcher* | RAINBOW KRIB | HI | 1984 |
| *Sarotherodon melanotheron* | BLACKCHIN TILAPIA | FL; HI | 1950s; >1970 |
| *Tilapia mariae* | SPOTTED TILAPIA | FL, NV | <1974 |
| *Tilapia rendalli* | REDBREAST TILAPIA | HI | 1957 |
| *Tilapia zilli* | REDBELLY TILAPIA | AZ, CA, HI, NC, TX | 1960s |
| Mugilidae: | | | |
| *Valamugil engeli* | KANDA | HI | 1955 |
| Channidae: | | | |
| *Channa striata* | CHEVRON SNAKEHEAD | HI | <1900 |
| Blenniidae: | | | |
| *Parablennius thysanius* | | HI | 1971 |

TABLE 1. (cont.)

| FAMILY & SPECIES | COMMON NAME[1] | STATE[2] | YEAR OF RELEASE |
|---|---|---|---|
| GOBIIDAE: | | | |
| *Acanthogobius flaimanus* | YELLOWFIN GOBY | CA | <1963 |
| *Neogobius melanostomus* | ROUNDNOSE GOBY | MI | 1988 |
| *Proterorhinus marmoratus* | TUBENOSE GOBY | MI | 1988 |
| *Tridentiger trigonocephalus* | CHAMELEON GOBY | CA | <1965 |
| ANABANTIDAE: | | | |
| *Trichopsis vittata* | CROAKING GOURAMI | FL | 1970s |

[1] After Courtenay et al. (1991)
[2] State abbreviations follow system of U.S. Postal System.
[3] One unidentified but distinct morphological species in each state.
[4] Hybrid between *Xiphophorus helleri* and *X. maculatus*.

Most Hawaiian data from Maciolek (1984), Randall (1987), and Springer (1991).

Man's introduction of fishes beyond natural barriers to dispersal has been so pervasive that original geographic distributions of several fishes may never be ascertained. Within the U.S. and other nations, fishes are suspected as having been moved into adjacent drainages, many as excess bait or unintentionally introduced with sport species, and others released intentionally. Undocumented transfers of tilapias within Africa and the presence of some Asian inland fishes in offshore Western Pacific island nations, for example, present similar problems to zoogeographers. If many of these fishes were introduced invasives and the approximate dates of their introduction and the introducers are unknown, it follows that it would be impossible to determine what their impacts on resident species may have been; additionally it could not be determined what resident fishes are no longer present because they were impacted or eliminated by the invasives.

Developed, or industrialized, nations have spent a lot of time, effort, and money in modifying and destroying natural ecosystems. Efforts to conserve relatively undisturbed areas were advanced later, as have governmental policies that ironically allow certain agencies to continue to modify habitats while other agencies of the same government are trying to save them. Developing nations should learn from these mistakes, but their desire to rise to industrial and economic prominence is usually aided by governments and international agencies of developed nations, almost assuring that the same, perhaps worse, environmental mistakes will occur. Introductions of invasive aquatic animals are a large part of these mistakes. The inherent biological qualities and resources of ecosystems targeted for introductions are rarely understood and even more rarely considered prior to making introductions. Because introduced species can reproduce and invade, they must be considered as potentially permanent biological pollutants.

## THE NORTH AMERICAN EXPERIENCE

Some 200 years following the first introduction (in the late 1600s) of an invasive fish, the goldfish, into the U.S., modified railcars were being towed behind steam locomotives, carrying eggs, larvae, and juveniles of eastern and

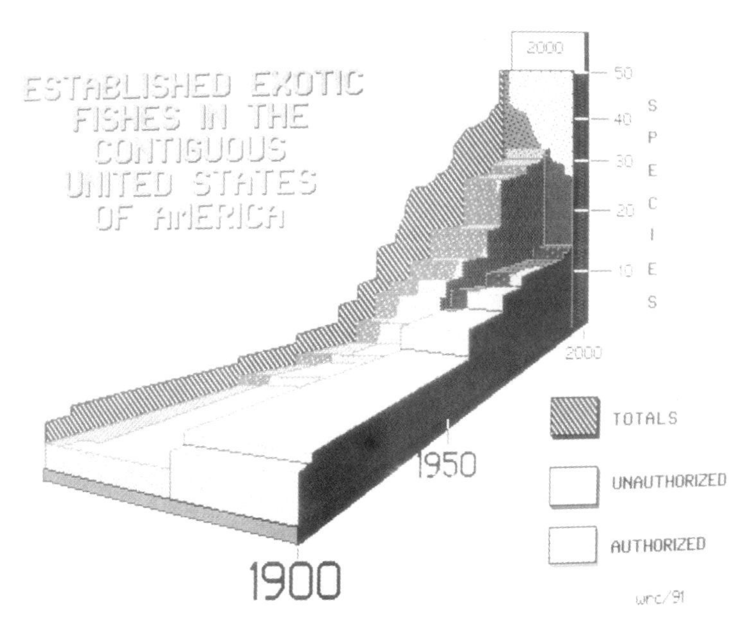


FIGURE 1.—Rates of establishment of exotic fishes in waters of the contiguous United States.

foreign fishes to be introduced to the American midlands and west (Fig. 1). On the return trip these trains hauled early life history stages of western fishes eastward for introductions there and occasionally for foreign export. At places where steam engines took on water, often at water tanks adjacent to streams or rivers, personnel in the "fish cars" sometimes found themselves on bridges, occasionally entertaining themselves by partially or totally dumping contents of fish containers into the waters below, probably thinking they were doing good things. Needless to say these events were not "officially" sanctioned or recorded (except in the notes of personnel in the fish cars, and localities of such releases were often omitted or inaccurate) but were events reported later. The U.S. Commissioner of Fisheries, Spencer F. Baird, was actively promoting introductions of fishes. Senators and congressmen from various states and representatives from territories where states were yet to be created clamored in the nation's capitol for stocks of the new "wonder fish" from Europe, common carp; many also wanted the Lochleven and/or the German trout, varieties of brown trout (*Salmo trutta*). In addition to carp, two other European cyprinids, ide (*Leuciscus idus*) and tench (*Tinca tinca*), were being cultured by the U.S. Fish Commission in ponds on the banks of the Potomac River in Washington, D.C.; a flood in 1889 washed these fishes into the Potomac (Baird 1893) where they became established. The ide and tench were introduced elsewhere, apparently with carp, into other U.S. waters. This was but a small part of what was taking place in North America during the second period of introductions.

Geographic North America encompasses three political entities: the U.S., Canada and Mexico. Rationale (or lack thereof, in some instances) for releases of invasive fishes into the waters of these nations has followed a pattern that can be categorized; the goals differed in some instances, but the pattern shown has been repeated and is being repeated internationally.

## INTRODUCTIONS FOR FOOD PURPOSES

Some have suggested that one of the original intents for introducing common carp and brown trout to North America was to provide new food resources (DeKay 1842; Baird 1879; Mather 1889). This may have been partly true, but it is more likely that a greater intent was introduction of sport species.

Although the U.S. Fish Commission accepted credit for bringing these fishes from Europe to North America, the first importation and introduction of common carp were made by an individual citizen in New York State in 1831 who apparently imported stock from Europe (DeKay 1842). In 1872 another citizen, in California's Sonoma Valley, made the second introduction with an importation from China (Moyle 1976a). Also in 1872 or 1873, Mexico obtained stocks of carp from Haiti and introduced them (Contreras and Escalante 1984).

After a few initial failures, the U.S. Fish Commission successfully imported 338 common carp from Germany in 1873 and distributed their progeny for culture and introductions until 1896 (Baird 1879; Laycock 1966; Scott and Crossman 1973). It remains unclear if the U.S. Fish Commission imported ide and tench separately from or together with carp from Europe, but later appearances of one or the other of these species in areas that had been stocked with carp suggest that shipments from Washington, D.C. were of intentionally or accidentally mixed foreign cyprinids.

The U.S. shipped carp to Canada in 1880, perhaps to both New Brunswick and Ontario; failure of a dam in 1896 led to release in Ontario, but there is evidence that carp were present in Ontario from an earlier introduction by 1891 (Crossman 1984, 1991). Other early introductions of carp into Canada were made by persons bringing stocks from adjacent U.S. localities and by migration of U.S.-introduced stocks into Canadian waters. Crossman (1984) stated that although carp were kept at a Canadian federal hatchery, there are no records that the Canadian government was directly responsible for any introductions of this species; private introductions, however, were allowed and assisted by the Canadian government until 1893.

In the U.S. and Canada, there are commercial fisheries for carp although this species is not a major factor in landings or value. There is also a limited sport fishery in both nations, which is now being encouraged by some (Sullivan 1988). Nevertheless, damage done by carp to aquatic vegetation and through competition with native fishes has been significant and reported by numerous authorities in many parts of the world for at least the past 50 years.

Introduction of common carp in Mexico was intended for food purposes, as has been the case with most purposeful fish releases there. In contrast to purposeful introductions of fishes in the U.S. and Canada, at least 30 species of fishes, most intended to be invasive, have been released into Mexican waters to supply new sources of protein (Contreras and Escalante 1984; Courtenay and Kohler 1986); some of these were also released to provide forage prey for other fishes (most of which are introduced) or for sport purposes. Within the past decade, tilapias, particularly blue tilapia (*Oreochromis aureus*), have been and continue to be widely introduced within Mexico by federal, state, and local governments. Many well-intentioned government officials, lacking knowledge of the biological consequences, are dedicated to spreading tilapias throughout the inland waters of Mexico. Several areas where these tilapias have been introduced contain endemic and often highly endangered species. The ability of tilapias to displace or replace native fishes is well-known, but mechanisms for this phenomenon are not fully understood. Although there have been few investigations of impacts of introduced fishes on native species in Mexico, there have been instances of hybridization and genetic swamping and severe predation (Contreras and Escalante 1984).

This scenario is being repeated in many developing nations with protein-deficient societies, maybe to the betterment of their people but not without substantial present or future losses to native species and perhaps habitat. Invariably in such situations, someone will suggest the "perfect fish" (or other organism) for introduction to solve one or another problem, but government officials often ignore the warning *caveat emptor*! It must be recognized that financial benefits will accrue to suppliers (and perhaps to some actually involved in the introduction) but not necessarily to the nation sanctioning and receiving the introduction and that there are no guarantees or reparations made by the suppliers or introducers. Developing nations and agencies within and without those nations that promote introductions must exercise extreme caution and concern for native aquatic resources. Creation and adoption of strong, detailed policies on introductions are required within all international agencies (such as FAO) and agencies of governments that assist other governments (e.g., U.S. Agency for International Development). Goals sought through introductions can be achieved and environmental damage minimized if done in a manner that recognizes and is designed to avoid the problems that are now being created. Tilapias or other introduced fishes in every body of water that will support them is not the answer, but they do carry a guarantee, namely the probability of becoming an environmental problem.

Brown trout were first introduced into the U.S. in 1883 by the U.S. Fish Commission. Canada separately introduced this species in 1884 or 1886 from Lochleven stock imported from Scotland; stock imported from Germany was released in 1892 (Crossman 1984, 1991). There are no records of this species having been introduced in Mexico, although both rainbow trout and brook trout (*Salvelinus fontinalis*) were introduced for food and sport purposes; only rainbow trout became established (Contreras and Escalante 1984). Of numerous fish introductions made in North America, the brown trout is generally regarded as an exemplary success. The many introductions of this species have not been without negative impacts. California has been attempting to eradicate the species from the Little Kern River, where it poses a demonstrated threat to native golden trout (*Oncorhynchus aguabonita*; E.P. Pister, pers. comm.). Similarly brown trout have been implicated in declines of native cutthroat trout (*O. clarki*) and the McCloud River population of Dolly Varden (*Salvelinus malma*) in California (Moyle, 1976a, b), and Gila trout (*O. gilae*) in New Mexico (J.N. Rinne, pers. comm.). Similarly there have been attempts to eradicate brown trout from Smoky Mountains National Park where they have negatively impacted native brook trout (Reiger 1981).

Some nonnative fishes were introduced in North America, at least in part, for commercial fishing purposes. Most occurred in Mexico, where aquaculture rarely implies pond or tank culture. A major example in the U.S., with "spillover" into Canada, was the stocking of the western Great Lakes with salmonids. Following decades of overfishing and poor management and substantial damage done to many native fishes by the invasive sea lamprey (*Petromyzon marinus*), some Pacific salmons (*Oncorhynchus* spp.) were introduced. Part of the rationale for these introductions was control (as forage) of the alewife (*Alosa pseudoharengus*), which had undergone a tremendous population explosion following the loss of most of their predators; the alewife may have been introduced into Lake Ontario prior to its later invasion of the western Great Lakes (Miller 1957), although Smith (1985) believes it occurred naturally in Lake Ontario. Some felt that these introductions might restore the Great Lakes to their former prominence in commercial freshwater fisheries, in addition to creating large sport fisheries. It was certainly not an attempt to restore the original fish fauna of the Great Lakes. These desires were tempered substantially when it was found that the introduced salmons were absorbing pollutants, particularly polychlorinated biphenyls (PCBs). Therefore, commercial fisheries, as anticipated, never materialized beyond local markets. What happened from these introductions, however, was development of a strong sport fishery, which may now be impaired. It appears that introduced salmons, perhaps coupled with some climatic changes, have overachieved desired control of alewives (Wells 1985; Eck and Brown 1988) and show little evidence of altering their diets to utilize the remaining native forage fish base. Native lake trout (*Salvelinus namaycush*) are reported to be making a recovery, and they do utilize the native forage base. Unfortunately, the concentration of organic toxins in lake trout has been measured at higher levels than in introduced Pacific salmonids (DeVault et al. 1985; Eshenroder, 1988). Adding insult to injury, lake trout being cultured for restocking the Great Lakes in several hatcheries became infected by a virus, termed "epizootic epitheliotropic virus disease" (EEVD), resulting in the purposeful destruction of 2.5 million lake trout in hatcheries in 1988 (Anonymous 1988). These events will doubtless result in conflicts between those who believe new introductions are needed and those who prefer to let nature take its course, with or without new releases of native lake trout but with man's help in reducing pollutant levels.

Two instances of introductions apparently made to create commercial fisheries occurred in Florida. The blue tilapia was introduced to west-central Florida in the early 1960s, initially for vegetation control purposes (Courtenay et al. 1974), was subsequently released elsewhere, and a commercial haul-seine fishery was permitted about a decade later. It has been suggested that the sudden appearance of populations of blue tilapia in other central Florida waters may have been caused by persons hoping to create their own local fisheries (Harris 1978). Similarly the blackchin tilapia (*Sarotherodon melanotheron*) escaped from aquarium fish culture into Tampa Bay in the late 1950s (Springer and Finucane 1963), where it became established and soon became part of the commercial fishery (Finucane and Rinckey 1964). About two decades later this tilapia appeared in an east coast estuary, the Banana River, immediately south of Cape Canaveral (Courtenay et al. 1984) where it is part of the commercial fishery there. There is no evidence of this species having been cultured on Florida's east coast. The only known aquarium culture, or aquarist release, of a tilapia in this area is the Mozambique tilapia, which also became established in the brackish waters of the Banana River (Courtenay et al. 1974). This suggests that the blackchin tilapia may have been a deliberate release to create a new commercial fishery on Florida's east coast.

## INTRODUCTIONS FOR SPORT PURPOSES

This category is complicated. It includes not only direct releases of potential sport fishes but also fishes introduced as forage for sport species and as released bait. Most introduced sport fishes are typically high-level predators, whereas a majority of forage and bait species are primary or secondary consumers. That all trophic levels of introduced fishes can become established, and often invasive, is of significant environmental concern.

Most introductions of foreign fishes in Canada, many in the U.S. and some in Mexico were made to establish new sport fisheries (Courtenay et al. 1984; Crossman 1984; Contreras and Escalante 1984) (Fig. 2). There has been only one intercontinental introduction of a forage fish into U.S. waters, an osmerid, the wakasagi (*Hypomesus nipponensis*), introduced into several California reservoirs (Wales 1962; Moyle 1976a).

There have been 48 records of introduced fishes in Canada; 12 were inter-

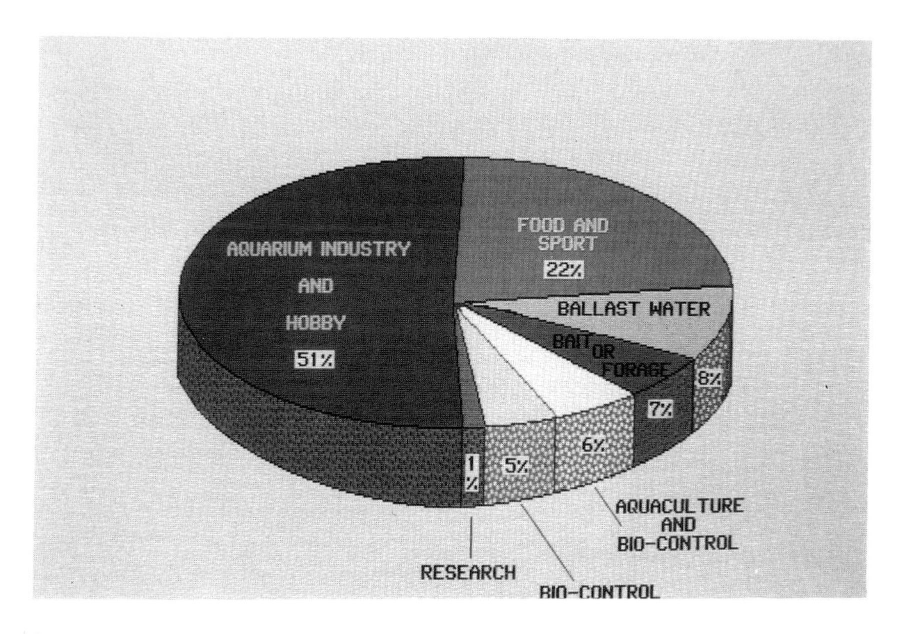


FIGURE 2.—Sources of exotic fishes established in waters of the United States.

continental transfers (with only 2 introduced for sport purposes: brown trout and common carp). The remainder were intracontinental introductions, most made for sport purposes (Crossman and McAllister 1986; Underhill 1986; McPhail and Lindsey 1986; Lindsey and McPhail 1986). Contreras and Escalante (1984) did not indicate that any fish had been introduced to Mexican waters solely for sport purposes; they listed six species as released for food (aquaculture) and sport purposes, another six for aquaculture and forage, one as an introduced forage species, one for aquaculture and biological control use, and seven as released bait fishes.

Transfers of native fishes beyond their natural ranges for, or resulting from, sport fishing activities have resulted in the majority of introductions of invasive aquatic animals in the United States. Courtenay and Taylor (1984) suggested 53 species as having been moved as game fishes and another 58 as releases of live bait (and some perhaps as forage) out of an estimated total of 168 intracontinental fishes that became established within the contiguous 48 states. Additionally, many invasive aquatic invertebrates (crayfishes, freshwater shrimps, the Asiatic clam [*Corbicula fluminea*] and others) apparently became established in new habitats as bait releases or as introduced forage.

Sport fishing has dominated introductions throughout North America, north of the Rio Grande, and, as noted above, has been a factor in fish releases into Mexico. At the beginning of this century, one generally could say that sport fish introductions were made only where there were human immigrants from Europe; this is no longer true, because what used to be a sport in a few countries is now an activity that has expanded throughout developed and into many developing nations. Moreover, there is presently a renewed interest in introducing new sport fishes from extracontinental sources into U.S. waters (Courtenay and Robins 1989), a trend that is likely to spread to other nations.

The advent of sport fishing in nations has been accompanied by establishment of governmental fishery agencies to support and enhance this activity.

Sport fishing generates income to many businesses, some of which exist only because of this resource. Developed and developing nations have experienced portions of rivers converted into reservoirs behind dams that may or may not supply electrical power but provide water to domestic, agricultural, and industrial users. Fishery resource managers are usually quick to stock reservoirs with species that flourish in lentic conditions, sometimes before but usually after the damming reduced or eliminated fishes that depend on flowing waters. Similarly fishery managers have stocked tailwaters of rivers or streams below dams with species that tolerate the temperature and perhaps the managed flow regimes. Within the U.S., virtually all introduced tailwater fishes were nonnative.

With increasing (and disturbing) frequency, fishery resource managers have not been able to properly manage these impounded or tailwaters because other managers, those who regulate water levels for the primary reasons the reservoirs were created, keep altering water levels and flow regimes and sometimes water quality. In some instances new dams are constructed upstream which significantly lower nutrient supplies to the fish forage base behind downstream dams. Searches are often instigated to find some species that can supposedly survive these altered conditions and fill perceived needs of the sport fishing public. Such efforts, however, are mostly undertaken for economic and political reasons, and rationalization often replaces recognition of actual problems. The result is frequently a proposal to introduce another nonnative species into a system with a history of such introductions without a full understanding of the system itself (Courtenay and Robins 1989). Rarely do such searches consider impacts, real or potential, within or especially beyond the target area of the proposed introduction. Instead they are focused on achieving short-term goals. Clearly the second period of introductions, the period of introduction madness, is still with us!

## INTRODUCTIONS FOR BIOLOGICAL CONTROL PURPOSES

Resource managers have often falsely assumed that many species considered to be pests must have specific predators that can be used to control them. Perhaps because of several successes in biological control achieved by introducing certain insects (Norris 1970; Drea 1993; Hight 1993), aquatic resource managers have assumed that introductions of fishes can achieve similar results. They fail to recognize, however, that many insects are oligophagous (consuming only a few dietary items) or monophagous (feeding on one item), whereas fishes and vertebrates in general are not (Courtenay and Robins 1975; Courtenay 1979). Additionally fishes are somewhat more complex than other vertebrates in that they change dietary preferences during their life histories and frequently remain opportunistic as adults. Resource managers ignore or are probably unaware of the fact that invasive, introduced, problem-creating species evolved in other ecosystems where they were kept in check by many, not one or a few, co-evolved components of their native ecosystems.

Where fishes were first employed in biological control is uncertain, but one of the first were mosquitofishes native to the southern U.S. Two species are recognized, *Gambusia affinis* and *G. holbrooki* (Wooten et al. 1988). The second species has typically been the one introduced for mosquito control purposes. Mosquitofishes have been introduced into all continents except Antarctica and have been disseminated widely beyond their native range within North America (Courtenay and Meffe 1989).

The ability of mosquitofishes to survive and become invasive beyond their native range has been dramatic. Equally dramatic has been their ability to create serious environmental problems, particularly to resident native fishes in receiving ecosystems, by preying on other small fishes, or the young stages of larger species. Their success as a control for mosquitos, however, has been minimal (Welcomme 1981; Courtenay and Meffe 1989); even in situations where they have been reported as beneficial, no descriptions of criteria or measures, beyond opinion, used to assess reported successes were provided (Welcomme 1981, 1988).

A fish widely used throughout much of the world for control of problem aquatic vegetation has been the grass carp (*Ctenopharyngodon idella*). Welcomme (1981) reported its introduction to at least 47 nations; he also noted that the species, in 1981, was reported as established in Mauritius, Rwanda, the Danube River basin, the Tisza River of Hungary, Mexico, and the United States. The U.S. may have been the only nation where natural reproduction of grass carp was considered a cause for concern (Lachner et al. 1970). Proponents for the introduction claimed that spawning requirements for grass carp were so specific as to preclude natural reproduction in U.S. waters. This display of ignorance and lack of understanding of both fish biology and U.S. river systems was a factor that led to the introduction and establishment of this species. It has been reproducing in the lower Mississippi River (Conner et al. 1980) and parts of the Atchafalaya River system in Louisiana since about 1975. Recently it was reported that young grass carp, picked up on intake screens of some Louisiana electrical power plants, amount to as much as 20 percent of the total fish biomass on the screens; it has also apparently become established in the Missouri River (J. D. Williams, pers. comm.) and the lower Ohio River (R. Wallus, pers. comm.).

Suppliers of grass carp—fish farmers in Arkansas, the first state to utilize the species for experimentation and stocking—shipped grass carp to many states, where they were often received illegally, undetected by law enforcement officials. The species was subsequently introduced into many waters in violation of state laws and regulations. Moreover, individual fish farmers from Arkansas have illegally trucked grass carp into nearby states, distributing them to farmers and others who wanted to stock outdoor ponds (F. Cross, pers. comm.).

Concern has been and remains that overcontrol of vegetation by grass carp that can reproduce amounts to no control of a biological control agent. Because aquatic vegetation provides important cover for early life history stages of most native fishes and both substrate and protection for their food supplies, overcontrol represents a distinct threat. Those who touted the use of grass carp (mostly plant pathologists, agronomists, and others, with little if any training in fishery science or ichthyology) fail to understand why concern was expressed over these introductions; they remain of the narrow opinion that if sport fish populations (usually native centrarchids) are not impacted within the few years after grass carp introduction, why should anyone be concerned with the loss of such fishes as cyprinids and cyprinodontids? This myopic view does not consider the long-term consequences of modifications of fish population dynamics as a result of introductions. These and other deleterious effects may not be manifested until decades following initial introduction and establishment (Courtenay 1979).

A technique has been devised within recent years involving thermal shocking of grass carp eggs during artificial fertilization with grass carp milt, resulting in triploid offspring (Thompson et al. 1987); another method uses pressure shocking (Cassani and Caton 1985). To completely assure that fish produced by either method are indeed triploid requires examining blood samples from each individual (Wattendorf 1986); this appears complicated, mistakenly, and is worth the effort and expense to assure triploidy and environmental protection. The technique of inducing triploidy does not guarantee sterility, but it assures that if triploids produce progeny, the likelihood of their survival is extremely low (Allen et al. 1987). Had this technique been developed prior to the first importation of grass carp into the U.S., concern over open-water introductions would have been reduced substantially.

Finally, other fishes used for control of undesirable aquatic vegetation in U.S. and some Mexican waters have involved tilapias. Studies conducted in pond situations with certain tilapias (e.g., *Oreochromis aureus, O. mossambicus, Tilapia zilli*) indicated that these fishes would consume aquatic vegetation; such would be expected of omnivores if all they had to consume was vegetation, but that was not stated in the study results. They were promoted as biological control species, and aquatic resource managers in several states accepted the promotion without question. Some of these species have undergone explosive invasions, particularly in the lower Colorado River of Arizona/California and in the lower Rio Grande and several powerplant reservoirs in Texas, to the exclusion of native and introduced species (Courtenay et al. 1984; Taylor et al. 1984; Noble and Germany 1986). In some areas of introduction, tilapias have been moderately successful in vegetation control; they crowded out or otherwise excluded other fishes, thus resulting in their "return" to the farm pond situation where all they had to consume was vegetation.

## INTRODUCTIONS FROM THE AQUARIUM TRADE AND HOBBY

Prior to 1900 there was one, possibly two, aquarium fish species established in open waters of the U.S. The first, goldfish, apparently accompanied early colonists to North America in the 1680s (DeKay 1842). The second may have been the ide, although its primary reason for importation and introduction seems to have been sport fishing. Bitterling (*Rhodeus sericeus*) were found in open waters north of New York City in the early 1920s, apparently from release of aquarium specimens (Myers 1925; Courtenay et al. 1984). Myers (1925) also suggested an introduction of rudd into New Jersey may have been from aquaria. In the 1930s oriental weatherfish escaped from an aquarium fish farm in Michigan and became established (Schultz 1960). Thus, prior to the end of World War II, there were as many as five foreign fishes established in North American waters as a result of aquarium fish releases, one of which was an escape from an aquarium fish culture facility.

A little over a decade after World War II, the number of aquarium fishes occurring in North American waters began to escalate dramatically. During the 1950s goldfish were first found in Canadian waters (Crossman 1984). The 1950s also witnessed the advent of international jet cargo aircraft that could bring fishes rapidly from distant parts of the world into U.S. ports-of-entry.

Aquarium fishes rarely imported previously became common imports, as did a flood of species new to aquarium hobbyists. The U.S. aquarium fish culture industry, relatively unimportant previously, grew rapidly, with much of it centered in the southern half of peninsular Florida. The number of fishes imported by air into the U.S. annually has averaged well over 100 million individuals (Ramsey 1985); in addition to these imports, usually of fishes that could not be cultured easily, the Florida aquarium fish culture industry alone was supplying about 80 percent of the aquarium fishes sold in North America (Boozer 1973), a staggering figure when one realizes that the other 20 percent (over 100 million per year) were imports!

Neotropical and African cichlids, South American suckermouth catfishes (Loricariidae) and others began to appear in waters of west and southeastern Florida during the late 1950s and early 1960s, the result of escapes and releases from aquarium fish culture facilities (Courtenay et al. 1974). Some aquarium fish farmers blamed hobbyists for releases, but the highest incidence of these fishes was in drainage systems where fish farms were adjacent.

Careless practices by fish farmers were clearly the major source of these introductions. These included unscreened effluent pipes from both indoor and outdoor

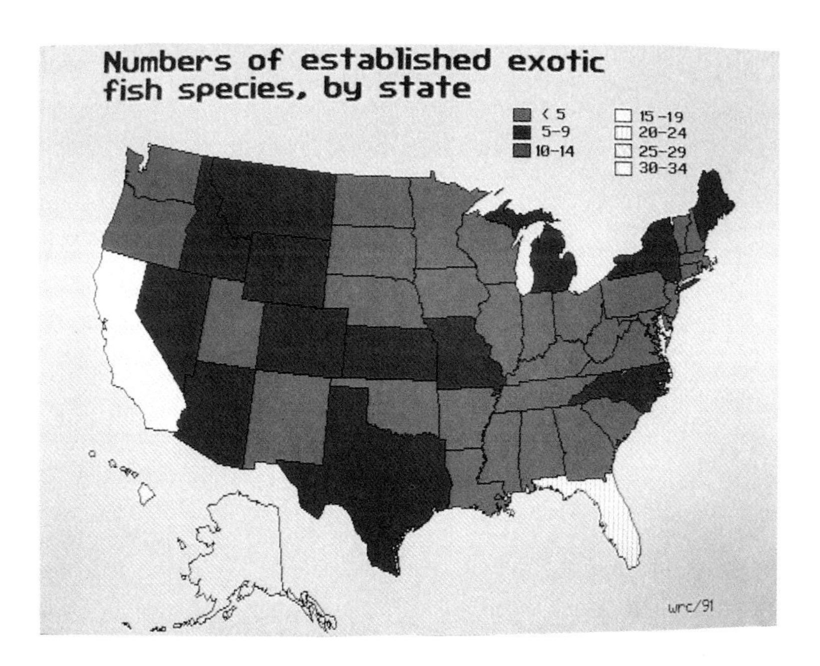


FIGURE 3.—Abundance of established exotic fishes in the United States.

culture tanks using flow-through water systems, the pumping of ponds containing aquarium fishes of different species mixed by birds (such as kingfishers and terns), and periodic flooding (most fish farms are constructed on low elevation land, and few are diked). It is likely that fish farm managers or employees periodically released excess, unwanted fishes into adjacent waters. They may also have released diseased fishes, but fortunately there is no evidence that this resulted in exotic diseases being introduced to native fishes as a result, at least in Florida. Escapes from aquarium fish farms also occurred in other states, California and Nevada in particular (Courtenay et al. 1984; Courtenay and Stauffer 1991).

Approximately 65 percent of the roughly 70 foreign fishes established in U.S. (including Hawaiian) waters are known or believed to have escaped from aquarium fish culture facilities or were introduced by hobbyists (Fig. 3). Most successful releases by hobbyists occurred in the American west where native fish diversity is comparatively low. More than 60 additional, nonestablished fishes, mostly aquarium species, have been collected in the wild. These events have not been restricted to subtropical or warm temperate North America but have occurred in northern states (Courtenay et al. 1984, 1986; Courtenay and Stauffer 1991) and one Canadian province (Crossman 1984, 1991; Crossman and McAllister 1986); establishment in northern waters has been successful only in thermal springs and their effluents. Contreras and Escalante (1984) recorded six species of aquarium fishes as established in Mexico and another two that failed to become established.

Aquarium fish hobbyists, tiring of their specimens or not being able to transport them when moving to another home, often release their fishes into open waters. College students do the same at the end of the academic year. Researchers have been suggested as being responsible for at least one introduction, the pike killifish (*Belonesox belizanus*) in Florida (Belshe 1961; Miley 1978) and may have been involved in the release of medaka (*Oryzias latipes*) on Long Island, New York. These persons believe they are doing something humane, but they are unwittingly contributing to a growing environmental problem when their releases become established. In addition, there have been releases of aquarium fishes to remote thermal springs, far from the nearest towns (Courtenay et al. 1988); these appear to have been deliberate introductions to establish tropical fishes in such waters. Thermal waters near tourist attractions and within national parks have also been targets of similar releases (Nelson 1984; Courtenay et al. 1988). Unfortunately, there has been at least one deliberate introduction of an aquarium fish, convict cichlid (*Cichlasoma nigrofasciatum),* into a thermal spring containing a federally listed, endemic endangered fish (Baugh et al. 1985); this introduction may have been in negative response to a successful reintroduction of the endangered fish into its type locality, located on private property.

To illustrate the magnitude of releases of tropical aquarium fishes in the U.S., such fishes as snakeheads (*Channa*) have been collected from waters in Maine and Rhode Island, piranhas (*Serrasalmus*) in Florida, Michigan, and Ohio, and pacus (*Colossoma*) in many states. These are but a few examples of aquarium fish introductions made by hobbyists that could have disastrous results in waters warm enough to allow their establishment.

## INTRODUCTIONS FROM AQUACULTURE

To date, most releases of invasive, nonnative fishes from aquaculture, except in Mexico (see above under INTRODUCTIONS FOR FOOD PURPOSES), have been from aquarium fish culture facilities. Past U.S. aquaculture has mostly involved native species, typically channel catfish (*Ictalurus punctatus*), usually cultured within its native range of distribution, and rainbow trout (cultured beyond its range). This, however, is changing as aquaculturists become interested in utilizing fishes of intercontinental origin (tilapias, Asiatic carps, etc.). Adults of bighead carp (*Hypophthalmichthys nobilis*) and silver carp (*H. molitrix*) have escaped (or were released from) aquaculture facilities into open waters in at least three states. Bighead carp is now established in the Missouri River (Pflieger 1989). There is increasing interest in culturing tilapias in Arizona, California, Florida, Idaho, Texas and Utah. More recently Atlantic salmon (*Salmo salar*) are being cultured in cages in Pacific waters of the northwestern U.S. (Courtenay and Robins 1989) where native pinnipeds have ripped cages apart, freeing these fish into open waters (T. W. Pietsch, pers. comm.).

Aquaculture, if not properly regulated, represents perhaps the greatest source for future introductions of invasive fishes into North American waters, and probably to most nations (Courtenay and Williams, 1992). Aquaculturists seem enamored with species of foreign origin, probably thinking they will provide higher financial returns on investments because of the mystique of marketing something exotic. Promoters who culture such fishes in hopes that aquaculturists will buy them in large numbers are the primary instigators in this movement to use foreign species. Unfortunately, politicians have become involved in this situation, which virtually assures that it will happen. Management of aquaculture, at least in the U.S., is being assigned to agriculture agencies rather than natural resource or fisheries agencies; politicians view this procedure as avoiding conflicts while assisting economics (which really translates to helping a few vocal constituents attempting circumvention of existing laws and regulations).

Aquaculture has a long way to go to reach the status of modern agriculture. Nutritional requirements and disease treatments are known for some species but not for most. Water quality for aquaculture is not well-understood by many fish farmers. Perhaps more importantly, aquaculture deals with natural stocks of animals that have the potential to survive on their own should they escape or the facility is deserted. At best it is a high risk for venture investors and especially for the environment.

Some of these problems, particularly nutrition, disease and parasite treatment, and water quality requirements, can be overcome. A more important step, however, is to prevent escape into open waters. This can be achieved in large part by requiring diking of the facilities and sand and gravel filtration of all effluents. Because many aquaculture facilities are located in areas subject to flooding (which assures escape), the ideal situation is to develop technology assuring either triploidy of aquaculture stock or (preferably) sterilization of all but the brood stock. Triploidy does not guarantee rapid growth, but it sharply reduces reproduction potential and essentially eliminates the threat of stunting (Shelton 1986). Sterilization assures maximum growth and no energy expended in reproduction, as cattle ranchers learned long ago; with fishes, however, gonadectomy would be impossible, so other means of achieving this goal must be developed and utilized. Those who accomplish sterilization techniques for aquaculture stocks will reap great rewards for a variety of reasons.

## INTRODUCTIONS FOR SPECIES PRESERVATION OR RESTORATION

Before the passage of the Endangered Species Act of 1973, a few fishes that were in danger of extinction were introduced beyond their native ranges in hopes of saving them. Among these early attempts was the release of two cyprinid species native to southern California; unfortunately, in their new environment in extreme northwestern Mexico, the result was massive hybridization (Hubbs and Miller 1943; Contreras and Escalante 1984). Other species were salvaged, at least temporarily, with varying degrees of success through transplants.

After 1973, however, emphasis changed to attempting to keep endangered species in refuges near or within their historic ranges and culturing in hatcheries, awaiting opportunities, should they occur, to reintroduce these species into their home waters. One would think that such refuges or hatcheries would be free from problems of invasive biota, but in two instances this proved otherwise. In one case the Asiatic tapeworm (*Bothriocephalus acheilognathi*), brought to the U.S. with grass carp, was found in endangered Colorado squawfish (*Ptychocheilus lucius*) being held in a federal endangered fish culture facility in New Mexico (Hoffman and Schubert 1984). In another, introduced mosquitofish posed a distinct threat to the endangered Clear Creek gambusia (*Gambusia heterochir*) through hybridization in its last remaining refuge, impounded headwater springs of Clear Creek, Texas (Johnson and Hubbs 1989).

The impact of the Asiatic tapeworm was also felt by an endangered fish in the Virgin River of southeastern Nevada, northwestern Arizona, and southwestern Utah. Populations of the endemic woundfin (*Plagopterus argentissimus*) along with two other endemic cyprinids have been on the decline for several decades, the result of habitat modification and introduction of nonnative fishes (Deacon 1979). It had been noted that an introduced cyprinid, red shiner (*Cyprinella lutrensis*), apparently established from baitfish releases in the Overton Arm of Lake Mead, a large Colorado River reservoir into which the Virgin River empties, was causing declines of woundfin. One of the mechanisms for severe declines in woundfin and the other endemic cyprinids of the Virgin River as a result of this introduction, other than spatial competition, is now known—Asiatic tapeworms, doubtless originating from grass carp being cultured with red shiners on fish farms in Arkansas (Heckman et al. 1986; Deacon 1988). Thus a problem apparently originating in the south-central U.S. has now become a problem in the southwest through a chain of events, from unconcerned fish farmers to uncaring bait dealers to unaware fishermen who released bait.

Introductions made for species preservation and restoration are perhaps the only introduction activities where there is genuine concern expressed against dissemination beyond historic ranges. This is because (1) ecologists and conservationists and particularly rare resource managers with a strong conservation ethic are involved in these efforts and (2) species recovery plans are developed, published, and circulated for commentary to agencies and organizations having interest in, but not actually conducting, the introductions. Whenever suggestions are made to introduce an endangered species beyond its historic range, except into special refuges or hatcheries, negative reaction is typically expressed, and the plan is adjusted accordingly. If all proposed intentional introductions, regardless of the purpose, followed this pattern, introductions would be fewer and made with a far greater margin of safety.

## INTRODUCTIONS FROM CANAL/AQUEDUCT CONSTRUCTION

To what extent the relatively few interdrainage connections from canals in North America have resulted in translocations of fishes is yet to be determined. Courtenay and Taylor (1984) indicated five translocations of U.S. fishes due to canal and aqueduct construction, and Smith (1985), Underhill (1986), and others reported other such instances.

Construction of the Welland Canal in 1829 provided commercial ships with access to ports throughout the Great Lakes beyond Lake Ontario. The previous barrier to ships, and to fishes in Lake Ontario and the downstream St. Lawrence River basin, was Niagara Falls. Dymond (1922) reported the first capture of the sea lamprey (Fig. 4) in Lake Erie, and Hubbs and Brown (1929) indicated Welland Canal as the probable vector. Creaser (1932) and Hubbs and Pope (1937) recorded the progressive spread of this predator into the western Great Lakes. Its introduction was disastrous to many fishes previously stressed by inadequate commercial fishing regulations and played a major role in extinctions of some native ciscos (*Coregonus* spp.). But what delayed invasion by the sea lamprey until almost a century after opening the Welland Canal?

Until recently there was no explanation. Various theories never proven suggested that sea lampreys must have navigated downstream-flushing waters from the locks or had hitched rides on ship hulls passing upstream through the Welland Canal, becoming established in western Great Lakes after native fish populations in Lake Erie and westward were depleted by overfishing. Ashworth (1986) presented the most plausible answer, through the three phases of Welland Canal construction, the last of which opened direct upstream flow through the locks, allowing passage of biological invasives above Niagara Falls. The third phase of construction of the Welland Canal also permitted invasion by alewives, as mentioned earlier, and rainbow smelt (*Osmerus mordax*), both of which eventually changed most of the forage base of the western Great Lakes.

Recent construction of the Central Arizona Project (CAP) Canal, from the lower Colorado River into the south-central interior of Arizona, was recognized (and since proven) as a potential vector for introductions of tilapias from the Colorado River into inland Arizona drainages (Grabowski et al. 1984). The CAP canal could also provide a route for invasion of rainbow smelt into these same drainages if Utah carries out a proposed introduction into a major reservoir upstream in the Colorado system (Courtenay and Robins 1989).

Finally, during a severe drought over much of the United States in 1988, several portions of the Mississippi became too shallow to support barge traffic. The governor of Illinois proposed opening the Chicago River (more properly

FIGURE 4.—Former USFWS biologist Vernon Applegate examines lamprey attached to a lake trout, circa 1950s. (photo provided by the University of Wisconsin Sea Grant College Program)

called the Chicago Sanitary and Ship Canal) from Lake Michigan through the Illinois River into the Mississippi to alleviate downstream flow problems. Burr and Mayden (1980) documented the introduction of rainbow smelt by this route from Lake Michigan into the Mississippi River during high flow periods. This proposed "pulling the plug" in 1988 probably would have resulted in the introduction of the alewife into the Mississippi basin, perhaps with disastrous results to many native species. The only reason that this proposal was rejected is that it would have raised the level of the Mississippi only by about 1 percent; introducing fishes was not a consideration.

## INTRODUCTIONS FROM BALLAST WATER OF SHIPS

Brittan et al. (1963, 1970) stated that the yellowfin goby (*Acanthogobius flavimanus*) was probably introduced into San Francisco Bay and the San Joaquin River estuary system of California from ballast pumped from ships. By the end of the 1970s, this western Pacific goby had dispersed slightly north of San

Francisco and far southward near the Mexican border (Miller and Lea 1972; Usui 1981 and pers. comm.). It has been suggested that the chameleon goby (*Tridentiger trigonocephalus*) was introduced into San Francisco Bay in the same manner, but Hubbs and Miller (1965) stated its origin was likely in the form of fertilized eggs on introduced Japanese oysters (*Crassostrea gigas*). This goby, however, later appeared in Los Angeles harbor, well to the south, doubtless arriving there in ship ballast. Both gobies were more recently found to be established in and around Australian ports, apparently from pumped ship ballast (Hoese 1976; Middleton 1982). A live specimen of chameleon goby was found in ballast water of an Australian vessel (Paxton and Hoese 1985). A more recent introduction to Australian waters was the Japanese sea bass (*Lateolabrax japonicus*), also probably released from ship ballast (Paxton and Hoese 1985).

The European flounder (*Platichthys flesus*) has been collected from waters of the Great Lakes on several occasions but has not become established, apparently due to its inability to reproduce in fresh water (Emery and Teleki 1978). Recently, however, the ruffe (*Gymnocephalus cernuus*) was found to be established in Duluth harbor and elsewhere in Lake Superior, with several year classes present, and invading waters along the Minnesota and Wisconsin coasts (Selgeby 1988; Simon and Vondruska 1991). In addition to this European percid, several species of European invertebrates were found, indicating ship ballast water as the transfer vector (Stanley 1988). Among the introduced zooplankters is a large European cladoceran, *Bythotrephes cederstroemi,* first detected in Lake Huron in 1984 and now present in all Great Lakes. Although it has become an important dietary item to many native and introduced fishes, it feeds heavily on native zooplankters and is believed to have caused a dramatic drop in native cladocerans, particularly *Daphnia* spp., major food items of early life history stages of Great Lakes fishes (Bur 1988).

Lately two species of Caspian sea gobies, the round goby (*Neogobius melanostomus*) and tubenose goby (*Proterorhinus marmoratus*), became established in Lake St. Clair and the Detroit River, between lakes Erie and Huron, the probable result of ballast water introduction (Robins et al. 1991). One or both species feed on the previously introduced zebra mussel (*Dreissena polymorpha*) (G. R. Smith, pers. comm.) and might not have become established before zebra mussel despite previous releases.

It must never be assumed that only marine waters are affected by ballast water introductions. Because many ships store ballast water in river ports before transoceanic voyages, fresh- and brackish-water species are unwittingly brought aboard as stowaways and are later pumped as refugees into foreign ecosystems. Such transfers, in reverse, have been noted in the Baltic Sea and adjacent waters, with one third of such transfers having originated in North America (Leppakoski 1984).

## DISCUSSION

International and intranational transfers of potentially invasive fishes and other aquatic organisms continue to increase at an alarming rate. Those who purposefully move and introduce non-native organisms often claim that there has been relatively little environmental damage demonstrated from such releases. Such claims are invalid for a variety of reasons. Many introductions are discovered well after they were made, and often in waters for which there has been only minimal study of native aquatic faunas prior to introductions; when changes in native species abundance or composition are noted, even when habitat may or may not have been disturbed, introduced species are found to be present and often implicated as a cause for the changes (probably a correct as-

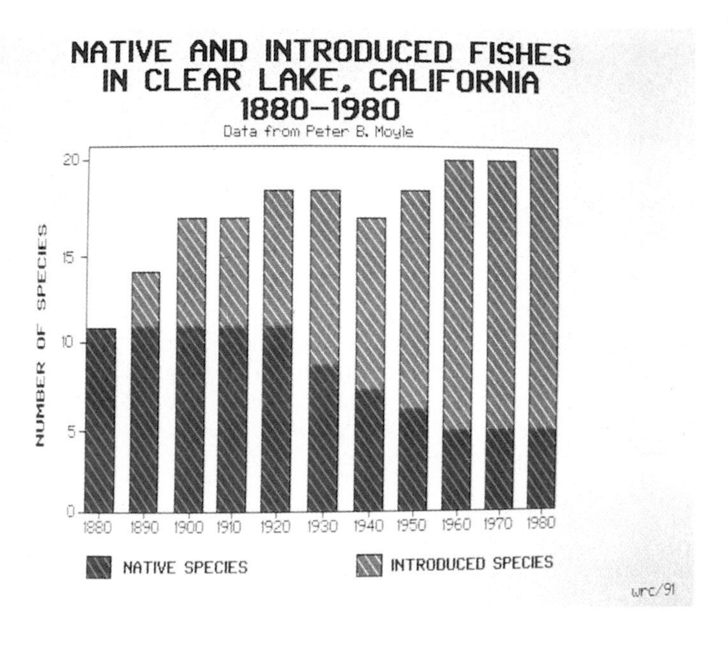


FIGURE 5.—Effects of introductions on biodiversity of native fishes in a lake with few other environmental perturbations.

sumption) (Fig. 5). Because we still know so little of the natural workings of native aquatic ecosystems, most potential impacts by introduced species are based on presumptions as to cause and effect; that is, because the changes were not seen, determined, and measured while they were occurring, their mechanics and sequencing have to become the hypothesis that introductions were causative. Even using a null hypothesis that introductions played no role, introduced species are too often found to have been major factors altering population dynamics in both terrestrial and aquatic habitats (Mooney and Drake 1986). An increasing number of studies have been and are being carried out recently that demonstrate negative effects on native species as a direct result of introduced fishes (e.g., Horwitz 1982; Phillip et al. 1983; Murphy and Terre 1984; Barel et al. 1985; Hocutt et al. 1986; Moyle et al. 1986; Noble and Germany 1986; Deacon 1988; Sheldon 1988; Echelle and Connor 1989; Marsh and Brooks 1989; He and Kitchell 1990; Waples et al. 1990; Brown and Moyle 1991; Minckley 1991; Minckley and Deacon 1991; Moyle and Leidy 1992). Only in recent years has the attention of professional ecologists and experimentalists been directed toward investigating introduction impacts and mechanics, long a fertile but ignored area for detailed studies.

Untold billions of dollars are expended annually by agencies and industry to import, culture, and often directly or indirectly introduce non-native fishes, yet few funds are expended to examine pre- or post-introduction impacts. Similarly well over 100 million federal dollars are provided to states annually, on a 3:1 matching basis, to improve sport fishing, while only a small fraction of other funds is expended on protection or recovery of endangered and threatened fishes (Wilcove, Bean and Lee 1992). And last, those who introduce fishes purposefully are more interested in impacts on fishes considered immediately useful to humans and not to the overall native fish fauna or ecosystem; that is, if it cannot be caught and used, it is of little or no importance. Therefore, whatever "impact" studies they might conduct have the same narrow focus as the rationale for making an introduction.

These last points are where some fishery biologists and many fishery managers differ strongly in opinion and approaches regarding intentional introductions with ecologists, limnologists, marine biologists, and ichthyologists. In fact, there is much dissatisfaction and consternation among many competent fishery biologists with what they are being directed to do by their agency managers and directors in terms of introductions. Unfortunately, this dichotomy also indicates which professionals have the better understanding of ecology and the importance of native biodiversity, by training, breadth of knowledge, and experience; it also emphasizes that fisheries managers in particular need far better training in ecology before they are placed in positions where they direct or influence introductions. The argument often posed by managers—that they live in the "real world" whereas ecologists live in a theoretical one—is biased and baseless and equally degrading to their often more knowledgeable employees who, should they speak out, lose their jobs or get transferred to other often lesser responsibilities or other regions. It is too easy for managers and directors to become political pawns rather than responsible conservation leaders.

Not all fish introductions are inherently bad, but every introduction will result in impacts to native biota, which may range from almost nil to major (including extinctions) with time. No natural system can accept a non-native species without adjustments. With aquatic biota, however, these impacts are more difficult to notice, observe, and measure than in terrestrial habitats where we live. Taylor et al. (1984) listed most impacts that can be expected to occur from fish introductions. Testing expected impacts prior to making an introduction is a far safer and more worthwhile approach than has characterized our history of introductions. Most answers will be provided by such research, but all possible ramifications will never be measurable. Agencies having short-term goals can be predicted to avoid this effort, willing to let future managers deal with correcting any problems created, often an impossible or very costly process by then. Perhaps only education and stricter regulation may curb future unintentional introductions from aquaculture and the aquarium fish culture industry and hobbyists. Because introduced species may not express an invasive nature beyond localized areas or negative impacts to native biota and/or habitat until years or decades following initial releases or ingress, every introduction must be viewed as a potential biological "time bomb" waiting to explode at some future time.

## ACKNOWLEDGMENTS

I thank M. N. Bruton, J. E. Deacon, A. R. Emery, D. P. Jennings, R. L. Lewis, P. J. Kailola, W. L. Minckley, P. B. Moyle, E. P. Pister, J. N. Rinne, C. R. Robins, R. J. Wattendorf, C. A. Usui, R. L. Welcomme, and J. D. Williams for supplying some information used in this paper, and several of them for their review of earlier manuscript drafts. The initial stimulus for this contribution was a Symposium on Invasive Aquatic Biota, held in Grahamstown, South Africa, in September 1987. Special thanks are due B. N. McKnight for having invited my participation in this current effort directed toward more responsible environmental management.

## REFERENCES

Allen, S. K., Jr., R. G. Thiery and N.T. Hagstrom. 1987. Cytological evaluation of the likelihood that triploid grass carp will reproduce. Trans. Amer. Fish. Soc. 115:841–48.

Anonymous. 1988. New virus plaguing lake trout hatcheries. Sport Fish. Inst. Bull. 396:5–6.

Ashworth, W. 1986. The late, Great Lakes; an environmental history. Wayne State Univ. Press, Detroit, Mich.

Baird, S. F. 1879. The carp. Rept. U.S. Fish Comm. 1876–77:40–44.

———. 1893. Report of the Commissioner. Rept. U.S. Fish Comm. 1890–91:1–96.

Balon, E. K. 1974. Domestication of the carp *Cyprinus carpio* L. Misc. Publ. Roy. Ont. Mus. Life Sci.:1–37.

Barel, C. D. N., R. Dorit, P. H. Greenwood, et al. 1985. Destruction of fisheries in Africa's lakes. Nature (London) 315:19–20.

Baugh, T. M., J. E. Deacon and D. Withers. 1985. Conservation efforts with the Hiko White River springfish. J. Aquar. Aquat. Sci. 4:49–53.

Belshe, J. F. 1961. Observations of an introduced tropical fish (*Belonesox belizanus*) in southern Florida. M.S. Thesis, Univ. Miami, Coral Gables.

Boozer, D. 1973. Tropical fish farming. Amer. Fish Farm. 4(8):4–5.

Brittan, M. R., A. B. Albrecht and J. D. Hopkirk. 1963. An oriental goby collected in the San Joaquin River Delta near Stockton, California. Calif. Fish & Game 49:302–04.

Brittan, M. R., J. D. Hopkirk., J. D. Conners, et al. 1970. Explosive spread of the oriental goby *Acanthogobius flavimanus* in the San Francisco Bay-Delta region of California. Proc. Calif. Acad. Sci. 38:207–14.

Brown, L. R. and P. B. Moyle. 1991. Changes in habitat and microhabitat partitioning with an assemblage of stream fishes in response to predation by Sacramento squawfish (*Ptycocheilus grandis*). Can. J. Fish. Aquat. Sci. 48:849–56.

Bruton, M. N. and J. van As. 1986. Faunal invasions of aquatic ecosystems in southern Africa, with suggestions for their management. Pages 47–61 *in* The ecology and management of biological invasions in southern Africa. I. A. W. Macdonald, F. J. Kruger and A. A. Ferrar (eds.). Oxford Univ. Press, Cape Town.

Bur, M. T. 1988. European cladoceran becomes a significant zooplankter in the Great Lakes. U.S. Fish Wild. Serv. Res. Inf. Bull. 88:41.

Burr, B. M. and R. L. Mayden. 1980. Dispersal of rainbow smelt, *Osmerus mordax*, into the upper Mississippi River (Pisces: Osmeridae). Amer. Midl. Nat. 104:198–201.

Cassani, J. R. and W. E. Caton. 1985. Induced triploidy in grass carp, *Ctenopharyngodon idella* Val. Aquacult. 46:37–44.

Conner, J. V., R. P. Gallagher and M. F. Chatry. 1980. Larval evidence for natural reproduction of the grass carp (*Ctenopharyngodon idella*) in the lower Mississippi River. Proc. Fourth Ann. Larv. Fish Conf., Biol. Serv. Prog., Nat. Power Plant Team, Ann Arbor, Mich. FWS/0bs-80/43:1–19.

Contreras-B. S. and M. A. Escalante-C. 1984. Distribution and known impacts of exotic fishes in Mexico. Pages 102–30 *in* Distribution, biology, and management of exotic fishes. W. R. Courtenay, Jr. and J. R. Stauffer, Jr. (eds.). Johns Hopkins Univ. Press, Baltimore, Md.

Courtenay, W. R., Jr. 1979. The introduction of exotic organisms. Pages 237–252 *in* Wildlife and America. H.P. Brokaw (ed.). Govt. Print. Off., Washington, D.C.

——— 1991. Pathways and consequences of the introduction of non-indigenous fishes in the United States. Unpublished rept. to Office of Technology Assessment, Congress of the United States.

Courtenay, W. R., Jr., D. A. Hensley, J. N. Taylor, et al. 1984. Distribution of exotic fishes in the continental United States. Pages 41–77 *in* Distribution, biology, and management of exotic fishes. W. R. Courtenay, Jr. and J. R. Stauffer, Jr. (eds.). Johns Hopkins Univ. Press, Baltimore, Md.

———. 1986. Distribution of exotic fishes in North America. Pages 675–98 *in* Zoogeography of North American freshwater fishes. C. H. Hocutt and E. O. Wiley (eds.). John Wiley & Sons, New York, N.Y.

Courtenay, W. R., Jr., D. P. Jennings and J. D. Williams. 1991. Appendix 2. Exotic fishes of the United States and Canada. Pages 97–110 *in* A list of common and scientific names of fishes from the United States and Canada, 5th ed., 1990. C. R. Robins et al. (ed.). Special Publication 20, Amer. Fisheries Soc.

Courtenay, W. R., Jr. and C. C. Kohler. 1986. Exotic fishes in North American fisheries management. Pages 401–13 *in* Fish culture in fisheries management. R. H. Stroud (ed.). Amer. Fish. Soc., Bethesda, Md.

Courtenay, W. R., Jr. and G. K. Meffe. 1989. Small fishes in strange places: a review of poeciliid introductions. Pages 319–31 *in* Evolution and ecology of livebearing fishes (Poeciliidae). G. K. Meffe and F. N. Snelson, Jr. (eds.). Prentice-Hall, New York, N.Y.

Courtenay, W. R., Jr. and C. R. Robins. 1975. Exotic organisms: an unsolved, complex problem. BioScience 25:306–13.

———. 1989. Fish introductions: Good management, mismanagement, or no management? Rev. Aquat. Sci. 1:159–72.

COURTENAY, W. R., JR., C. R. ROBINS, R. M. BAILEY, et al. 1988. Records of exotic fishes from Idaho and Wyoming. Great Basin Nat. 47:523–26.

COURTENAY, W. R., JR., H. F. SAHLMAN, W. W. MILEY, II, et al. 1974. Exotic fishes in fresh and brackish waters of Florida. Biol. Conserv. 6:292–302.

COURTENAY, W. R., JR. and J. R. STAUFFER, JR. 1991. The introduced fish problem and the aquarium fish industry. J. World Aquacult. Soc. 21:145–59.

COURTENAY, W. R., JR. and J. N. TAYLOR. 1984. The exotic ichthyofauna of the contiguous United States, with preliminary observations on intranational transplants. EIFAC Tech. Pap. 42:466–87.

COURTENAY, W. R., JR. and J. D. WILLIAMS. 1992. Dispersal of exotic species from aquaculture sources, with emphasis on freshwater fishes. Pages 49–81 *in* Dispersal of living organisms into aquatic ecosystems. A. Rosenfield and R. Mann (eds.). Maryland Sea Grant Program, College Park, Md.

CREASER, C. W. 1932. The lamprey *Petromyzon marinus* in Michigan. Copeia 1932:157.

CROSSMAN, E. J. 1984. Introduction of exotic fishes into Canada. Pages 78–101 *in* Distribution, biology, and management of exotic fishes. W. R. Courtenay, Jr. and J. R. Stauffer, Jr. (eds.). Johns Hopkins Univ. Press, Baltimore, Md.

———. 1991. Introduced freshwater fishes: A review of the North American perspective with emphasis on Canada. Can. J. Fish. Aquat. Sci. 48, Suppl. 1:46–57.

CROSSMAN, E. J. and D. E. MCALLISTER. 1986. Zoogeography of freshwater fishes of the Hudson Bay drainage, Ungava Bay and the Arctic archipelago. Pages 53–104 *in* Zoogeography of North American freshwater fishes. C. H. Hocutt and E. O. Wiley (eds.). John Wiley & Sons, New York, N.Y.

DEACON, J. E. 1979. Endangered and threatened fishes of the west. Great Basin Nat. Mem. 3:41–64.

——— 1988. The endangered woundfin and water management in the Virgin River, Utah, Arizona, Nevada. Fisheries (Bethesda) 13:18–24.

DEKAY, J. E. 1842. Zoology of New York-IV: Fishes. W. & A. White and J. Visscher (eds.), Albany, N.Y.

DEVAULT, D. S., W. A. WILLFORD and R. J. HESSELBERG. 1985. Contaminant trends in lake trout (*Salvelinus namaycush*) from the upper Great Lakes. U.S. Environ. Prot. Agency 905/3-85-001.

DI SYLVA, S. S. (ed.). 1989. Exotic aquatic organisms in Asia. Proc. workshop on introduction of exotic aquatic organisms in Asia. Asian Fish. Soc. Spec. Publ. 3.

DREA, J. J. 1993. Classical biocontrol: an endangered discipline? Pages 215–22 *in* Biological pollution: the control and impact of invasive exotic species. B. N. McKnight (ed.). Indiana Acad. Sci., Indianapolis.

DYMOND, J. R. 1922. A provisional list of the fishes of Lake Erie. Univ. Toronto Stud. Biol. Sci. Ser. 20, Publ. Ont. Fish. Res. Lab. 4:57–73.

ECHELLE, A. A. and P. J. CONNOR. 1989. Rapid, geographically extensive genetic introgression after secondary contact between two pupfish species (*Cyprinodon,* Cyprinodontidae). Evol. 43:717–27.

ECK, G. W. and E. H. BROWN, JR. 1988. Climatic events influence alewife populations in Lake Michigan. U.S. Fish Wildl. Serv. Res. Inf. Bull. 88-43.

ELTON, C. S. 1958. The ecology of invasions by animals and plants. Chapman and Hall, London.

EMERY, A. R. and G. TELEKI. 1978. European flounder (*Platichthys flesus*) captured in Lake Erie, Ontario, Canada. Can. Field-Nat. 92:89–91.

ESHENRODER, R. L. 1988. Socioeconomic aspects of lake trout rehabilitation in the Great Lakes. Trans. Amer. Fish. Soc. 116:309–13.

FINUCANE, J. H. and G. R. RINCKEY. 1964. A study of the African cichlid, *Tilapia heudeloti* Dumeril, in Tampa Bay, Florida. Proc. Ann. Conf. SE Assoc. Game, Fish Comm. 18:259–69.

GRAHAM, E. H. 1944. Natural principles of land use. Oxford Univ. Press, New York, N.Y.

GRABOWSKI, S. J., S. D. HIEBERT and D. M. LIEBERMAN. 1984. Potential for introduction of three species of nonnative fishes into central Arizona via the Central Arizona Project—a literature review and analysis. Mimeo. Rept., Environ. Sci. Sect., U.S. Bur. Rec., Denver, Colo.

HARRIS, C. 1978. Tilapia: Florida's alarming menace. Fla. Sportsman 9:12, 15, 17–19.

HE, X. and J. F. KITCHELL. 1990. Direct and indirect effects of predation on a fish community: a whole lake experiment. Trans. Amer. Fish. Soc. 119:825–35.

HECKMAN, R. A., J. E. DEACON and P. D. GREGER. 1986. Parasites of the woundfin minnow, *Plagopterus argentissimus* and other endemic fishes from the Virgin River, Utah. Great Basin Nat. 46:662–76.

HIGHT, S. D. 1993. Control of the ornamental purple loosestrife by exotic organisms. Pages 147–48 *in* Biological pollution: the control and impact of invasive exotic species. B. N. McKnight (ed.). Indiana Acad. Sci., Indianapolis.

HOCUTT, C. H., R. E. JENKINS and J. R. STAUFFER, JR. 1986. Zoogeography of the fishes of the central Appalachians and central Atlantic coastal plain. Pages 161–212 *in* The zoogeography of North American freshwater fishes. C. H. Hocutt and E. O. Wiley (eds.). John Wiley & Sons, New York, N.Y.

Hoese, D. F. 1973. The introduction of the gobiid fishes, *Acanthogobius flavimanus* and *Tridentiger trigonocephalus,* into Australia. Koolewong 2:3–5.
Hoffman, G. L. and G. Schubert. 1984. Some parasites of exotic fishes. Pages 233–61 *in* Distribution, biology, and management of exotic fishes. W. R. Courtenay, Jr. and J. R. Stauffer, Jr. (eds.). Johns Hopkins Univ. Press, Baltimore, Md.
Horwitz, R. J. 1982. The range and co-occurrence of the shiners *Notropis analostanus* and *N. spilopterus* in southeastern Pennsylvania. Proc. Acad. Nat. Sci. Philad. 134:178–93.
Hubbs, C. L. and D. E. S. Brown. 1929. Materials for a distributional study of Ontario fishes. Trans. Roy. Can. Inst. 17:1–56.
Hubbs, C. L. and R. R. Miller. 1943. Mass hybridization between two genera of cyprinid fishes in the Mohave Desert, California. Pap. Mich. Acad. Sci., Arts, Lett. 28:343–78.
———. 1965. Studies of cyprinodont fishes. XXII. Variation in *Lucania parva,* its establishment in western United States, and description of a new species from an interior basin in Coahuila, Mexico. Misc. Publ. Mus. Zool. Univ. Mich. 127:1–111.
Hubbs, C. L. and T. E. B. Pope. 1937. The spread of the sea lamprey through the Great Lakes. Trans. Amer. Fish. Soc. 66:172–76.
Johnson, J. E. and C. Hubbs. 1989. Status and conservation of poeciliid fishes in the United States. Pages 301–17 *in* Ecology and evolution of livebearing fishes (Poeciliidae). G. K. Meffe and F. N. Snelson, Jr. (eds.). Prentice-Hall, New York, N.Y.
Jordan, D. S. 1891. A reconnaissance of the streams and lakes of the Yellowstone National Park, Wyoming, for the purposes of the U.S. Fish Commission. Bull. U.S. Fish Comm. 9:41–63.
Kailola, P. J. 1990. Translocated and exotic fishes: Towards a cooperative role for industry and government. Pages 31–37 *in* Introduced and translocated fishes and their ecological effects. D. A. Pollard (ed.). Bur. Rur. Res. (Canberra, Australia) Proc. 8.
Lachner, E. A., C. R. Robins and W. R. Courtenay, Jr. 1970. Exotic fishes and other aquatic organisms introduced into North America. Smithsonian Stud. Zool. 59:1–29.
Laycock, G. 1966. The alien animals. Nat. Hist. Press, Garden City, N.Y.
Leppakoski, E. 1984. Introduced species in the Baltic Sea and its coastal ecosystems. Ophelia, Suppl. 3:123–25.
Lindsey, C. C. and J. D. McPhail. 1986. Zoogeography of fishes of the Yukon and Mackenzie basins. Pages 639–74 *in* Zoogeography of North American New York. C. H. Hocutt and E. O. Wiley (eds.). John Wiley & Sons, New York, N.Y.
Maciolek, J. A. 1984. Exotic fishes in Hawaii and other islands of Oceania. Pages 131–61 *in* Distribution, biology, and management of exotic fishes. W. R. Courtenay, Jr. (ed.). The John Hopkins Univ. Press, Baltimore, Md.
Marsh, P. C. and J. E. Brooks. 1989. Predation by ictalurid catfishes as a deterrent to re-establishment of hatchery-reared razorback suckers. Southwest. Nat. 34:188–95.
Mather, F. 1889. The brown trout in North America. Bull. U.S. Fish Comm. 7:21–22.
McKay, R. J. 1984. Introductions of exotic fishes in Australia. Pages 177–99 *in* Distribution, biology, and management of exotic fishes. W. R. Courtenay, Jr. and J. R. Stauffer, Jr. (eds.). Johns Hopkins Univ. Press, Baltimore, Md.
———. 1989. Exotic and translocated freshwater fishes in Australia. Pages 217–39 *in* Exotic aquatic organisms in Asia. Proceedings of the workshop on introduction of exotic aquatic organisms in Asia. S. D. De Silva (ed.). Asian Fish. Soc. Spec. Publ. 3.
McPhail, J. D. and C. C. Lindsey. 1986. Zoogeography of the freshwater fishes of Cascadia (the Columbia system and rivers north to the Stikine). Pages 615–37 *in* Zoogeography of North American freshwater fishes. C. H. Hocutt and E. O. Wiley (eds.). John Wiley & Sons, New York, N.Y.
Middleton, M. J. 1982. The oriental goby, *Acanthogobius flavimanus* (Temminck and Schlegel), an introduced fish in the coastal waters of New South Wales. J. Fish Biol. 21:513–23.
Miller, D. J. and R. N. Lea. 1972. Guide to the coastal marine fishes of California. Calif. Dept. Fish & Game Fish. Bull. 157:1–235.
Miller, R. R. 1957. Origin and dispersal of the alewife, *Alosa pseudoharengus,* and the gizzard shad, *Dorosoma cepedianum,* in the Great Lakes. Trans. Amer. Fish Soc. 86:97–111.
Miley, W. W., II. 1978. Ecological impact of the pike killifish, *Belonesox belizanus* Kner (Poeciliidae), in southern Florida. M.S. Thesis, Florida Atlantic Univ., Boca Raton.
Minckley, W. L. 1991. Native fishes of the Grand Canyon region: an obituary? Pages 124–77 *in* Colorado River ecology and management. Nat. Acad. Press, Washington, D.C.
Minckley, W. L. and J. E. Deacon (eds.). 1991. Battle against extinction. Univ. Ariz. Press, Tucson.
Misik, V. 1958. Biometrika dunajskeho kapra (*Cyprinus carpio carpio* L.) z dunajskeho systemu na Slovensku (Biometry of the Danube wild carp [*Cyprinus carpio carpio* L.] of the Danube basin in Slovakia). Biologike prace Slovenskej Akad. Vied 4:55–125.
Mooney, H. A. and J. A. Drake (eds.). 1986. Ecology of biological invasions of North America and Hawaii. Springer-Verlag, New York, N.Y.
Moyle, P. B. 1976a. Inland fishes of California. Univ. Calif. Press, Berkeley.
———. 1976b. Fish introductions in California: history and impact on native fishes. Biol. Conserv. 9:101–18.

MOYLE, P. B. and R.A. LEIDY. 1992. Loss of biodiversity in aquatic ecosystems: evidence from fish faunas. Pages 127–69 *in* Conservation biology: the theory and practice of nature conservation, preservation, and management. P. L. Fiedler and S. Jain (eds.). Chapman and Hill, New York, N.Y.

MOYLE, P. B., H. W. LI and B. A. BARTON. 1986. The Frankenstein Effect: impacts of introduced fishes on native fishes in North America. Pages 415–26 *in* Fish culture in fisheries management. R. H. Stroud (ed.). American Fisheries Society, Bethesda, Md.

MURPHY, B. R. and D. R. TERRE. 1984. Genetic conservation of North American walleye stocks: a review of problems and needs. Unpubl. abstract, 114th meeting Amer. Fish. Soc., Ithaca, N.Y.

MYERS, G. S. 1925. Introduction of the European bitterling (*Rhodeus*) in New York and of the rudd (*Scardinius*) in New Jersey. Copeia 140:20–21.

NELSON, J. S. 1984. The tropical fish fauna in Cave and Basin hotsprings drainage, Banff National Park, Alberta. Can. Field-Nat. 97:255–61.

NOBLE, R. L. and R. D. GERMANY. 1986. Changes in fish populations of Trinidad Lake, Texas, in response to abundance of blue tilapia. Pages 455–61 *in* Fish culture in fisheries management. R. H. Stroud (ed.). Amer. Fish. Soc., Bethesda, Md.

NORRIS, K. R. 1970. General biology. Pages 107–14 *in* The insects of Australia: a textbook for students and research workers. Melbourne Univ. Press, Carlton, Victoria, Australia.

PAXTON, J. R. and D. F. HOESE. 1985. The Japanese sea bass, *Lateolabrax japonicus* (Pisces, Percichthyidae), an apparent marine introduction into eastern Australia. Japan. J. Ichthyol. 31:369–72.

PFLIEGER, W. L. 1989. Natural reproduction of bighead carp (*Hypophthalmichthys nobilis*) in Missouri. Introd. Fish Sect., Amer. Fish. Soc. 9(4):9–10.

PHILIPP, D. P., W. F. CHILDERS and G. S. WHITT. 1983. A biochemical genetic evaluation of the northern and Florida subspecies of largemouth bass. Trans. Amer. Fish. Soc. 112:1–20.

RAMSEY, J. S. 1985. Sampling aquarium fishes imported by the United States. J. Alabama Acad. Sci. 56:220–45.

RANDALL, J. M. 1987. Introductions of marine fishes to the Hawaiian Islands. Bull. Marine Sci. 41:490–502.

REIGER, G. 1981. The wildness factor. Field & Stream 40:137–38.

ROBINS, C. R., R. M. BAILEY, C. E. BOND, et al. 1991. World fishes important to North America. Amer. Fish. Soc. Spec. Publ. 21. 243 p.

SCHULTZ, E. E. 1960. Establishment and early dispersal of a loach, *Misgurnus anguillicaudatus* (Cantor), in Michigan. Trans. Amer. Fish. Soc. 89:376–77.

SCOTT, W. B. and E. J. CROSSMAN. 1973. Freshwater fishes of Canada. Bull. Fish. Res. Bd. Canada 184:1–966.

SELGEBY, J. H. 1988. A European percid invades Lake Superior. U.S. Fish Wild. Serv. Res. Inf. Bull. 88–42.

SHELDON, A. L. 1988. Conservation of stream fishes: Patterns of diversity, rarity, and risk. Conserv. Biol. 2:149–56.

SHELTON, W. L. 1986. Reproductive control of exotic fishes—a primary requisite for utilization in management. Pages 427–34 *in* Fish culture in fisheries management. R. H. Stroud (ed.). Amer. Fish. Soc., Bethesda, Md.

SIMON, T. P. and J. T. VONDRUSKA. 1991. Larval identification of the ruffe, *Gymnocephalus cernuus* (Linnaeus) (Percidae: Percini), in the St. Louis River Estuary, Lake Superior drainage basin, Minnesota. Can. J. Zool. 69:436–42.

SMITH, C. L. 1985. The inland fishes of New York state. N.Y. Dept. Conserv., Albany. 522 p.

SPRINGER, V. G. 1991. Documentation of the blenniid fish *Parablennius thysanius* from the Hawaiian Islands. Pacific Sci. 45:72–75.

SPRINGER, V. G. and J. H. FINUCANE. 1963. The African cichlid, *Tilapia heudeloti* Dumeril, in the commercial fish catch of Florida. Trans. Amer. Fish. Soc. 92:317–18.

STANLEY, J. G. 1988. Ballast-transported organisms pose a threat to Great Lakes. U.S. Fish Wildl. Serv. Res. Inf. Bull. 88–45.

SULLIVAN, C. R. 1988. More on carp. Fisheries 13: inside front cover.

TAYLOR, J. N., COURTENAY, W. R., JR. and J. A. MCCANN. 1984. Known impacts of exotic fishes in the continental United States. Pages 322–73 *in* Distribution, biology, and management of exotic fishes. W. R. Courtenay, Jr. and J. R. Stauffer, Jr. (eds.). Johns Hopkins Univ. Press, Baltimore, Md.

THOMPSON, B. Z., R. J. WATTENDORF, R. S. HESTAND, et al. 1987. Triploid grass carp production. Prog. Fish-Cult. 49:213–17.

UNDERHILL, J. C. 1986. The fish fauna of the Laurentian Great Lakes, the St. Lawrence lowlands, Newfoundland, and Labrador. Pages 105–36 *in* Zoogeography of North American freshwater fishes. C. H. Hocutt and E. O. Wiley (eds.). John Wiley & Sons, New York, N.Y.

USUI, C. A. 1981. Behavioral, metabolic, and seasonal size comparisons of an introduced gobiid fish, *Acanthogobius flavimanus* and a native cottid, *Leptocottus armatus*, from upper Newport Bay, California. M.S. Thesis, Calif. St. Univ., Fullerton.

Wales, J. B. 1962. Introduction of pond smelt from Japan into California. Calif. Fish & Game 48:141–42.

Waples, R. S., G. A. Winans, F. M. Utter, et al. 1990. Genetic approaches to the management of Pacific salmon. Fisheries (Bethesda) 15(5):19–25.

Wattendorf, R. J. 1986. Rapid identification of triploid grass carp with a Colter Counter and channelyzer. Prog. Fish-Cult. 48:125–32.

Welcomme, R. L. 1981. Register of international transfers of inland fish species. FAO Fish. Tech. Pap. 213.

———. 1988. International introductions of inland aquatic species. FAO Fish. Tech. Pap. 294. 318 p.

Wells, L. 1985. Changes in Lake Michigan's prey fish populations with increasing salmonid abundance, 1962 to 1984. Pages 13–25 *in* Presented papers from the Council of Lakes Committee plenary session on Great Lakes predator-prey issues. R. L. Eshenroder (ed.). Great Lakes Fish. Comm. Spec. Publ. 85-3.

Wilcove, D., Bean, M. and P. C. Lee. 1992. Fisheries management and biological diversity: problems and opportunities. Trans. 57th N. Amer. Wildl. Nat. Res. Conf. p. 373–83.

Wooten, M. C., K. T. Scribner and M. H. Smith. 1988. Genetic variability and systematics of *Gambusia* in the southeastern United States. Copeia 1988:283–89.

# Biology of Recent Invertebrate Invading Species in the Great Lakes: The Spiny Water Flea, *Bythotrephes cederstroemi*, and the Zebra Mussel, *Dreissena polymorpha*

David W. Garton[1*], David J. Berg[2],
Ann M. Stoeckmann[1] and Wendell R. Haag[1]

## INTRODUCTION

Environmental alterations in the Laurentian Great Lakes during the 19th and 20th centuries have caused significant changes in species richness and aquatic community structure (Dambach 1969; Thomas 1981). The combined effects of commercial exploitation, loss of spawning habitat and/or overall environmental degradation have endangered or extirpated many native species or caused significant changes in the abundance and distribution of other species (reviewed in Moyle 1986). Continuous environmental perturbation has led to long-term community instability, thus rendering Great Lakes aquatic communities vulnerable to invasion. Deliberate or incidental introductions and natural invasions by exotic species have also contributed to continuous instability of native aquatic communities. During the past two centuries, dozens of exotic plant and animal species have established themselves in the Great Lakes. Of 115 species identified as nonindigenous to the Great Lakes, 28 percent are plants, 19 percent fish, 23 percent algae, 10 percent oligochaetes and 9 percent molluscs (Mills et al. 1991).

Most invading species have had minimal or nondetectable impacts in Great Lakes ecosystems. However, some species have had significant impacts on invaded communities by displacing native species, altering trophic interactions (food-web energy-flow patterns), or changing species composition of communities (i.e., biodiversity). Examples of species that have caused significant changes include purple loosestrife (*Lythrum salicaria,* an aquatic macrophyte), rainbow smelt (*Osmerus mordax*), sea lamprey (*Petromyzon marinus*), alewife (*Alosa pseudoharengus*), and white perch (*Morone americana*) (Mills et al. 1991). The serious ecological and economic consequences resulting from the introduction of these exotic species fuels concern regarding effects of recent invading species in the Great Lakes.

The appearance of non-indigenous species has been documented continuously since the mid-1800s. More recently, during the 1980s, a number of European freshwater species have successfully invaded the Great Lakes, notably three species of fish (river ruffe, *Gymnocephalus cernuus*; tubenose goby, *Proterorhinus marmoratus*; and round goby, *Neogobius melanostomus*), a crustacean (spiny water flea, *Bythotrephes cederstroemi*) (Fig. 1), and a mollusc (zebra mussel, *Dreissena polymorpha*) (Fig. 2). These recent successful invading species probably

*Present Address: Department of Biological and Physical Sciences, Indiana University at Kokomo, 2300 S. Washington St., Kokomo, IN 46904

1. Department of Zoology, Ohio State University, 1735 Neil Ave., Columbus, OH 43210
2. Department of Entomology, Ohio State University, 1735 Neil Ave., Columbus, OH 43210

FIGURE 1.—Spiny water flea, *Bythotrephes cederstroemi,* parthenogenetic female with brood, total length approximately 1 cm.

shared a common method of introduction, unregulated discharge of ballast water from trans-Atlantic shipping (Sprules et al. 1990). Ballast water transport has been implicated in the transoceanic dispersal of many marine, estuarine, and freshwater organisms on a global scale (Carlton 1985; Locke et al. 1991).

The introduction of *Bythotrephes cederstroemi* and *Dreissena polymorpha* into the Great Lakes generated considerable concern because both species were recognized as having the potential to alter pelagic and benthic food webs, with possible detrimental consequences for native species. *Bythotrephes cederstroemi* is a predatory cladoceran, feeding primarily on smaller cladocerans and rotifers, with the potential to compete with planktivorous fish and invertebrates in pelagic food webs. *Dreissena polymorpha* is an efficient filter-feeding bivalve mollusc and can alter the dynamic plankton-benthos relationship by diverting energy from pelagic to benthic food webs by consuming large amounts of phytoplankton. Therefore, studies on the natural history, ecology, physiology, genetics, and reproductive biology of these two species were initiated immediately following their discovery in the Great Lakes. The F. T. Stone Laboratory, Ohio State University's field station in western Lake Erie (Fig. 3), has been the focal site for several recent studies on these invading species. The biology of *Bythotrephes cederstroemi* has been studied since 1986, whereas studies on the biology of *Dreissena polymorpha* were begun in the fall of 1988. This paper summarizes the results of studies on the physiology, genetics, and ecology of these two species, presented as independent case histories. Each case history also assesses the relative impact of these two species on native communities. As invasion success can be considered to be an intrinsic property of the invading species, the stability of the invaded system, or a combination of the two, these long-term studies on *B. cederstroemi* and *D. polymorpha* allow conclusions to be drawn regarding important biological criteria for determining success during an invasion event.

FIGURE 2.—Zebra mussel, *Dreissena polymorpha* colony. Note many small, juvenile mussels attached to shell of adult zebra mussels. Total shell length of adult mussels approximately 2.0 cm; juvenile mussels 1–5 mm. (photo by David Dennis)

## CASE HISTORY: *BYTHOTREPHES CEDERSTROEMI*

### Potential Impacts

As is the case with many exotic species, food web alterations represent the most important potential consequences of the appearance of *Bythotrephes cederstroemi* in Nearctic lakes. Any alterations to food webs will be the result of interactions between the exotic and native species. In the case of *B. cederstroemi*, these alterations include predation on native zooplankton species, competition with other planktivores that may result from this predation, and utilization of the exotic organism as a food source by planktivorous fishes.

*Bythotrephes cederstroemi* and its congener, *B. longimanus*, are voracious predators on a variety of small zooplankton, especially rotifers and other cladocerans (Mordukhai-Boltovskaia 1958, 1960; Monakov 1972). Reported feeding rates are all derived from laboratory studies, but a few European studies have noted effects of these predatory cladocerans on zooplankton dynamics. In Lago Maggiore, Italy, predation by *B. longimanus* is a major factor accounting for seasonal declines in *Daphnia* abundance (de Bernardi et al. 1987), and competition with *Leptodora kindti* may limit *B. longimanus* abundance at times when predation pressure on the latter is low (de Bernardi and Guissani 1975).

The importance of *Bythotrephes* (both species) as prey for planktivorous fishes and the role of planktivores in regulating the population dynamics of these large cladocerans are fairly well-documented. *Bythotrephes* are consumed by percids (Guma'a 1978) and are preferred prey of coregonids (Berg and Grimaldi 1966; de Bernardi and Guissani 1975; Hamrin and Persson 1986) and salmonids (Stenson 1972; Fitzmaurice 1979). In Sweden, *Bythotrephes* are present in fishless lakes and those with low numbers of planktivorous fishes and absent from lakes with large numbers of vertebrate planktivores (Stenson 1972, 1978).

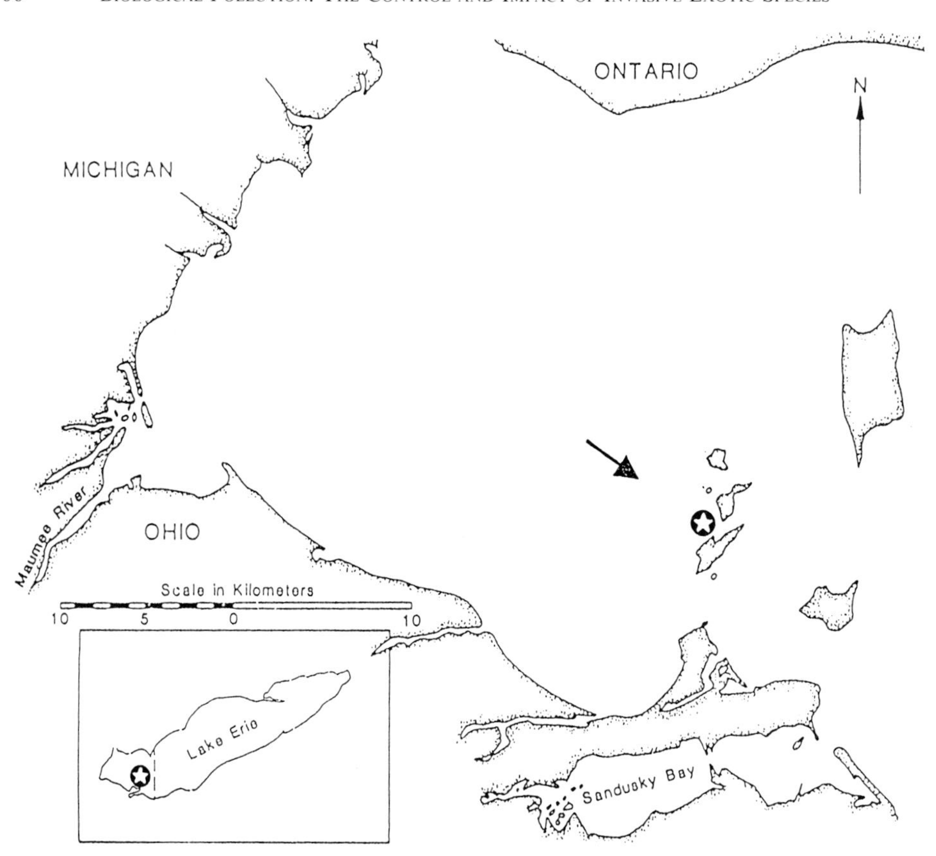


FIGURE 3.—Western Lake Erie indicating location of Bass Island region (arrow) and Ohio State University's F. T. Stone Laboratory (star).

Because of its relatively large size, significant predation on *Bythotrephes* is probably limited to adult fish. Young European perch (*Perca fluviatilis*) show negative electivity for *B. longimanus* (Guma'a 1978), while small rainbow trout (*Onchorhynchus mykiss*) have great difficulty handling *B. cederstroemi* due to its large caudal spine (Fig. 1) (Barnhisel 1991). In summary, fish predation on *Bythotrephes* is a function of the size of the predator.

The potential impacts of *Bythotrephes cederstroemi* on Nearctic food webs are quite variable. As a predator, this species may affect population dynamics of smaller zooplankton, especially rotifers and cladocerans, while being a potential competitor with other planktivores, including *Leptodora kindti* and fishes. Because this may include competition with young-of-year fishes, *B. cederstroemi* has the potential for indirectly affecting fish recruitment. It will certainly be a new food source for larger planktivores. The European literature indicates that *B. cederstroemi* may have a variety of food-web impacts and that these may be both direct and indirect effects on native species.

## Introduction and Spread in North America

The first record of *Bythotrephes cederstroemi* in the Laurentian Great Lakes is from southern Lake Huron in December 1984 (Bur et al. 1986). By the end of 1985 this species had spread throughout all three basins of Lake Erie (Bur et al. 1986) and was seasonally abundant by 1987 (Berg and Garton 1988). First records from Lake Ontario were also from 1985 (Lange and Cap 1986). The species was abundant in 1987 but present at only very low densities in 1988 (Makarewicz and Jones 1990). In 1986 *B. cederstroemi* was found at a number of stations across Lake Michigan, and it was a conspicuous member of the zooplankton community by 1987 (Lehman 1987, 1988; Evans 1988). Initially reported from stomachs of Lake Superior fishes in 1987 (Cullis and Johnson 1988), high densities of *B. cederstroemi* were present by 1988 (Garton and Berg 1990). The first records of its presence outside of the Great Lakes are from the Muskoka Lakes of southern Ontario (1989) and lakes in the Cloquette River watershed of northern Minnesota (1990) (Yan et al. 1992). Thus, in less than ten years this exotic has expanded its range to include all of the Great Lakes and several inland watersheds.

Several hypotheses have been advanced to account for the sudden appearance of *Bythotrephes cederstroemi* in North America, including transport in ballast water of ocean-going freighters (Bur et al. 1986), atmospheric transport (Lange and Cap 1986), and transport of resting eggs in mud adhering to nautical or fishing equipment (Lehman 1987; Evans 1988). The currently accepted hypothesis invokes ballast water transport in ships moving from St. Petersburg (formerly Leningrad), Russia to Great Lakes seaports (Sprules et al. 1990). Although normally brackish or saline, St. Petersburg harbor is relatively dilute, and at certain times of the year ships may take on freshwater ballast. Under this scenario the source of North American populations would be Lake Ladoga. This lake contains only *B. cederstroemi* (Zozulya and Mordukhai-Boltovskoi 1977), accounting for the fact that *B. longimanus* has not been introduced into North America even though it is more prominent than its congener throughout much of Europe (not considering any unsuccessful introductions). Sprules et al. (1990) estimated that in the late 1970s and early 1980s multiple seedings by freighters occurred throughout the Great Lakes, and that these, combined with interbasin transfer by currents, shipping and other anthropogenic factors (that is, fishermen) account for the current distribution of *B. cederstroemi* in North America.

## Reproductive Biology

Like most cladocerans, members of the genus *Bythotrephes* are cyclic parthenogens. These organisms generally reproduce asexually throughout most of the growing season, with environmental stress (extreme water temperature, low food abundance) inducing the production of sexual resting eggs (Herzig 1985). Parthenogenesis is followed by direct development of young in the female brood pouch. Length of development time in the brood pouch is a function of water temperature (Yurista 1992), while brood size may vary from one to greater than ten individuals (Berg 1991). Resting eggs are released into the water column, allow survival through winter, and provide the initial "inoculum" as water temperature warms in the spring (Herzig 1985). Sexual broods may contain up to seven resting eggs (Berg and Garton 1988).

Under favorable environmental conditions, when parthenogenesis is the primary mode of reproduction, populations are nearly 100 percent female (Fig. 3) (Berg 1991). During times when sexual eggs are present in females, the proportion of males increases, exceeding 30 percent of the population (Garton and Berg 1990; Berg 1991). Thus, sex ratio is an indicator of the condition of the population, with significant proportions of males associated with sexual reproduction. Normally, sexual reproduction would be expected to occur in late autumn, as water temperature declines and prey items become scarce. However, in the western basin of Lake Erie, a shallow basin that experiences water temperatures as high as 30°C throughout the water column, a maximum proportion of males and sexual reproduction also occur in mid-summer (Berg 1991). This indicates that both cold and hot temperatures may provide the stress necessary to induce sexual reproduction in *Bythotrephes cederstroemi.*

## Environmental Physiology

In Europe, *Bythotrephes cederstroemi* is typically found in cold, oligotrophic lakes in Scandinavia and Russia but is absent from central and southern Europe. In the Laurentian Great Lakes it is present throughout the summer and autumn in Lakes Ontario (Makarewicz and Jones 1990) and Michigan (Lehman 1991) and the Muskoka Lakes of Canada (Yan et al. 1992). Populations in Lake Ontario, however, are very unpredictable, presumably a combination of intense predation and unfavorable environmental conditions (Makarewicz and Jones 1990). In both lakes Superior and Huron we have collected *B. cederstroemi* in July and September (Berg and Garton, unpubl. data).

In Lake Erie, seasonal presence of this species varies between basins. In the deeper, cooler central basin, *B. cederstroemi* is present from June through October (Bur and Klarer 1991), while in the shallower, warmer western basin, its presence is highly variable between years (Fig. 5) (Berg 1991). The high summer water temperatures of western Lake Erie are the most likely cause of the unpredictable presence of *Bythotrephes cederstroemi* in this basin (Garton et al. 1990). The western basin is shallow (average depth of 7 m), warm (maximum water temperature of 27–30°C), and isothermal. Laboratory experiments indicate that mortality of *B. cederstroemi* in summer and autumn is positively correlated with water temperature (Fig. 6). *Leptodora kindti*, a native predatory cladoceran, exhibits a seasonal acclimation response in which minimum mortality occurs at ambient temperatures. This response is completely lacking in *B. cederstroemi.* Field observations corroborate the findings of these laboratory studies. Not only is *B. cederstroemi* present for only short periods of time in western Lake Erie, but populations are least fit in summer. Average size of individuals is smaller in summer, and reproductive effort is well below the potential maximum due to low fecundity (Berg 1991). Such observations would be consistent with a scenario in which *B. cederstroemi* are forced to maintain abnormally high basal metabolic rates, shunting energy away from growth and reproduction. On the other hand *L. kindti* is present from April through December and exhibits maximum abundance and fitness during summer (Andrews 1948; Berg 1991). The thermal acclimation response may indicate that this species is capable of lowering basal metabolism and therefore continuing to grow and reproduce. While the native species appears to be well-adapted to the conditions of western Lake Erie, summer water temperatures create an inhospitable environment for the exotic species. As a result it is only marginally successful in western Lake Erie. Little information on demographics of *B. cederstroemi* in other North American lakes has been published. However, a comparison of samples from western Lake Erie and Lake Superior show that Lake Superior animals are larger, presumably due to the more favorable thermal regime of this deep, cold lake (Garton and Berg 1990). Thus, *B. cederstroemi* can best be described as a cold-water, stenothermic organism that has experienced limited success in its colonization of western Lake Erie due, at least in part, to thermal constraints.

## Population Genetics

A potential explanation for lack of a thermal acclimation response in *Bythotrephes cederstroemi* may be low genetic variability due to passage through a

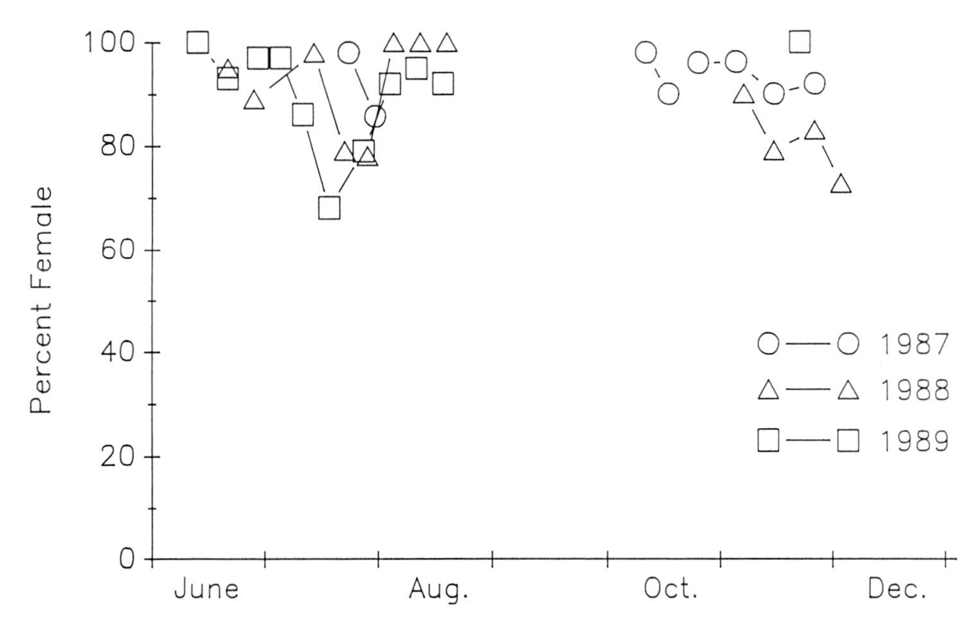


FIGURE 4.—*Bythotrephes cederstroemi*. Percent females in summer and autumn populations, 1987–1989. Gaps during late summer represent period when *B. cederstroemi* is absent from western Lake Erie. Bouts of sexual reproduction occur when percent females in population declines (late July and November). Data from Berg (1991).

genetic bottleneck during the invasion process. This lack of genetic variability may inhibit the ability of a population to respond to environmental challenges. North American populations of *B. cederstroemi* have low genetic variation when compared with other cladoceran species, but this variation is comparable to that found in European populations (Berg 1991; Weider 1991). It is unlikely that *B. cederstroemi* passed through a genetic bottleneck during its invasion of North America (Berg 1991). Therefore, the inability of *B. cederstroemi* to acclimate to warm temperatures is most likely characteristic of the species itself and not a function of reductions in genetic variability during the invasion process.

Comparisons of genetic structure among *Bythotrephes cederstroemi* populations in North America and Europe lead to a number of conclusions (Berg 1991). Populations within the Laurentian Great Lakes are not genetically distinct over space or time. This indicates that a single European population was the source of all Nearctic populations. North American populations are more similar to populations from Finland than from Sweden or Germany, further evidence that Lake Ladoga may be the ultimate source of the invasion. Because of the inherently low variability in populations of *B. cederstroemi*, the small absolute sizes of founding populations may not have represented loss of significant amounts of genetic variation.

## Ecological Impacts

As stated earlier, the food-web effects of the invasion of North America by *Bythotrephes cederstroemi* may be considered from two perspectives: the exotic species as a predator competing with native planktivores and the exotic species as a source of nutrition for planktivorous fishes. The majority of North American studies have focused on addressing these issues.

The effects of predation by *Bythotrephes cederstroemi* on population dynam-

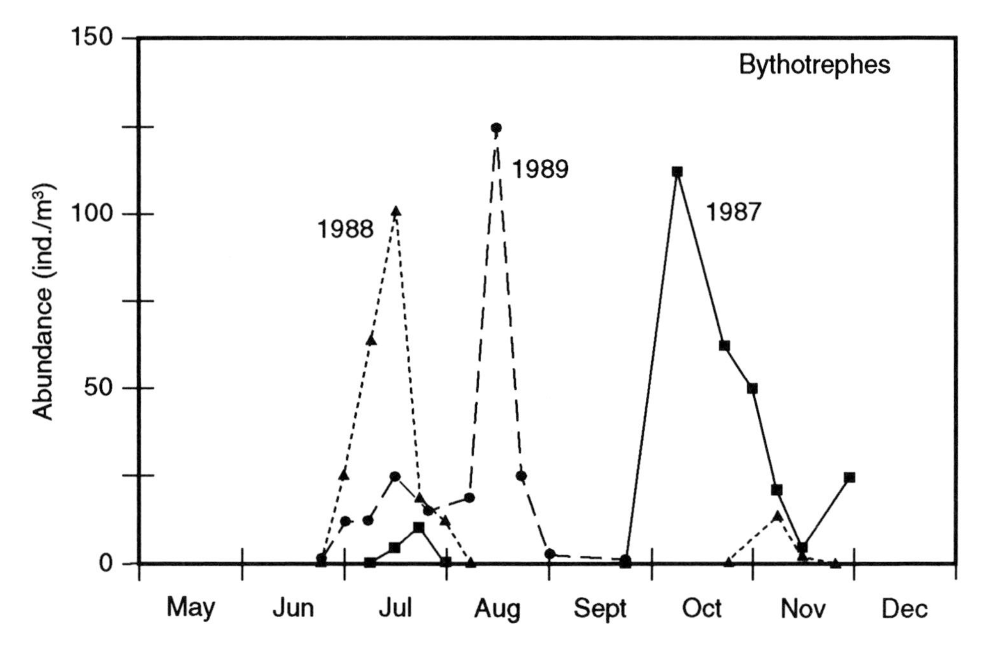


FIGURE 5.—*Bythotrephes cederstroemi*. Seasonal abundance in western Lake Erie, 1987–1989, data from early May to early December. Data from Berg and Garton (1988) and Berg (1991).

ics of potential prey species (small zooplankters) are unclear. Early food web models developed for Lake Michigan predicted that predation by the carnivore would drive down abundances of *Daphnia,* allowing copepods to become the dominant herbivores (Scavia et al. 1986). This prediction was supported by work showing large changes in zooplankton community composition following the appearance of *B. cederstroemi* in Lake Michigan (Lehman 1988, 1991). Based on estimated feeding rates and low relative abundance of *B. cederstroemi* in Lake Michigan, other studies have concluded that *B. cederstroemi* is unable to significantly affect population dynamics of *Daphnia* in Lake Michigan. Rather, recent declines in abundance of these herbivores may be due to planktivory by the exotic alewife (*Alosa pseudoharengus*) (Sprules et al. 1990).

If predation by *Bythotrephes cederstroemi* is a major factor controlling *Daphnia* population fluctuations, one would predict that *Daphnia* death rates would be positively correlated and *Daphnia* abundance would be negatively correlated with abundance of *B. cederstroemi*. In eastern Lake Erie, studies conducted from 1989 through 1991 found no relationships between abundance of *B. cederstroemi* and *Daphnia* population dynamics (H. P. Riessen, unpubl. data). Patterns of small zooplankton abundance are similar in preinvasion and postinvasion years (1986 and 1988–89, respectively) in western Lake Erie (Phipps 1987; Berg 1991). Seasonal patterns of zooplankton abundance are similar in 1988 and 1989 even though *B. cederstroemi* is present at different times during the two years (Fig. 4) (Berg 1991). Thus, the impacts of *B. cederstroemi* as a predator are still unclear and deserving of further research.

Little is known of competitive interactions between *Bythotrephes cederstroemi* and planktivorous fishes. Declines in abundance of larval bloater (*Coregonus hoyi*) in Lake Michigan coincided with the appearance of *B. cederstroemi.* This has been attributed to food-limited competition between fish larvae and the exotic cladoceran (Warren and Lehman 1988). Several studies have examined relationships between *B. cederstroemi* and *Leptodora kindti*. In 1987 western

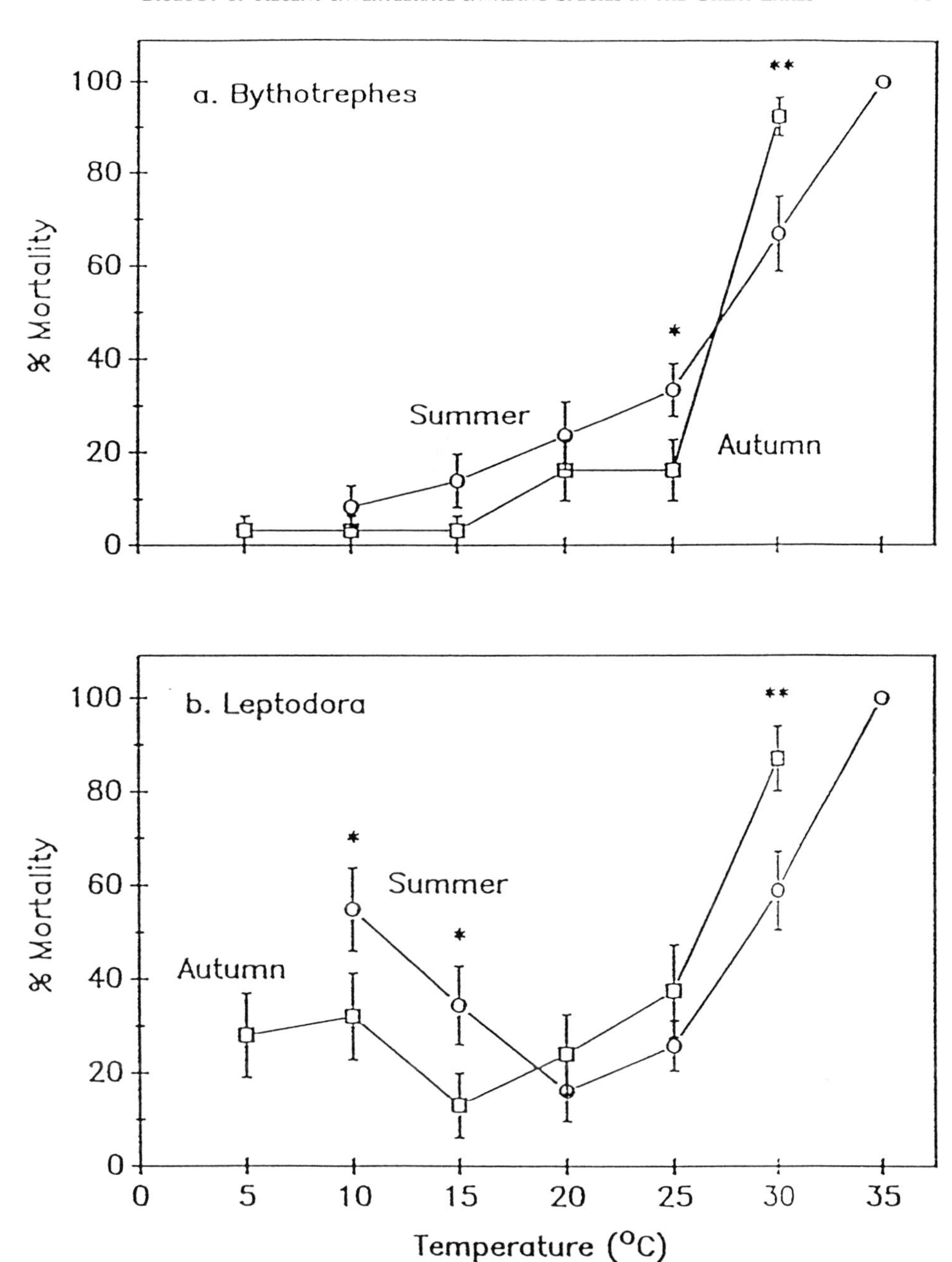


FIGURE 6.—*Bythotrephes cederstroemi* and *Leptodora kindti*. Percent mortality after 12 h at each exposure temperature, between season comparison for each species. Note lack of seasonal acclimation response for *B. cederstroemi* relative to *L. kindti*. Ambient temperature during the summer, 25°C; autumn, 7.1–10.5°C. Vertical bars represent 1 SD of sample mean. Results for test of equality of percentages comparing summer and autumn mortality at each temperature: * $p < 0.05$; ** $p < 0.01$. Data from Garton et al. (1990).

Lake Erie abundances of these species were negatively correlated (Berg and Garton 1988). Similar results were also reported from 1986–1990 in Lake Michigan (Lehman 1991). However, in 1988 and 1989 abundance and fitness of *B. cederstroemi* were independent of *L. kindti* abundance and fitness (Berg 1991). Furthermore, there was no evidence of food limitation (the key component of competition between these species) of predatory cladocerans in western Lake Erie. Although it is safe to conclude that western Lake Erie populations of carnivorous cladocerans do not compete for food, it is not yet clear whether the same is true for other populations in the Great Lakes.

A wide variety of Great Lakes fishes consume *Bythotrephes cederstroemi*, including yellow perch (*Perca flavescens*), walleye (*Stizostedion vitreum*), deepwater sculpin (*Myxocephalus thompsoni*), and chinook salmon (*Onchorhynchus tshawytscha*) (Bur et al. 1986; Cullis and Johnson 1988; Evans 1988; Keilty 1990; Schneeberger 1991; Baker et al. 1992). In the central basin of Lake Erie, *B. cederstroemi* can make up to 80 percent of the diet by volume for a variety of fish species (Bur and Klarer 1991). In laboratory experiments, yellow perch consistently detected *B. cederstroemi* before detecting identically sized *Leptodora kindti* (Berg 1991). In western Lake Erie, *B. cederstroemi* is present at high abundance for very short periods of time (Fig. 5). The precipitous declines in abundance are consistent with a scenario of high predation by planktivorous fishes (Berg 1991). While the long caudal spine of *B. cederstroemi* provides protection from juvenile fishes (Barnhisel 1991), it does not deter consumption by adults (Berg 1991). Results from the Great Lakes are consistent with European findings that *B. cederstroemi* abundance is consistently high only when fish predation is low (Stenson 1972, 1978).

In summary *Bythotrephes cederstroemi* seems to have had little impact on the food-web structure of western Lake Erie. The combination of high summer water temperatures (without a hypolimnion to act as a thermal refuge) and high levels of planktivory results in *B. cederstroemi* being present for only short periods in the summer and autumn. In deeper, colder basins of the Great Lakes, this may not be the case. Summer water temperatures are not nearly as warm, the hypolimnion forms a thermal refuge, and fish predation is lower. In these systems *B. cederstroemi* may play a more prominent role in determining food web dynamics.

## CASE HISTORY: *DREISSENA POLYMORPHA*

### Potential Impacts

The zebra mussel, *Dreissena polymorpha*, is well-known as a significant biofouling organism in Europe and, in many aquatic communities, the dominant benthic species (reviewed in Mackie et al. 1989). Detrimental impacts resulting from the introduction of *D. polymorpha* include direct and indirect effects on humans and native biota. In Europe, zebra mussels attain densities often exceeding 10,000 individuals/m$^2$ (Wiktor 1963; Lewandowski 1982; and many others). Direct effects of high mussel densities are well-documented; e.g., fouling of water intake pipes, hydroelectric installations and navigation locks; overgrowth of available substrate for other organisms, and submerging of aquatic macrophytes (reviewed in Mackie et al. 1989). Indirect effects result from filter feeding by *D. polymorpha* on pelagic and benthic food webs. Zebra mussels are efficient filter feeders, and studies of nutrient cycles in small Polish lakes have shown that zebra mussel consumption of algae and production of waste products can "turn over" the entire pool of nitrogen (N) and phosphorus (P) every 2–20 days, depending upon the particular lake (Stancyzkowska and Planter 1985). Numerous European studies have documented the significant role of *D. polymorpha* in aquatic communities and its economic and ecological impacts.

## Introduction and Spread in North America

Through the 18th century the zebra mussel was apparently restricted to the Ponto-Caspian Basin in the southern part of the former Soviet Union (drainage basins of the Aral, Black, and Caspian seas) (Morton 1969). Construction of canals and increased commerce among Russia and other European nations in the 1800s allowed *Dreissena polymorpha* to rapidly expand its range. This range expansion even crossed a saltwater barrier, the English Channel, as zebra mussels were discovered in the Thames River at London in the 1830s (Morton 1969). One common name for *D. polymorpha*, "die Wandermuschel," or "the wandering mussel," recognized the ability of this species to invade novel habitats. Presently *D. polymorpha* is widespread throughout Europe, including Great Britain, extending from Scandinavia and the Baltic Sea southward to the Aral, Black, and Caspian seas, and eastward to the Ural Mountains (Zhadin 1952).

In North America the first specimens of *Dreissena polymorpha* were collected in June 1988 from Lake St. Clair, a lake connecting southern Lake Huron with western Lake Erie (Hebert et al. 1989). The age of these first specimens, estimated from growth rate studies in European mussel populations, was between one and two years. Benthic surveys conducted in Lake St. Clair between 1983 and 1985 did not find any specimens, therefore it is likely that *D. polymorpha* was successfully introduced in 1986 or 1987 (Hebert et al. 1989). Shipping channels through Lake St. Clair are narrow, and coupled with the shallow lake waters the risk of grounding is greater than in other sections of the St. Lawrence Seaway (Griffiths et al. 1991). Release of ballast water to avoid grounding or to refloat a grounded vessel may have been the proximal cause of the introduction of *D. polymorpha* into southern Lake St. Clair rather than some other Great Lakes seaport (Griffiths et al. 1991).

Initial reproductive success of *Dreissena polymorpha* in southern Lake St. Clair was extremely high, and zebra mussels soon spread downstream into western Lake Erie from the Detroit River (Griffiths et al. 1991). The range of the zebra mussel in North America has expanded rapidly during the past few years. The first specimens of *D. polymorpha* collected at Stone Laboratory were found in October 1988 (Garton and Haag 1991). By the end of 1989 *D. polymorpha* were present throughout Lake Erie and into western Lake Ontario. In 1990 zebra mussels were transported into the upper Great Lakes and were present in lakes Superior, Huron, and Michigan as well as further downstream in the St. Lawrence River and across New York from the New York State Barge Canal (Griffiths et al. 1991). By the end of 1991 *D. polymorpha* had expanded its range beyond the Great Lakes drainage basin and was present in the Hudson, Illinois, Mississippi, Ohio, and Tennessee rivers and inland lakes, and reservoirs in Indiana, New York, Ohio, and Ontario (Schloesser et al. 1992). The spread of *D. polymorpha* from a central source (that is, Lake St. Clair) is supporting evidence for a single (successful) introduction event. Within the first three years since its discovery, *D. polymorpha* has successfully entered most of the major drainage basins in eastern North America (Fig. 7).

## Reproductive Biology

One of the important factors contributing to the rapid spread of *Dreissena polymorpha* is its free-swimming veliger larva, a feature unique among all freshwater molluscs. Sexes are separate, and fertilization occurs in the water column. Within two to three days the fertilized egg develops into a planktotrophic veliger, followed by pediveliger and plantigrade stages. The total duration of planktonic

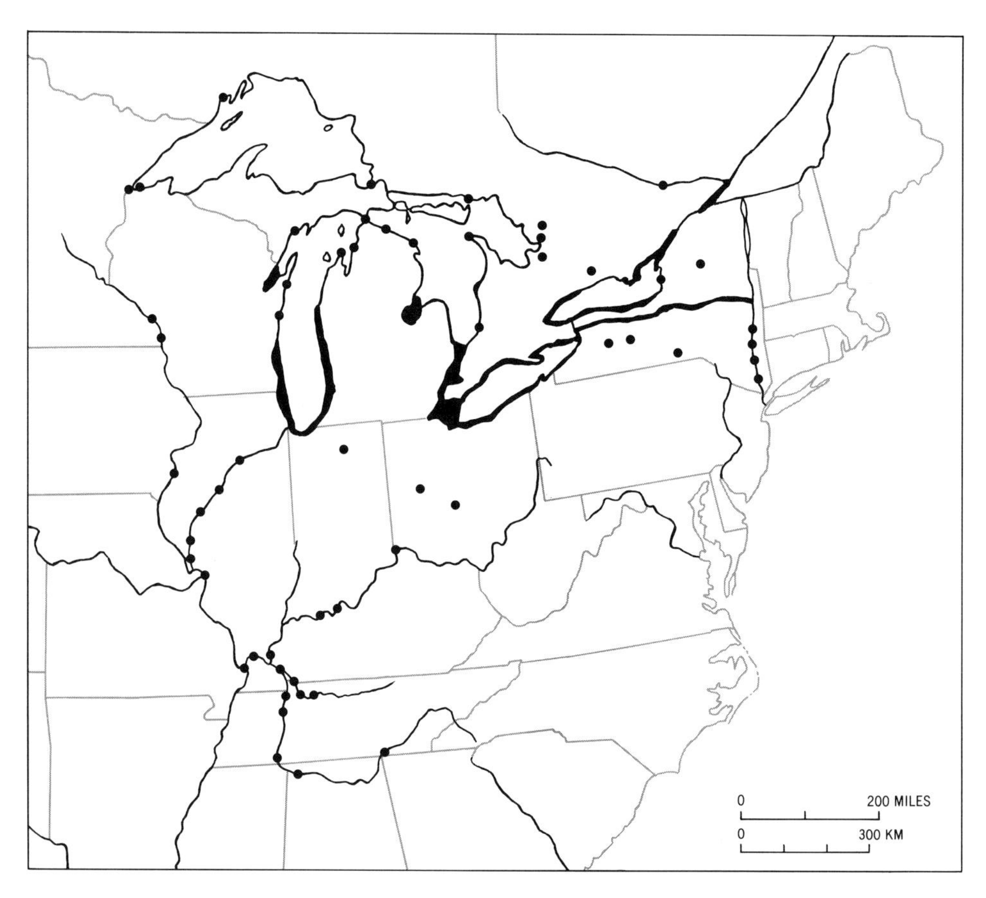


FIGURE 7.—Present distribution of *Dreissena polymorpha* in North America. Thick lines in the Great Lakes indicate continuous populations, points indicate locations along rivers or within lakes where adult specimens have been collected and hence breeding populations have been assumed to be established. Data compiled from NOAA Sea Grant programs (Illinois-Indiana, Michigan, Minnesota, New York, Ohio, and Wisconsin).

development has been estimated as lasting from 8 to 33 days (Sprung 1989). During the course of a spawning season a mature female will produce up to 300,000 eggs (Walz 1978), although perhaps only 1 percent will survive to the settling stage (Garton, unpubl. data).

Seasonal variation in water temperature has been identified as one of the primary factors regulating the timing of reproduction of *Dreissena polymorpha* (Galperina 1978; Sprung 1989; Borcherding 1991). In Europe the occurrence of veligers is highly variable; they normally appear in late spring, peak in midsummer and disappear in early fall (reviewed in Sprung 1989). Temperatures above 12°C have been reported as necessary for the initiation of spawning; in heated lakes, spawning begins earlier and persists later in the year than in nearby nonheated lakes (Sprung 1989).

The reproduction of *Dreissena polymorpha* in Lake Erie follows an overall cycle similar to that reported in European studies, but significant differences do exist (Haag and Garton 1992). In western Lake Erie, veliger abundances are highly seasonal and peak in midsummer (Fig. 8). However, spawning in local populations around Stone Laboratory does not begin until water temperature reaches 20–22°C, considerably higher than reported in European studies. Although water temperature acts as a rate-regulating factor for reproductive pro-

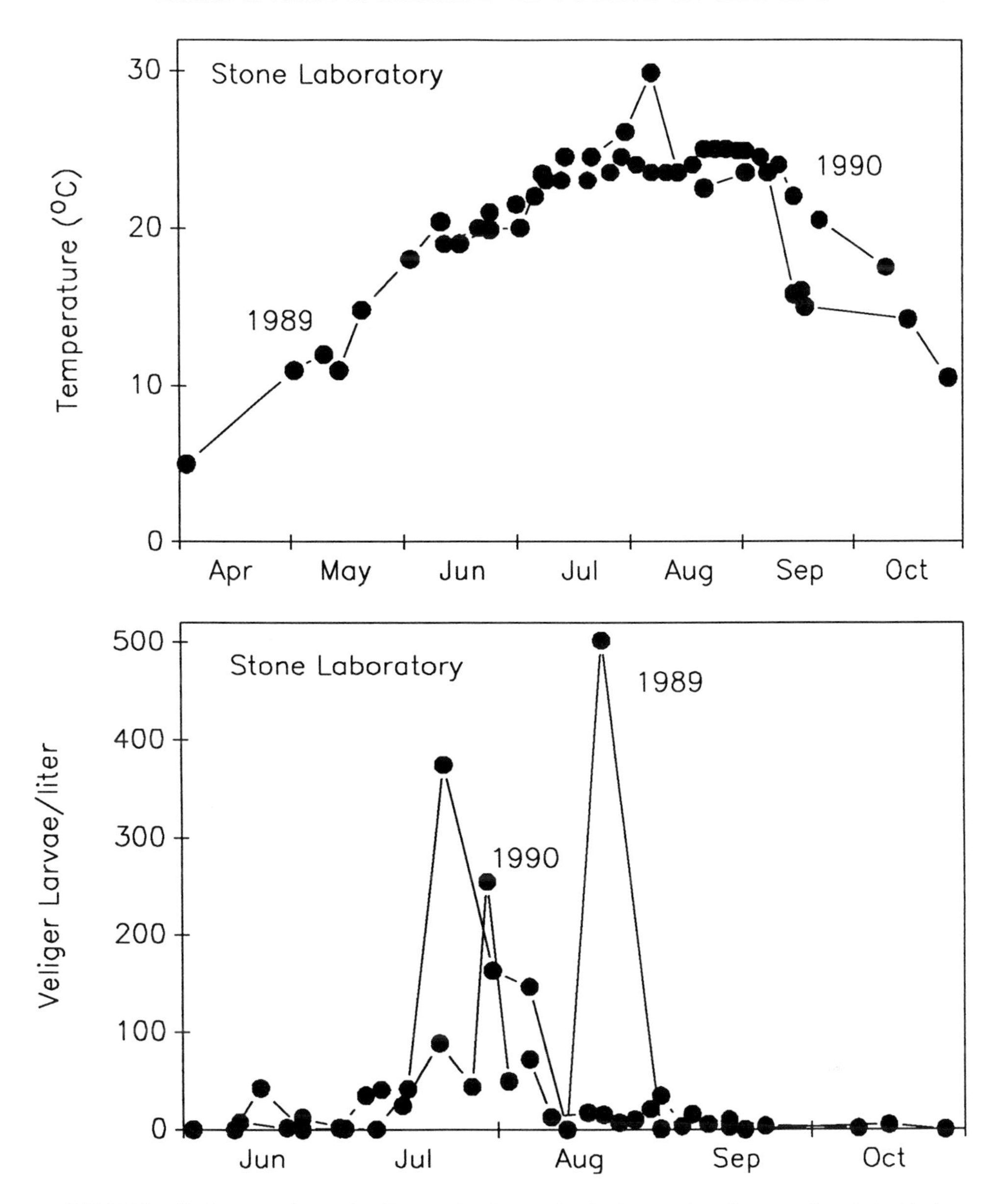


FIGURE 8.—*Dreissena polymorpha*. Seasonal patterns of water temperature (top panel) and planktonic veliger abundance (bottom panel) at Stone Laboratory during 1989 and 1990. Data from Garton and Haag (1992).

cesses (gonadal development and gametogenesis), triggering and synchronization of spawning within mussel beds may depend on other environmental cues, such as abundance of phytoplankton (Haag and Garton 1992).

The number and timing of spawning events also varied each year from 1989 to 1991. In the summer of 1989 a single, highly synchronized spawning event occurred in late August; in 1990 two distinct spawning events occurred (late July and late August); while in 1991 three spawning periods were observed (mid-May, early and late July) (Garton and Haag 1992; Garton and Stoeckmann) (Table 1). Early spawning in 1991 was the result of unusually warm temperatures in the

TABLE 1. Annual variation in spawning and total settlement of *Dreissena polymorpha* in the Bass Island region, near F. T. Stone Laboratory, western Lake Erie, 1989–1991. Data for 1989–1990 from Garton and Haag (1992); 1991, from Garton and Stoeckmann (unpubl. data).

| Year | Spawning period | # of spawns | Peak veliger density (#/L) | Temperature range (°C) | Settling period | Total settled Larvae (#/m²) | Temperature range (°C) |
|---|---|---|---|---|---|---|---|
| 1989 | late Aug | 1 | 500 | 22–23 | Jul–late Aug | 30,000 | 22–25 |
| 1990 | mid Jun–Jul | 2–3 | 300 | 20–25 | Jul–early Sep | 23,000 | 24–25 |
| 1991 | mid May–Jul | 3 | 20 | 18–25 | Jul–late Aug | 5,000 | 23–25 |

spring. These studies indicate that reproduction of *Dreissena polymorpha* is flexible, responding to geographic and temporal variation in temperature, phytoplankton abundance, water currents, and depth. This variation is responsible for variable spawning events in local populations of zebra mussels in Lake Erie (Haag and Garton 1992).

## Environmental Physiology

Water temperature represents an important factor limiting the distribution of aquatic organisms. In Europe, *Dreissena polymorpha* is found in habitats that experience broad temperature extremes on a seasonal basis. In western Lake Erie, water temperature ranges from less than 4°C in winter (underneath ice cover) to 30°C in late summer. Clearly, zebra mussels readily tolerate and acclimate to seasonal changes in water temperature, characteristic of western Lake Erie and many other temperate lakes.

The ability of zebra mussels to acclimate to fluctuating temperatures is reflected in acute thermal tolerance and acclimation to seasonal changes in water temperature. The resistance of *Dreissena polymorpha* to higher temperatures increases with increasing ambient temperature. In spring at an ambient temperature of 17°C, acute 24-hr $TL_{50}$ is 31.3°C, while in midsummer (ambient temperature 26.0°C), $TL_{50}$ increases to 34.5°C (Stoeckmann and Garton 1991). Metabolic rates, measured as oxygen consumption, are fairly insensitive to seasonal changes in water temperature. As water temperature declines from 25°C (midsummer) to 15°C (early fall), zebra mussels maintain constant metabolic rates (Fig. 9) (Garton and Stoeckmann 1990). Elevated metabolic rates observed in early summer are the result of reproductive activities (discussed below). *Dreissena polymorpha* acclimates rapidly following seasonal changes in water temperature.

## Costs of Reproduction

High larval mortality of *Dreissena polymorpha* is offset by high fecundity, with each female capable of producing in excess of several hundred thousand eggs during the spawning season (Walz 1978). Spawning is stressful and associated with rapid loss of body mass and elevated metabolic rates. In western Lake Erie, mussel dry tissue mass declines rapidly during spawning periods, and mussels lose 50 percent of their tissue body mass (Haag and Garton 1992; Garton and Haag 1992). Similarly, metabolic rates are elevated two- to threefold during the spawning season, further evidence of physiological stress associated with spawning (Stoeckmann and Garton 1991).

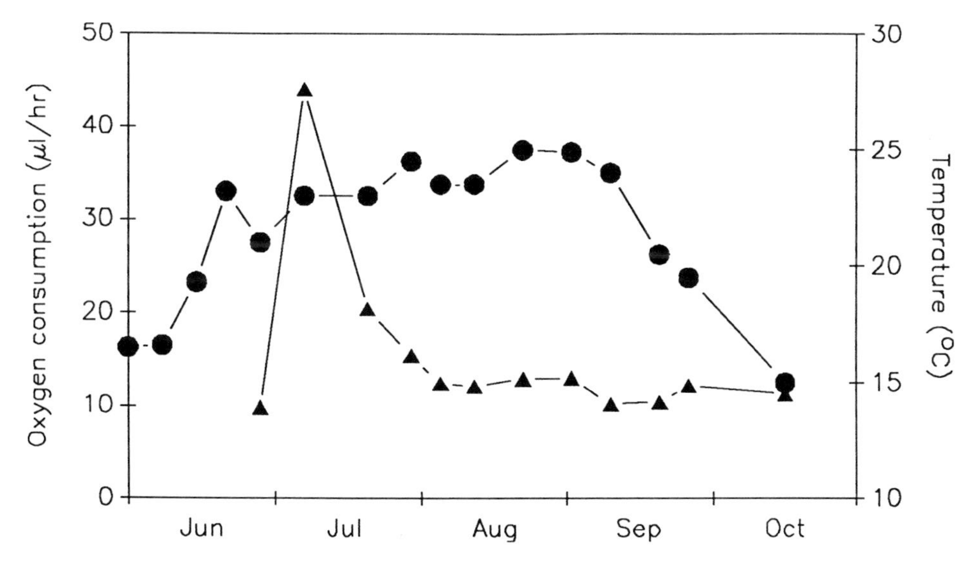


FIGURE 9.—*Dreissena polymorpha*. Seasonal patterns of oxygen consumption measured at ambient lake temperature (triangles) and water temperature (circles) at Stone Laboratory during 1990. Peak in metabolism in early summer is associated with spawning in the local population (see text). Data from Garton and Stoeckmann (1991).

Following spawning, mussels are vulnerable to additional environmental stress and subsequent mortality. Post-spawning mortality is reflected in the population demographics of zebra mussels in western Lake Erie. A zebra mussel is sexually mature within 11 months of settling, at a shell length of approximately 15 mm (settlement peaks in August, while spawning peaks the following July). Most mussels die following their first spawning season, therefore during the winter and spring months the population is dominated by new recruits. Those mussels surviving into a second spawning season are in the size range of 20–25 millimeters. To date, the largest mussel collected near Stone Laboratory was 33.5 mm in length and likely three years of age (Stoeckmann and Garton, unpubl. data).

## Population Genetics

Newly established populations often pass through genetic bottlenecks and suffer reduced levels of intra-population genetic variability relative to the source population. However, populations of *Dreissena polymorpha* in North America show no evidence of reduced genetic variability; rather, populations in Lake St. Clair and Lake Erie possess high levels of intra-population genetic variation (Hebert et al. 1989; Garton and Haag 1991). However, as a result of larval dispersal, zebra mussel populations throughout western Lake Erie are genetically similar; that is, the frequency of genotypes across populations is homogeneous (Haag 1991). Unfortunately, little is known of population genetic structure of *D. polymorpha* in Europe, either for indicating potential source populations or providing insight on population genetic structure (Hebert et al. 1989).

A positive relationship between size, and hence fitness, and individual genetic variability has been reported for several species of marine molluscs. Likewise, individual genetic variability is correlated with increasing size in *Dreissena polymorpha* (Garton and Haag 1991). In many marine bivalves, individual genetic variability is correlated with increased growth and fitness and viability in stressful environments (reviewed in Mitton and Grant 1984). Relationships between genetic variability, physiological plasticity (that is, acclimation to fluctuating temperature) and high fecundity may have contributed to the success of this species as a colonizer in diverse freshwater habitats in Europe and North America.

## Ecological Impacts

The mechanisms of the primary ecological impacts of *Dreissena polymorpha* can be placed into two broad categories: removal of suspended material by efficient filter feeding and biofouling, and the encrustation of solid substrate by dense beds of mussels. In Europe, the deliberate introduction of *D. polymorpha* has been proposed in order to improve water clarity and quality in small eutrophic lakes (Reeders et al. 1989). In western Lake Erie, water clarity increased and phytoplankton abundance decreased markedly following the introduction of zebra mussels (Leach 1992; Nicholls and Hopkins 1992). Attributing changes in water clarity and phytoplankton abundance to filter feeding by zebra mussels is confounded by a historical trend in declining nutrient inputs to Lake Erie, especially phosphorous (P). During the past two decades P levels have declined steadily, resulting in a corresponding decline in phytoplankton abundance. However, rates of phytoplankton decline and water clarity increase accelerated more rapidly following the introduction of *D. polymorpha* than could be accounted for by P decline (Nicholls and Hopkins 1992). In addition, changes in water clarity and phytoplankton abundance followed the invasion front of zebra mussels as it moved from west to east in Lake Erie (Nicholls and Hopkins 1992).

Initial observations that *Dreissena polymorpha* readily colonized live native unionids (Bivalvia) led to concerns regarding negative impacts (Mackie 1991; Nalepa et al. 1991; Schloesser and Kovalak 1991). Native unionids in Lake St. Clair and Lake Erie heavily encrusted by zebra mussels did experience high mortality (Gillis and Mackie 1992). A similar die-off of unionids was reported for Lake Balaton, Hungary, following the introduction of *D. polymorpha* in the 1930s (Sebestyen 1938). Unionid species with thinner shells (subfamilies Anodontinae and Lampsilinae) are more vulnerable than thick-shelled species (subfamily Ambleminae) (Haag et al. 1993). Therefore unionid communities will likely experience significant changes in species richness following heavy encrustation by zebra mussels (Haag et al. 1992). As a result of severe encrustation, native unionids have been essentially extirpated from areas of high zebra mussel density in lakes St. Clair and Erie. In Lake St. Clair, unionids declined from approximately 15 individuals/$m^2$ in 1986 to less than 0.05 individuals/$m^2$ in 1991 (Gillis and Mackie 1992). There are over 150 species of native unionids in the United States; nearly a third of which are included on the Federal Endangered Species List, leading to concern regarding the ultimate impact of zebra mussels on this fauna as *Dreissena* spreads throughout the Mississippi drainage basin.

In addition to exclusion of infaunal bivalves, the high density and biomass of local zebra mussel populations will restructure benthic communities. Filter-feeding *Dreissena polymorpha* deposit feces and pseudofeces on the surrounding substrate, which in turn provides a food source for scavenging epibenthic invertebrates. Certain benthic invertebrates, such as oligochaetes and amphipods, may use this new resource, whereas species requiring firm substrates for attachment (sponges, bryozoans and cnidarians) may colonize *D. polymorpha* shells. The accumulation of feces/pseudofeces may also alter the characteristics of the substrate (particle size, stability, pH, chemical composition, bacterial community, etc.). Clearly, *D. polymorpha* will cause significant changes in the benthic community.

## SUMMARY

### Implications for Predicting Invasion Success

*Bythotrephes cederstroemi* and *Dreissena polymorpha* represent two successful invading species in the Great Lakes. Both species were introduced in the mid-1980s, and both spread rapidly throughout all five Great Lakes within three years of introduction (Berg 1991; Griffiths et al. 1991). In contrast, the three fish species introduced during the same period in the 1980s (perhaps even simultaneously with *B. cederstroemi* and *D. polymorpha*) have spread more slowly, with the river ruffe still limited to extreme western Lake Superior and the two goby species found only in the St. Clair River-Lake St. Clair-Detroit River system (Jude 1992; Selgeby and Ogle 1992). However, as in other invasions, it may be many years before the full impact of the three fish species may be apparent.

The success of *B. cederstroemi* and *D. polymorpha* can be attributed to several factors, including pelagic dispersal stages; transport in ballast water of inter-lake freighters, pleasure craft, fishing tackle and/or nets; environmental conditions similar to native habitats in Europe, and lack of competition and/or predation sufficient to limit rapid population growth.

Ehrlich (1986) listed possible characteristics determining the potential for a species to invade novel habitats. These important traits included:

—abundant in original range
—polyphagous diet
—short generation times
—high genetic variation
—fertilized female able to colonize alone
—larger than most closely related species
—associated with *Homo sapiens*
—wide environmental tolerance

Life history studies of *Bythotrephes cederstroemi* and *Dreissena polymorpha* have revealed that while these species possess some "invader" characters, some traits of these two species are not likely indicators of invasion potential.

Both species are distributed widely throughout Europe, but *Bythotrephes cederstroemi* occurs only seasonally in the plankton, primarily during summer months. *Dreissena polymorpha* occurs year round in the adult stage, but like *B. cederstroemi,* the larval pelagic stage is also highly seasonal. As the pelagic forms are most likely to be transported in ballast water, the risk of invasion increases during periods when pelagic stages are abundant (adult *Bythotrephes*; larval *Dreissena*).

*Bythotrephes cederstroemi* and *Dreissena polymorpha* are both polyphagous. *Bythotrephes cederstroemi* consumes a wide-range of zooplankton prey (rotifers, cladocerans, and nauplii), while *D. polymorpha* filters a wide size range of phytoplankton, detritus, and small zooplankton (MacIsaac et al. 1991). Likewise both species have short generation times, on the order of days for *B. cederstroemi* and less than one year for *D. polymorpha* (Yurista 1992; Haag and Garton 1992).

Population genetic variability of these two species present strong contrasts. Populations of North American and European *Bythotrephes cederstroemi* and European *B. longimanus* have little genetic variability and show evidence of restricted gene flow between drainage basins (Berg 1991). However, there is no evidence of a genetic bottleneck during the invasion of North America by *B. cederstroemi*, as European and North American populations have similar levels of genetic variability. Although genetic variation in European populations of *Dreissena polymorpha* has not been studied, given the high genetic variability of North American populations of *D. polymorpha,* it is unlikely this species passed through a bottleneck during the invasion of North America. Therefore, population genetic variation *per se* does not accurately predict likelihood of invasion, as a less-genetically variable species (*B. cederstroemi*) and highly variable species (*D. polymorpha*) have both successfully invaded North America. However, the genetically variable species has spread more rapidly and over a greater geographic range than the less variable species.

These two invading species differ in the number of founders required to initiate viable populations. Populations of *Bythotrephes cederstroemi* can theoretically be established by a single female reproducing parthenogenetically (which mimics the situation of winter-resting eggs hatching during the spring). However, *Dreissena polymorpha* is dioecious, and fertilization is external. For a species with external fertilization, a minimum density of founders must be present for successful reproduction and recruitment. *Dreissena polymorpha* were introduced as larvae, adults, or both. In the case of larval stages, the number of veligers in the founding population must have been fairly large to compensate for dilution when ballast was discharged into receiving waters. Similarly, if zebra mussels were introduced in the adult stage (attached to an anchor, for example), then veligers resulting from reproduction by these adults would also be widely dispersed by water currents. A viable population will be established only if veliger larvae settle at a density sufficient for external fertilization to occur.

*Bythotrephes cederstroemi* and *Dreissena polymorpha* represent freshwater fauna unique to central Eurasia. The Cercopagidae (*Bythotrephes* and related genera) and Dreissenacea (*Dreissena*) are endemic to the Ponto-Caspian Basin (Zhadin 1952; Mordukhai-Boltovskoi 1966). The success of these two species in spreading across Europe and to North America may be a lack of closely related, and possibly competing, species in the invaded communities. *Leptodora kindti* is a large predatory cladoceran ecologically similar to *B. cederstroemi* and is present in Europe and North America, however there is little evidence of inter-specific competition limiting the distribution of either of these two species (Berg 1991). *Dreissena polymorpha* occupies a unique niche in freshwater communities as an epibenthic, filter-feeding bivalve. Other species of filter-feeding bivalves are infaunal, burrowing forms.

Human activity has assisted the spread of both of these species, acting as an agent of direct transport and by altering the environment. Both species were likely transported to North America in ballast water. Construction of canals and reservoirs in Europe provided both habitat and routes of invasion for pelagic stages of *Bythotrephes cederstroemi* and *Dreissena polymorpha*. For example, both species were absent or rare in the Volga River prior to the construction of locks and dams for navigation, flood control, and electric power generation (Dzyuban 1978; Mordukhai-Boltovskoi 1978). Following the completion of these projects in the 1950s, which formed large reservoirs (Rybinsk and Kuibyshev), and completion of canals connecting Lake Ladoga and Lake Onega with the Volga River, *D. polymorpha* spread northward while *Bythotrephes* spread southward throughout the Volga River system (Dzyuban 1978).

The two species differ markedly in tolerance and response to environmental factors such as temperature. *Bythotrephes cederstroemi* is a cold-water stenotherm, and in western Lake Erie, populations are ephemeral and limited by warm summer water temperatures (Garton et al. 1990). Since its discovery in 1984, *B. cederstroemi* has not been reported further south than western Lake Erie and has only spread out of the Great Lakes into a few inland lakes in Minnesota and Ontario. In contrast *Dreissena polymorpha* is eurytopic and has spread rapidly southward from the Great Lakes.

In summary, these two species possess some but not all of the traits favoring invasion reviewed by Ehrlich (1986). *Bythotrephes cederstroemi* is widespread and seasonally abundant, polyphagous, parthenogenetic, possesses a short generation time, and is large relative to other zooplankton. Invasion potential is limited by being stenothermic and having low levels of intrapopulation genetic variability. *Dreissena polymorpha* is abundant and widespread in Europe, polyphagous, possesses a short generation time, eurytopic, and genetically variable. Invasion potential is limited by external fertilization requiring a large founding population. Ehrlich (1986) concluded that chance played a large role in species invasion, and that no one particular trait or combination of traits could accurately predict which species would invade while another would not. Although a species may not possess all traits predicting a high risk of invasion, the more "invader" traits, the greater the risk of a successful invasion.

## Risk for Future Invasions in the Great Lakes

In 1989 the Canadian Coast Guard developed the voluntary Great Lakes Ballast Water Control Guidelines in order to reduce the risk of additional species introductions into the St. Lawrence Seaway and Great Lakes. The guidelines require that ships must exchange freshwater ballast with mid-ocean seawater prior to entering the St. Lawrence Seaway. To examine compliance with voluntary guidelines, Locke et al. (1991) sampled ships entering the Great Lakes from May to December 1990. Although compliance was fairly high (89 percent of ships with ballast; 95 percent for all ships), live freshwater organisms were present in the few ships that did not exchange ballast, as well as some ships that had exchanged part of their ballast water. Locke et al. (1991) concluded that risk of invasion still exists from ships not in compliance with voluntary guidelines and from live organisms remaining in ballast water after exchange has occurred. Although risk has been reduced, it is likely new species will continue to be introduced into the Great Lakes through ballast water discharge.

# ACKNOWLEDGMENTS

Over the years many individuals have made valuable contributions to our research described in this paper, assisting in field collections and laboratory experiments. To all, our heartfelt thanks and appreciation. We would like to especially acknowledge the continued support by John Hageman, Laboratory Manager, and his staff at F. T. Stone Laboratory. Without his assistance much of our work would not have been accomplished. Ms. Kathy Bruner kindly reviewed an earlier version of this manuscript. David Dennis, Jose Diaz, and Matt Meadows assisted in the preparation of figures. Research described in this paper was supported fully or in part by the Ohio Sea Grant College Program, Projects R/ZM-10, R/ER-10, R/ER-15, and R/ER-20–PD, under Grants NA88A-D-SG094, NA89AA-D-SG132 and NA90AA-D-SG496 of the National Sea Grant College Program, National Oceanic Atmospheric Administration in the U.S. Department of Commerce, and from the State of Ohio, to DWG. Additional support was provided by a Sigma Xi Grant-in-Aid of Research and an Ohio State University Presidential Fellowship to DJB.

# REFERENCES

ANDREWS, T. F. 1948. The life history, distribution, growth and, abundance of *Leptodora kindti* (Focke) in western Lake Erie. Ph.D. Thesis, Ohio State Univ., Columbus.

BAKER, E. A., S. A. TOLENTINO and T. S. MCCOMISH. 1992. Evidence for yellow perch predation on *Bythotrephes cederstroemi* in southern Lake Michigan. J. Great Lakes Res. 18:190–93.

BARNHISEL, D. R. 1991. The caudal appendage of the cladoceran *Bythotrephes cederstroemi* as defense against young fish. J. Plankton Res. 13:529–37.

BERG, A. and E. GRIMALDI. 1966. Ecological relationships between planktophagic fish species in the Lago Maggiore. Verh. Internat. Verein. Limnol. 16:1065–73.

BERG, D. J. 1991. Genetics and ecology of an invading species: *Bythotrephes cederstroemi* in western Lake Erie. Ph.D. Thesis, Ohio State Univ., Columbus.

BERG, D. J. and D. W. GARTON. 1988. Seasonal abundance of the exotic predatory cladoceran, *Bythotrephes cederstroemi*, in western Lake Erie. J. Great Lakes Res. 14:479–88.

BORCHERDING, J. 1991. The annual reproductive cycle of the freshwater mussel *Dreissena polymorpha* Pallas in lakes. Oecologia (Berlin) 87:208–18.

BUR, M. T. and D. M. KLARER. 1991. Prey selection for the exotic cladoceran *Bythotrephes cederstroemi* by selected Lake Erie fishes. J. Great Lakes Res. 17:85–93.

BUR, M. T., D. M. KLARER and K. A. KRIEGER. 1986. First records of a European cladoceran, *Bythotrephes cederstroemi*, in lakes Erie and Huron. J. Great Lakes Res. 12:144–46.

CARLTON, J. T. 1985. Transoceanic and interoceanic dispersal of coastal marine organisms: the biology of ballast water. Ocean. Mar. Biol. Ann. Rev. 23:313–71.

CULLIS, K. I. and G. E. JOHNSON. 1988. First evidence of the cladoceran *Bythotrephes cederstroemi* Schoedler in Lake Superior. J. Great Lakes Res. 14:524–25.

DAMBACH, C. A. 1969. Changes in the biology of the lower Great Lakes. Pages 1–17 *in* Proceedings conference on changes in the biota of lakes Erie and Ontario. R. Sweeney (ed.). Bull. Buffalo Soc. Nat. Sci. Vol. 25(1).

DE BERNARDI, R. and G. GUISSANI. 1975. Population dynamics of three cladocerans of Lago Maggiore related to predation pressure by a planktophagous fish. Verh. Internat. Verein. Limnol. 19:2906–12.

DE BERNARDI, R. G. GUISSANI and M. MANCA. 1987. Cladocera: predators and prey. Hydrobiologia 145:225–43.

DZYUBAN, N. A. 1978. Zooplankton of the Volga. Pages 195–231 *in* The river Volga and its life. Ph.D. Mordukhai-Boltovskoi (ed.). Junk bv Publishers, The Hague.

EHRLICH, P. R. 1986. Which species will invade? Pages 79–95 *in* Ecology of biological invasions of North America and Hawaii. H. Mooney and J. Drake (eds.). Springer-Verlag, New York, N.Y.

EVANS, M. S. 1988. *Bythotrephes cederstroemi*: its new appearance in Lake Michigan. J. Great Lakes Res. 14:234–40.

FITZMAURICE, P. 1979. Selective predation on Cladocera by brown trout *Salmo trutta*. J. Fish Biol. 15:521–26.

GALPERINA, G. E. 1978. The relationship of spawning of some north Caspian bivalves to the distribution of their larvae in the plankton. Malac. Rev. 11:108–09.

GARTON, D. W. and D. J. BERG. 1990. Occurrence of *Bythotrephes cederstroemi* (Schoedler 1877) in Lake Superior, with evidence of demographic variation within the Great Lakes. Great Lakes Res. 16:148–52.

GARTON, D. W. and W. R. HAAG. 1991. Heterozygosity, shell length and metabolism in the European mussel, *Dreissena polymorpha*, from a recently established population in Lake Erie. Comp. Biochem. Physiol. 99A:45–48.

———. 1992. Seasonal reproductive cycles and settling patterns of *Dreissena polymorpha* in western Lake Erie. Pages 111–28 *in* Zebra mussels: biology, impact and control. T. Nalepa and D. Schloesser (eds.). Lewis Publishers, Ann Arbor, Mich. (in press).

GARTON, D. W. and A. STOECKMANN. 1990. Temperature-dependent metabolism of zebra mussels: seasonal and short-term acclimation experiments. J. Shellfish Res. 10:249.

GARTON, D. W., D. J. BERG and R. J. FLETCHER. 1990. Thermal tolerances of the predatory cladocerans *Bythotrephes cederstroemi* and *Leptodora kindti*: relationship to seasonal abundance in western Lake Erie. Can. J. Fish. Aquat. Sci. 47:731–38.

GILLIS, P. L. and G. L. MACKIE. 1992. The effect of exotic zebra mussel (*Dreissena polymorpha*) on native bivalves (Unionidae) in Lake St. Clair. North American Benthological Society, Annual Meeting Abstracts, Louisville, Ky.

GRIFFITHS, R. W., D. W. SCHLOESSER, J. H. LEACH, et al. 1991. Distribution and dispersal of the zebra mussel (*Dreissena polymorpha*) in the Great Lakes region. Can. J. Fish. Aquat. Sci. 48:1381–88.

GUMA'A, S. A. 1978. The food and feeding habits of young perch, *Perca fluviatilis*, in Windermere. Freshw. Biol. 8:177–87.

HAAG, W. R. 1991. Genetic variation during ontogeny of a freshwater bivalve. M.S. Thesis, Ohio State Univ., Columbus.

HAAG, W. R. and D. W. GARTON. 1992. Synchronous spawning in a recently established population of the zebra mussel, *Dreissena polymorpha*, in western Lake Erie, USA. Hydrobiologia 234:103–10.

HAAG, W. R., D. J. BERG, D. W. GARTON, et al. 1993. Reduced survival and fitness in native bivalves in response to fouling by the introduced zebra mussel (*Dreissena polymorpha*) in western Lake Erie. Can. J. Fish. Aquat. Sci. 49:(in press).

HAMRIN, S. F. and L. PERSSON. 1986. Asymmetrical competition between age classes as a factor causing population oscillations in an obligate planktivorous fish species. Oikos 47:223–32.

HEBERT, P. D., B. W. MUNCASTER and G. L. MACKIE. 1989. Ecological and genetic studies on *Dreissena polymorpha* (Pallas): a new mollusc in the Great Lakes. Can. J. Fish. Aquat. Sci. 46:1587–91.

HERZIG, A. 1985. Resting eggs—a significant stage in the life cycle of crustaceans *Leptodora kindti* and *Bythotrephes longimanus.* Verh. Internat. Verein. Limnol. 22:3088–98.
JUDE, D. 1992. Impact of the tubenose and round gobies on Great Lakes ecosystems. North American Benthological Society, Annual Meeting Abstracts, Louisville, Ky.
KEILTY, T. J. 1990. Evidence for alewife (*Alosa pseudoharengus*) predation on the European cladoceran *Bythotrephes cederstroemi* in northern Lake Michigan. J. Great Lakes Res. 16:330–33.
LANGE, C. and R. CAP. 1986. *Bythotrephes cederstroemi* (Schoedler) (Cercopagidae: Cladocera): a new record for Lake Ontario. J. Great Lakes Res. 12:142–43.
LEACH, J. 1992. Population dynamics of larvae and adult zebra mussels in Lake Erie and changes in water quality in western Lake Erie. Second Annual Zebra Mussel Conference Meeting Abstracts, Toronto, Canada.
LEHMAN, J. T. 1987. Palearctic predator invades North American Great Lakes. Oecologia (Berlin) 74:478–80.
———. 1988. Algal biomass unaltered by food-web changes in Lake Michigan. Nature 332:537–38.
———. 1991. Causes and consequences of cladoceran dynamics in Lake Michigan: implications of species invasion by *Bythotrephes.* J. Great Lakes Res. 17:437–45.
LEWANDOWSKI, K. 1982. The role of early developmental stages in the dynamics of *Dreissena polymorpha* (Pall.) (Bivalvia) populations in lakes. I. Occurrence of larvae in the plankton. Ekol. Pol. 30:81–110.
LOCKE, A., D. M. REID, W. G. SPRULES, et al. 1991. Effectiveness of mid-ocean exchange in controlling freshwater and coastal zooplankton in ballast water. Can. Tech. Rpt. of Fish. Aq. Sci. No. 1822.
MACISAAC, H. J., W. G. SPRULES and J. H. LEACH. 1991. Ingestion of small-bodied zooplankton by zebra mussels (*Dreissena polymorpha*): can cannibalism on larvae influence population dynamics? Can. J. Fish. Aquat. Sci. 48:2051–60.
MACKIE, G. L. 1991. Biology of the exotic zebra mussel, *Dreissena polymorpha*, in relation to native bivalves and its potential impact in Lake St. Clair. Hydrobiologia 219:251–68.
MACKIE, G. L., W. N. GIBBONS, B. W. MUNCASTER, et al. 1989. The zebra mussel, *Dreissena polymorpha*: a synthesis of European experiences and a preview for North America. Ontario Ministry of the Environment.
MAKAREWICZ, J. C. and H. D. JONES. 1990. Occurrence of *Bythotrephes cederstroemi* in Lake Ontario offshore waters. J. Great Lakes Res. 16:143–47.
MILLS, E. L., J. LEACH, C. SECOR, et al. 1991. Species invasions in the Great Lakes. Abstract of presented paper, *in* Ecology and management of the zebra mussel and other introduced aquatic nuisance species. J. Yount (ed.). USEPA, Washington, D.C.
MITTON, J. B. and R. C. GRANT. 1984. Associations among protein heterozygosity, growth rate and developmental homeostasis. Ann. Rev. Ecol. Syst. 15:479–99.
MONAKOV, A. V. 1972. Review of studies on feeding of aquatic invertebrates conducted at the Institute of Biology of Inland Waters, Academy of Sciences, USSR. J. Fish. Res. Bd. Can. 29:363–83.
MORDUKHAI-BOLTOVSKAIA, E. D. 1958. Preliminary notes on the feeding of the carnivorous cladocerans *Leptodora* and *Bythotrephes.* Doklady Akad. Nauk. SSSR, Biol. Sci. Sect. 122:828–30.
———. 1960. The nutrition of the carnivorous Cladocerae *Leptodora* and *Bythotrephes.* Biull. Inst. Biologie Vodokr. 6:21–22.
MORDUKHAI-BOLTOVSKOI, PH.D. 1966. On the taxonomy of the Polyphemidae. Crustaceana 14:197–209.
——— (ed). 1978. Zooplankton and other invertebrates living on substrata in the Volga. Pages 235–68 *in* The river Volga and its life. Ph.D. Mordukhai-Boltovskoi (ed.). Junk bv Publishers, The Hague.
MORTON, B. S. 1969. Studies on the biology of *Dreissena polymorpha* (Pall.). II. Population dynamics. Proc. Malacological Soc. London 38:471–82.
MOYLE, P. B. 1986. Fish introductions into North America: patterns and ecological impact. Pages 27–43 *in* Ecology of biological invasions of North America and Hawaii. H. Mooney and J. Drake (eds.). Springer-Verlag, New York, N.Y.
NALEPA, T. F, B. A. MANNY, J. C. ROTH, et al. 1991. Long-term decline in freshwater mussels (Bivalvia: Unionidae) of the western basin of Lake Erie. J. Great Lakes Res. 17:214–19.
NICHOLLS, K. H. and G. J. HOPKINS. 1992. Recent changes in Lake Erie (N. shore) phytoplankton: the relative effects of phosphorus loading and zebra mussels. Second International Zebra Mussel Conference Meeting Abstracts, Toronto, Canada.
PHIPPS, T. 1987. Genetics of seasonal changes in adult size in *Daphnia* (Cladocera: Daphnidae). Ph.D. Thesis, Ohio State Univ., Columbus.
REEDERS, H. H., A. BIJ DE VAATE and F. J. SLIM. 1989. The filtration rate of *Dreissena polymorpha* (Bivalvia) in three Dutch lakes with reference to biological water quality management. Freshw. Biol. 22:133–41.
SCAVIA, D., G. L. FAHNENSTIEL, M. S. EVANS, et al. 1986. Influence of salmonid predation and weather on long-term water quality trends in Lake Michigan. Can. J. Fish. Aquat. Sci. 43:435–43.
SCHLOESSER, D. W. and W. KOVOLAK. 1991. Infestation of unionids by *Dreissena polymorpha* in a power plant canal in Lake Erie. J. Shellfish Res. 10:355–59.

SCHLOESSER, D. W., J. R. FRENCH, S. J. NICHOLLS, et al. 1992. Zebra mussels in North America: distribution, biology, impacts and control. North American Benthological Society, Annual Meeting Abstracts, Louisville, Ky.

SCHNEEBERGER, P. J. 1991. Seasonal incidence of *Bythotrephes cederstroemi* in the diet of yellow perch (ages 0–4) in Little Bay de Noc, Lake Michigan, 1988. J. Great Lakes Res. 17:281–85.

SEBESTYEN, O. 1938. Colonization of two new fauna-elements of Pontus-origin (*Dreissena polymorpha* Pall. and *Corophium curvispinum* G.O. Sars) in Lake Balaton. Verh. Internat. Verein. Limnol. 8:169–81.

SELGEBY, J. H. and D. H. Ogle. 1992. Impacts of the invading ruffe (*Gymnocephalus cernuus*) on the native fish community of the St. Louis River. North American Benthological Society, Annual Meeting Abstracts, Louisville, Ky.

SPRULES, W. G., H. R. RIESSEN and E. H. JIN. 1990. Dynamics of the *Bythotrephes* invasion of the St. Lawrence Great Lakes. J. Great Lakes Res. 16:346–51.

SPRUNG, M. 1989. Field and laboratory observations of *Dreissena polymorpha* larvae: abundance, growth, mortality and food demands. Arch. Hydrobiol. 115:537–61.

STANCYZKOWSKA, A. and M. PLANTER. 1985. Factors affecting nutrient budget in lakes of R. Jorka watershed (Masurian lakeland, Poland) X. Role of *Dreissena polymorpha* (Pallas) in N and P cycles in a lake ecosystem. Ekol. Pol. 33:345–56.

STENSON, J. 1972. Fish predation effects on the species composition of the zooplankton community in eight small forest lakes. Inst. Freshwater Res. Drottningholm Rep. 52:132–48.

———. 1978. Relations between vertebrate and invertebrate zooplankton predators in some Arctic lakes. Astarte 11:21–26.

STOECKMANN, A. M. and D. W. GARTON. 1991. Metabolic responses to increased food supply and induced spawning. Second International Zebra Mussel Research Conference Abstracts, Rochester, N.Y.

THOMAS, N. A. 1981. Ecosystem changes in lakes Erie and Ontario. Pages 1–20 *in* Proceedings conference on changes in the biota of lakes Erie and Ontario. R. Cap and V. Frederick (eds.). Bull. Buffalo Soc. Nat. Sci. Vol. 25(4).

WALZ, N. 1978. The energy balance of the freshwater mussel *Dreissena polymorpha* Pallas in laboratory experiments and in Lake Constance. II. Reproduction. Arch. Hydrobiol. (Suppl.) 55:106–19.

WARREN, G. J. and J. T. LEHMAN. 1988. Young-of-the-year *Coregonus hoyi* in Lake Michigan: prey selection and influence on the zooplankton community. J. Great Lakes Res. 14:420–26.

WEIDER, L. J. 1991. Allozymic variation in *Bythotrephes cederstroemi*: a recent invader of the Great Lakes. J. Great Lakes Res. 17:141–43.

WIKTOR, J. 1963. Research on the ecology of *Dreissena polymorpha* (Pall.) in the Szczecin Lagoon (Zalew Szczecinski). Ekol. Pol. 11:275–80.

YAN, N. D., W. I. DUNLOP, T. W. PAWSON, et al. 1992. *Bythotrephes cederstroemi* (Schoedler) in Muskoka lakes: first evidence of the European invader in inland lakes in Canada. Can. J. Fish. Aquat. Sci. 49:422–26.

YURISTA, P. 1992. Embryonic and post-embryonic development in *Bythotrephes cederstromii.* Can. J. Fish. Aquat. Sci. 49:1118–25.

ZHADIN, V. I. 1952. Mollusks of fresh and brackish waters of the USSR. Akad. Nauk USSR. (Translated from Russian, U.S. Department of Commerce, 1965).

ZOZULYA, S. S. and PH.D. MORDUKHAI-BOLTOVSKOI. 1977. Seasonal variability of *Bythotrephes longimanus* (Crustacea, Cladocera). Doklady Akad. Nauk. SSSR, Biol. Sci. Sect. 232:75–77.

# Cut-Leaved and Common Teasel (*Dipsacus laciniatus* L. and *D. sylvestris* Huds.): Profile of Two Invasive Aliens

Mary Kay Solecki[1]

## INTRODUCTION

Two species of teasel, cut-leaved teasel (*Dipsacus laciniatus*) and common teasel (*Dipsacus sylvestris*), are invasive plant species alien to North America that can be problematic in Midwestern natural areas. Though both species are somewhat similar in appearance, cut-leaved teasel is distinguished by deeply lobed or laciniate leaves, white flowers, and involucral bracts that do not surpass the head length, while common teasel has undivided leaves, pink or purple flowers, and involucral bracts that are longer than the head (Fernald 1950; Werner 1979) (Fig. 1). In Illinois, cut-leaved teasel is more aggressive than common teasel and has severely threatened the natural quality of several natural areas in northern and central counties. If left unchecked, cut-leaved teasel quickly can form large monocultures excluding most native vegetation.

## DISTRIBUTION

Teasel is an Old World genus native to Eurasia and northern Africa (Gleason 1952). Like many exotic plant immigrants it arrived in North America prior to 1900 (Werner 1979). Early settlers probably introduced both teasel species deliberately as ornamentals or coincidentally with decorations or toys made from the flowering heads (Werner 1979). Another species, the cultivated teasel (*Dipsacus fullonum*), was introduced for the use of its bristly heads in raising or "teasing" the nap of wool cloth (Mullins 1951). Possibly cut-leaved and common teasel were introduced with *D. fullonum* or introduced accidentally with other plant material from Europe.

Common teasel is distributed throughout most of the United States, except for the north-central United States (Fig. 2) (Fernald 1950; Hitchcock and Cronquist 1973; Werner 1979; Elmore 1989). Cut-leaved teasel, which has a more restricted distribution, primarily occurs in the northeastern and Midwestern United States (Fig. 2) (Fernald 1950; Ferguson 1965; Salamun and Cochrane 1974; Wherry et al. 1979; Seymour 1982; Barkley 1986). Common teasel was already widely distributed in the northeastern United States by 1913, while cut-leaved teasel was only reported from Albany, New York (Britton and Brown 1913).

Cut-leaved teasel has spread rapidly in the Midwest during the last 10 to 30 years. This rapid range expansion probably was aided by the interstate highway system which acts as a dispersal corridor. For example, cut-leaved teasel was first recorded from Missouri in 1980 in the vicinity of Kansas City. It is now recorded from 24 Missouri counties and is particularly prevalent along Interstate 70 and

1. Illinois Nature Preserves Commission, P.O. Box 497, Sidney, IL 61877

FIGURE 1.—Inflorescence of *Dipsacus sylvestris* with characteristic bracts in Champaign Co., Illinois. (photo by B. N. McKnight)

the counties bordering it (George Yatskievych, pers. comm.). In Illinois, cut-leaved teasel is most abundant along roadsides and interstate highways. Large cut-leaved teasel populations can now be found along Illinois interstates and state highways that had no or only small scattered populations 5 or 10 years ago (pers. obser.). Presently, many of these populations consist of dense colonies containing several hundred to several thousand individuals.

Horticultural use of common teasel has aided in expansion of its North American range. For example, teasel used in floral decorations at grave sites has resulted in dispersal in and around cemeteries. Swink and Wilhelm (1979) mention that common teasel "often escapes near cemeteries, florist establishments, and greenhouses."

## HABITAT

The preferred habitat of both species of teasel can be characterized as open sunny conditions with few or no trees or shrubs. Optimal conditions seem to be mesic habitats although it may occasionally be found in dry sites. The largest

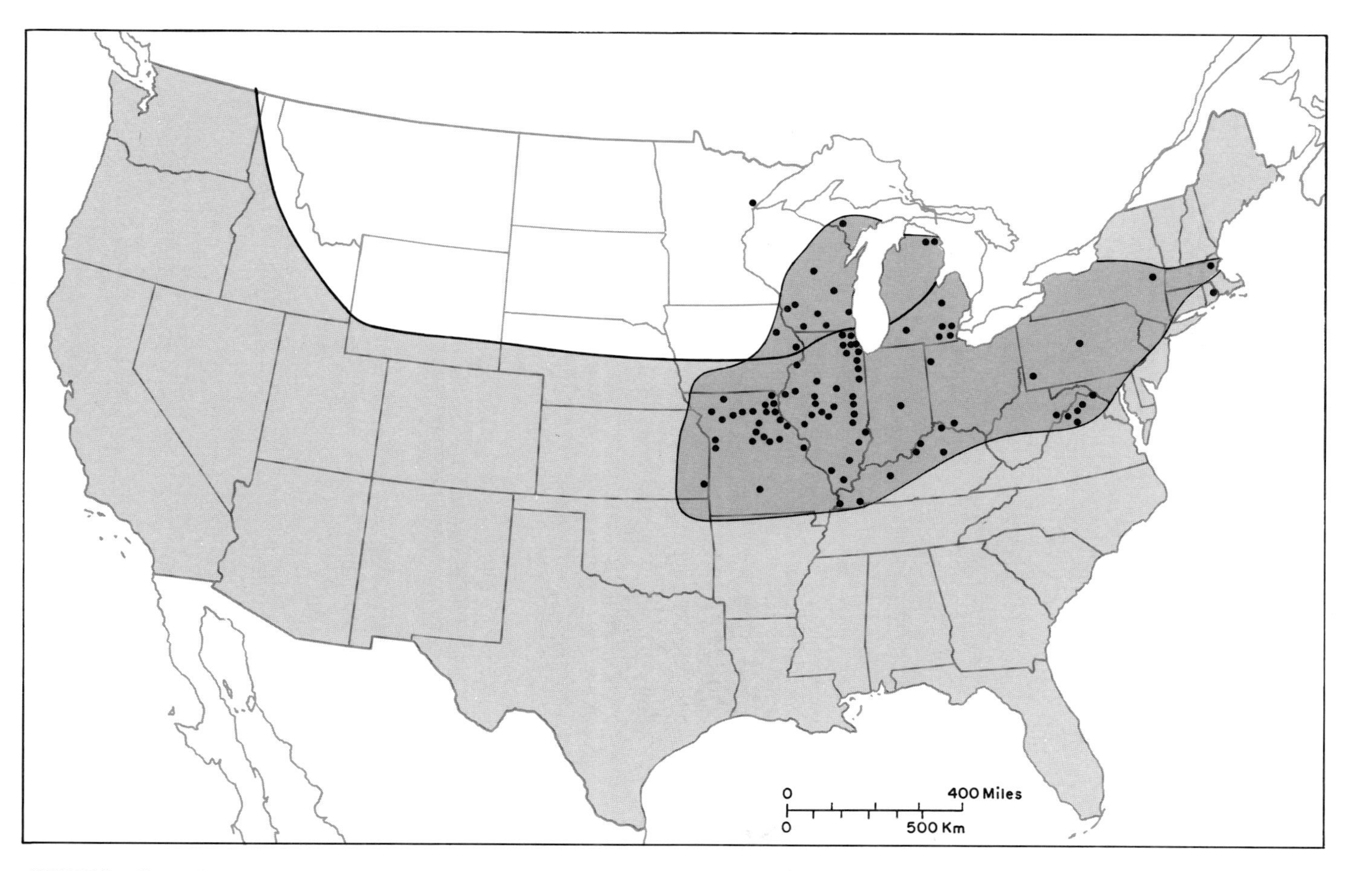


FIGURE 2.—Generalized range map of *Dipsacus sylvestris* (light gray) within the United States, adapted from Elmore (1989) and generalized range map of *Dipsacus laciniatus* within the United States. Counties from which herbarium specimens or confirmed reports are known (see acknowledgments) appear in black.

teasel plants are found when soil moisture is maintained relatively high throughout the growing season. Cut-leaved teasel often occurs in wetter habitats than common teasel (Werner 1979). Common teasel occurs on a variety of soils, from sandy soils with abundant available moisture to heavy clay soils. Both teasel species tolerate saline conditions found along roadsides. Each species typically occurs as a monospecific colony, although the two species occur together occasionally.

Roadsides and cemeteries act as refuges from which teasel can invade nearby natural communities. Illinois natural areas infested with large teasel populations are usually adjacent to or near cemeteries or roads harboring teasel populations. Natural communities that have been invaded by teasel include high quality prairies, savannas, seeps, and sedge meadows. Cut-leaved and common teasel typically are not invasive in forests or dense, shrubby areas, although they can inhabit moist forest openings (Werner 1979).

## LIFE HISTORY

Although often considered biennials, common and cut-leaved teasel are actually monocarpic perennials. The plant grows as a basal rosette for a minimum of one year, although this rosette period frequently is longer. After an adequate rosette is formed, the plant forms a tall flowering stalk. It dies after flowering. The period of time in the rosette stage apparently varies depending on the amount of time needed to acquire adequate resources for flowering to occur (Werner 1975b). Typically, common teasel rosettes form a flowering stalk only after reaching a critical size of about 30 cm in diameter (Werner 1979), although flowering does occur rarely with plants as small as 7 cm in diameter (Werner 1975b).

Several attributes enable teasel to establish large populations. Common teasel possesses a thick, well-developed taproot that may reach over 75 cm deep (Werner 1979). When mature, cut-leaved teasel generally is a larger, more robust plant than common teasel and probably has a deeper taproot. These large, deep taproots provide substantial water and carbohydrate storage. Moreover, both common and cut-leaved teasel are heavily armed with numerous spines on the leaves and stems. The flower heads themselves have sharp barbs and stiff bracts that present an effective defense against grazing by herbivores. Finally, both species are photosynthetically active for a longer period than most native species occupying the same habitats, becoming green earlier in the spring and remaining green longer in autumn. These collective attributes appear to give both teasel species a competitive advantage over many native plant species in open habitats.

Seed production is high for both species, and seeds readily germinate at high rates. Werner (1979) found that a single common teasel plant can be expected to produce approximately 3,333 seeds. No pre-germination requirements of freezing or cold have been found for common teasel (Werner 1979). This is in contrast to many native herbaceous perennials that often require cold stratification before germination will occur. In addition, common teasel seeds stored under dry conditions remain viable at least six years (Werner 1979). Werner also observed that common teasel seeds harvested in October from naturalized populations germinate at rates exceeding 90 percent on soil or moist filter paper in a laboratory or greenhouse. Likewise, mature cut-leaved teasel seeds germinate at rates above 90 percent on moist filter paper (Solecki 1989).

Under a variety of field conditions, common teasel seed germination ranges from 28 to 86 percent (Werner 1979). The presence of plant litter inhibits germination of common teasel, and successful teasel establishment is apparently related to the amount and type of plant litter and cover on the site (Werner 1975a; Werner 1979). Germination was lowest in a field with a thick cover of quack grass and its litter and highest in the field with the highest amount of bare ground and the least quack grass (Werner 1979).

When mature, both common and cut-leaved teasel seeds fall passively from the parent plant. Ninety-nine percent of common teasel seeds are deposited less than 1.5 m from the parent plant (Werner 1979). At times, however, animals, including humans, bump the plants or heads causing seed to scatter farther. Ants, goldfinches, and blackbirds are reported to eat the seeds (Werner 1979) and may act as dispersal agents. Long-distance seed dispersal probably is aided, at least in part, by the ability of teasels to float and subsequently germinate. In a laboratory experiment, common teasel seeds germinated after floating 16 days in distilled water. This ability to germinate after floating allows waterways such as streams and ditches to provide dispersal corridors (Werner 1979).

Parent plants often provide an optimal nursery site for new teasel plants after the adult dies. The mature rosettes are wide with leaves that spread horizontally over the ground. Dead adult plants leave a relatively large area of bare ground, formerly occupied by their own basal leaves, that new plants readily occupy.

The combination of these life history attributes, including the high seed and seedling production, the lack of natural predators, and the provision of nursery sites by parent plants contribute to the ability of teasels to invade and establish large populations. Interestingly, teasel apparently does not reproduce vegetatively, an attribute common to many aggressive exotic plants.

## CASE HISTORY

Loda Cemetery Prairie, a prairie owned by The Nature Conservancy in east-central Illinois (Iroquois Co.), provides an interesting case history of cut-leaved teasel spread and control efforts. Cut-leaved teasel invaded the 1.4 ha (3.4 acres) Loda Cemetery Prairie from an adjacent cemetery and dramatically increased in numbers from about 1982 to 1988 when it was one of the dominant plants in the prairie. For seven years, efforts to control cut-leaved teasel at this prairie consisted of annually cutting stems at the base after flower buds developed (prior to peak flowering) and leaving the cut stems lying on the ground. Stems that resprouted were cut again the same year. It was thought that repeated cutting of cut-leaved teasel would prevent on-site seed set and eventually decrease the population size and the seed bank.

In addition, the prairie was periodically burned in spring, typically on a rotation basis with each half of the prairie burned on an alternating, biennial regime. This spring burning regime did not effectively control cut-leaved teasel in part because dense colonies of green rosettes did not burn well. Although isolated rosettes usually suffered fire damage, in many instances the central core of the rosette remained undamaged, enabling the rosette to recover.

By 1988 the cut-leaved teasel population was not noticeably reduced in size, and it seemed apparent that control efforts were not preventing establishment of new seedlings. Monitoring plots contained up to 1,926 cut-leaved teasel seedlings in a single 3 $m^2$ plot (Solecki 1989). It was thought that cut flowering stems left lying in the prairie were producing viable seed.

Germination studies showed that cut-leaved teasel seeds harvested at Loda Cemetery Prairie from inflorescences that were green and only partially flowering at the time of harvest were subsequently viable, and germination success increased with post-harvest seed age (Fig. 3) (Solecki 1989). Germination rates of seven-month-old seed stored in the laboratory and germinated on moist filter paper are comparable to rates reported by Werner (1979) for mature common teasel seed (over 90 percent) germinated under similar conditions.

After documenting the ability of immature cut-leaved teasel seeds to germinate, control measures at Loda Prairie were modified. Currently, foliage of individual teasel rosettes is sprayed with 1.5–2 percent glyphosate during early spring. Any cut-leaved teasel stems that form flowering stalks are cut, usually in late July or early August, and the cut flowering stems are removed from the prairie and disposed of off-site to prevent on-site seed dispersal. Following three years of this modified control regime, the cut-leaved teasel population was markedly reduced. Many new seedlings, however, have emerged each year. Several years of continued control effort likely will be needed before the cut-leaved teasel seed bank is depleted.

## CONTROL MEASURES

An effective control strategy for common and cut-leaved teasel must concentrate on eliminating or limiting the number of plants established and on preventing the dispersal of viable seed. Cutting, removal, burning and/or herbicides offer solutions for control in natural areas. The following control information taken from Glass (1991) is pertinent for both species.

Individual rosettes can be extracted using a dandelion digger or similar tool, taking care to remove as much of the root as possible to prevent resprouting. Alternatively, flowering stalks of flowering plants can be cut once flowering has initiated. Normally the plant will not reflower and will die at the end of the growing season. To prevent dispersal of viable seed, cut-flowering stalks should be removed from the site. Cutting the flowering stalk before flower buds develop should be avoided because the plant will frequently send up new flowering stalk(s). Cutting of flowering stems may need to be repeated for several years to achieve control over both species of teasel. Teasel in nearby areas should also be eliminated to prevent introduction of new seed.

Periodic late spring burns may be useful for controlling teasel before it becomes dense. However, once an area is densely covered with teasel rosettes, fire does not carry well through the teasel-infested area. Prescribed burns probably work best in conjunction with other control methods.

Foliar application of glyphosate (available under the trade name Roundup) or 2,4-D amine herbicide (available under a number of trade names) is effective and useful where other treatments mentioned above are not feasible. Although glyphosate is most effective during summer when the plant is actively growing, it is also effective in late autumn or early spring. Application during late autumn or early spring will result in less potential harm to non-target species because most native plants are not photosynthetically active at these times. Roundup should not be used to control teasel in natural areas during the active growing season of most native plants. This will result in unnecessary injury to native species because glyphosate is nonselective (it will kill any plants it contacts).

In high-quality natural areas, Roundup should be applied carefully by hand-sprayer to individual teasel rosettes at a 1.5 percent solution according to label instructions during late fall or early spring. Spray coverage should be uniform and complete. Do not spray so heavily that herbicide drips off the target species.

Application of 2,4-D amine is selective to broadleaf plants; it will not harm most grasses. The herbicide, 2,4-D amine, should be applied in early spring when the rosettes are young. The 2,4-D amine should be applied by hand-sprayer at the recommended application rate on the label for spot-spraying weeds. Application should be uniform with the entire leaf being wet. As with Roundup, do not spray

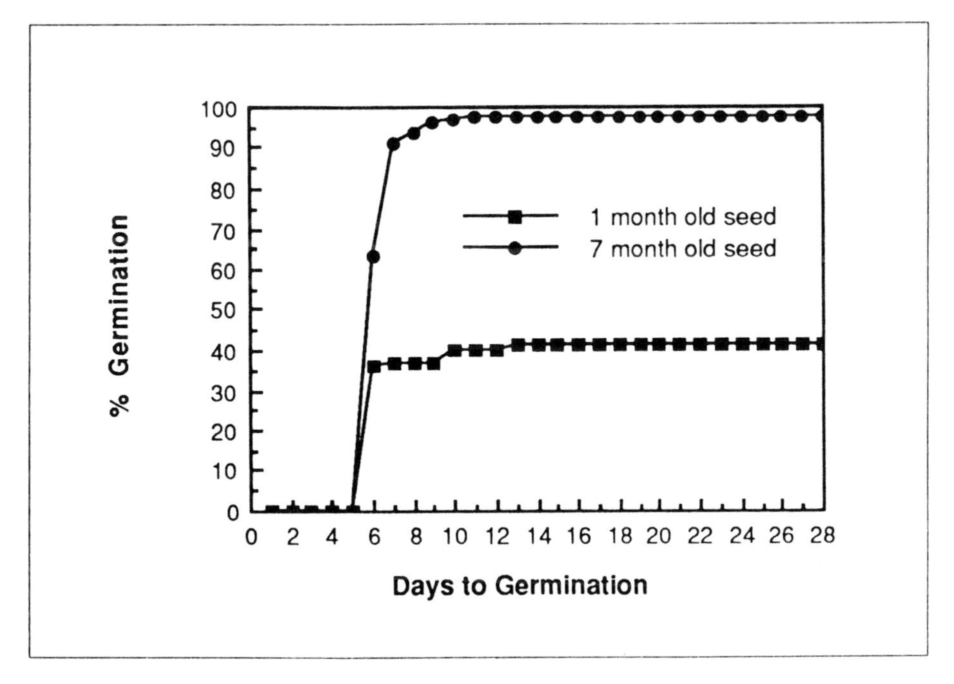


FIGURE 3.—Percent germination of cut-leaved teasel (*Dipsacus laciniatus* L.) seed harvested from stems cut prior to full flowering. Seed ages are months after harvest. (reprinted with permission from the Natural Areas Journal)

so heavily that herbicide drips off the target species. The amine formulation of 2,4-D should be used, rather than the ester formulation, to reduce vapor drift. Using Roundup or 2,4-D amine on the rosettes may have to be followed with cutting of any flowering stalks that survive spraying.

The above control measures can reduce population levels of teasel but rarely result in complete eradication of well-established, large populations. Biological control measures may offer alternate control techniques eventually. Currently, however, this has received little attention, and no biological controls have been reported that are feasible in natural areas.

Finally, any potential area of invasion should be monitored periodically for the presence of teasel. As with many invasive exotics, it is easiest to control teasel when it is at low population levels and before it has an opportunity to establish a large soil seed bank. Given teasel's ability to rapidly build up its population to large levels, early detection is an essential component of an effective control strategy.

## ACKNOWLEDGMENTS

I thank John Taft and two anonymous reviewers for their thoughtful review of the manuscript. I also thank curators and personnel of the following herbaria and agencies for providing distributional information: Illinois Natural History Survey, Kentucky Nature Preserves Commission, Indiana Nature Preserves Commission, Indiana University, The Nature Conservancy, University of California—Berkeley, University of Iowa, University of Michigan, University of Minnesota, University of Oregon, University of Pennsylvania, University of Tennessee, University of Wisconsin—Madison, and Virginia Polytechnic Institute.

## REFERENCES


Barkley, T. M. (ed.). 1986. Flora of the Great Plains. Univ. Press of Kansas, Lawrence.

Britton, N. and A. Brown. 1913. An illustrated flora of the northeastern United States and Canada, Vol. III. Dover reprint ed., 1970. Dover Publications, New York, N.Y.

Elmore. C. D. (ed.). 1989. Weed identification guide. Southern Weed Science Society, Champaign, Ill.

Fernald, M. L. 1950. Gray's manual of botany. 8th ed. D. Van Nostrand Co. New York, N.Y.

Ferguson, I. K. 1965. The genera of Valerianaceae and Dipsacaceae in the southeastern United States. J. Arnold Arboretum. 46:218–31.

Glass, B. 1991. Vegetation management guideline: cut-leaved teasel (*Dipsacus laciniatus* L.) and common teasel (*Dipsacus sylvestris* Huds.). Nat. Areas J. 11:213–14.

Gleason, H. A. 1952. The new Britton and Brown illustrated flora of the northeastern United States and adjacent Canada. Vol. 3. The sympetalous dicotlyedoneae. Hafner Press, New York, N.Y.

Hitchcock, C. L. and A. Cronquist. 1973. Flora of the Pacific Northwest. Univ. Washington Press, Seattle.

Mullins, D. 1951. Teasel growing, an ancient practice. World Crops 3:146–47.

Reed, C. F. 1970. Selected weeds of the United States. USDA-ARS Agricutural Handbook No. 366. U.S. Government Printing Office, Washington, D.C.

Salamun, P. J. and T. S. Cochrane. 1974. Preliminary reports on the flora of Wisconsin No. 65. Dipsacaceae—teasel family. Trans. Wisconsin Acad. Sci., Arts and Let. 62:253–60.

Seymour, F. C. 1982. The flora of New England. Moldenke, Plainfield, N.J.

Solecki, M. K. 1989. The viability of cut-leaved teasel (*Dipsacus laciniatus* L.) seed harvested from flowering stems: management implications. Nat. Areas J. 9:102–05.

Werner, P. A. 1975a. The effects of plant litter on germination in teasel, *Dipsacus sylvestris* Huds. Amer. Midl. Nat. 94:470–76.

———. 1975b. Prediction of fate from rosette size in teasel (*Dipsacus fullonum* L.). Oecologia 20:197–201.

———. 1979. The biology of Canadian weeds. 12. *Dipsacus sylvestris* Huds. Pages 134–45 *in* G. Mulligan (ed.). The biology of Canadian weeds, contributions 1–32. Information Services, Agriculture Canada, Ottawa, Ont.

Wherry, E. T., J. Fogg and H. Wahl. 1979. Atlas of the flora of Pennsylvania. Univ. Pennsylvania, Morris Arboretum, Philadelphia.

# Native and Exotic Earthworms in Deciduous Forest Soils of Eastern North America

Paul J. Kalisz[1]

## INTRODUCTION

It is difficult to obtain accurate information about organisms that live underground. This generalization applies to earthworms (Annelida: Oligochaeta) not only because they live underground but also because the 100 or so species found in the eastern United States are superficially similar. Earthworms are widespread and common; they seem familiar to us, when in fact we know little concerning the food, reproductive systems, competitive relations, and other life history characteristics of most species. Frank Smith (1925), a zoologist at the University of Illinois, described this lack of accurate information by referring to " . . . earthworms of two quite different kinds, those that are found out of doors, . . . and those that are found in textbooks . . . ," and commented that "It is no more reasonable to talk about 'the common earthworm' than it would be to talk about the common species of bird or the common kind of fish."

Although much has been learned about earthworm taxonomy and anatomy in the 66 years since Smith's comments, we are still steeped in ignorance regarding basic ecological and biogeographical relationships. This ignorance extends to our understanding of the distribution of earthworm species prior to the wholesale disturbance of ecosystems, and introduction and extirpation of species that occurred over the last 300 to 500 years in conjunction with European colonization of North America and other parts of the world. Put simply—we are still not absolutely sure which of the earthworm species now inhabiting eastern North America are exotic.

## BIOGEOGRAPHICAL CONSIDERATIONS

The two most important events affecting the present distribution of earthworms in eastern North America were glaciation and European settlement. During the most recent (Wisconsinian) glacial advance, which peaked about 18,000 years ago, ice extended southward into the present deciduous forest region of eastern North America. Wisconsinian glacial ice extended as far south as southern Indiana and south-central Ohio (to about 39° N latitude); earlier glacial advances had extended slightly further south in the deciduous forest region and substantially further south in Illinois and in the prairie region to the west (Fig. 1). Everyone agrees that earthworms were extirpated from ice-covered land areas (e.g., Lee 1985; Sims and Gerard 1985). There is disagreement, however, concerning the nature of the earthworm fauna that occurred in the glaciated portion of eastern North America prior to glaciation, and concerning the possibility that remnants of the pre-glacial earthworm fauna persisted throughout the Quaternary on ice-free nunataks and coastal refugia north of the glacial limits (see Pielou 1991, for a discussion of the occurrence of ice-free land areas during gla-

1. Department of Forestry, University of Kentucky, Lexington, KY 40546-0073

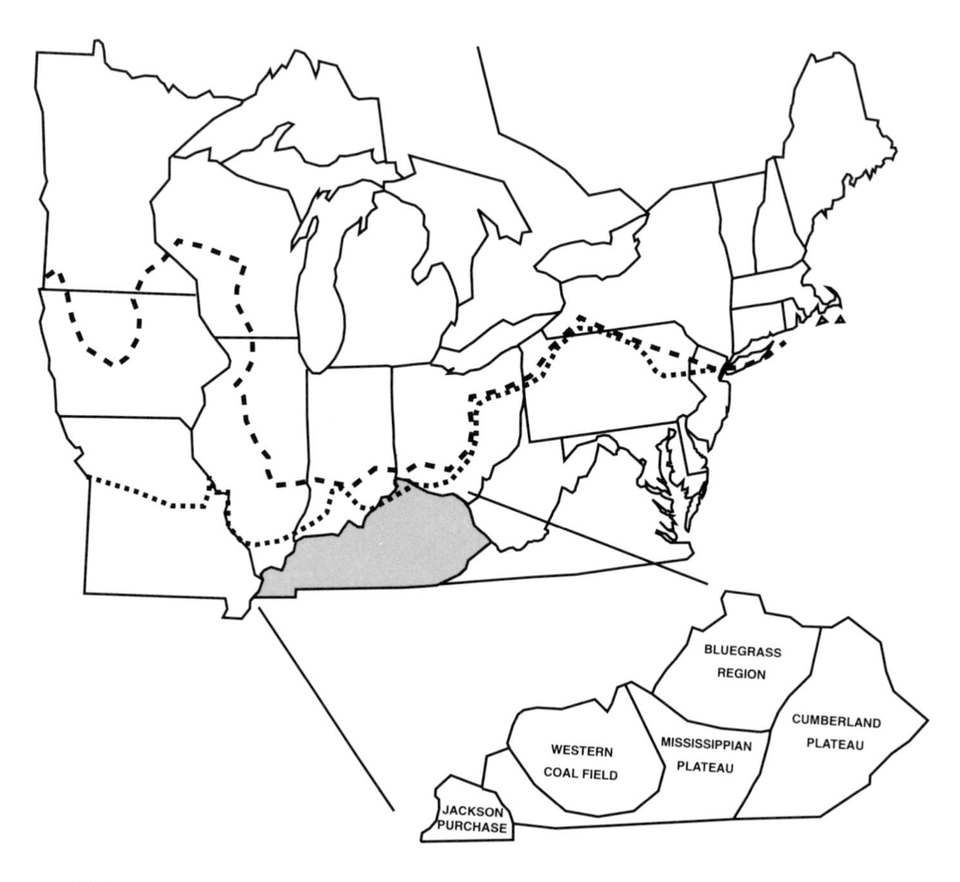


FIGURE 1.—Map of a portion of the eastern United States showing state boundaries, the southern limits of the Wisconsinian glaciation (heavy dashed line), earlier glaciations (light dashed line), and the physiographic divisions of Kentucky.

cial advances). This disagreement is fundamental to defining "exotic" and "native" as applied to earthworm species in eastern North America and is most clearly expressed in the conflicting ideas of G. E. Gates (1966, 1976) and P. Omodeo (1963).

With only minor differences among taxonomists, 5 genera and over 60 species of earthworms in 5 families (Table 1) are considered endemic to eastern North America (Sims and Gerar 1985; James 1990; Schwert 1990). There is disagreement, however, concerning whether the 20 or so species of the family Lumbricidae that are now dominant north of the glacial limits (Table 2) are exotic or native. These are "peregrine" species that are adapted to prosper in disturbed habitats (Lee 1985; Sims and Gerard 1985) and have spread from the Palearctic region throughout the world in association with humans. Gates (1966, 1970) considered the endemic taxa (Table 1) to have survived Quaternary glaciations by taking refuge in the southeastern United States. He pictured glaciated North America as devoid of earthworms after the Wisconsinian glaciation and believed that earthworms now inhabiting this region (Table 2) were introduced by European colonists during the last 400 to 500 years. This interpretation was based chiefly on the fact that Gates found the endemic taxa mostly south of the glacial limits while the "exotics" were primarily northern. Gates explained occasional reports of native earthworms, especially *Bimastos* spp. and *Sparaganophilus* spp., far to the north in locations such as Michigan and Massachusetts due to the redistribution of endemics by European colonists. He also called upon the results of his own surveys of greenhouses (Gates 1963) and on the records of the U.S. Bureau of Quarantine (Gates 1976) to show that stowaways in potted plants, packing material, and ship ballast could account for earthworm introductions to North America. Gates found over 12 exotic genera in greenhouse soils and found records that over 40 exotic species, including all but two or three of the "exotic" taxa (Table 2), had been intercepted at ports of entry.

TABLE 1. Endemic families and genera of earthworms in eastern North America. Approximate numbers of species are given in parentheses. Taxonomy according to Sims and Gerard (1985).

|  |
|---|
| AILOSCOLECIDAE [KOMAREKIONIDAE] |
| *Komarekiona* (1) |
| LUMBRICIDAE |
| *Bimastos* (9) |
| *Eisenoides* (2) |
| LUTODRILLIDAE |
| *Lutodrilus* (1) |
| ACANTHODRILIDAE [MEGASCOLECIDAE] |
| *Diplocardia* (>40) |
| SPARGANOPHILIDAE |
| *Sparganophilus* (10) |

Omodeo (1963), based chiefly on biogeographical considerations, argued that many of the species of Lumbricidae now found north of the glacial limits in North America were remnants of a pre-glacial (Tertiary) Holarctic fauna that somehow survived glaciation *in situ* and then re-colonized parts of their former range. Thus, according to Omodeo, some of Gates' "exotics" occurred north of the glacial limits long before European colonization of North America. A fossil cocoon of *Dendrodrilus rubidus* discovered in marly lake sediments in southern Ontario and dated at > 10,000 years before present (Schwert 1979) seems to support Omodeo's contention that some species of Lumbricidae occurred north of the glacial limits long before European contact. Omodeo also referred to a number of ecological relationships that suggested that earthworms had occurred north of the glacial limits in North America for relatively long periods of time. For example, although earthworms were recognized as important agents in the genesis of soils, there was no evidence that soils of glaciated North America differed from soils elsewhere that were known to have developed under the influence of earthworms. Similarly, animals specialized to feed on earthworms, such as moles (Talpidae) and woodcock (Scolopacidae; *Philohela minor*), occur far north of the glacial limits in North America; this can only be explained by either the long-term presence of earthworms in the north or else major changes in the ranges or diets of these animals over the last 400 to 500 years.

Earthworms are soft-bodied and composed of about 90 percent water on a fresh mass basis and about 75 percent protein on a dry mass basis (Lee 1985). Therefore, dead earthworms decompose rapidly and are not preserved as fossils in rocks or as skeletal or other remains in dry burial sites such as caves or rock-shelters. Thus, we cannot generally rely on dated remains to interpret temporal and spatial patterns of earthworm occurrence in the past. Given the absence of a fossil record, the fact that natural ecosystems have been drastically altered, and that many types of organisms, including earthworms, have been introduced,

TABLE 2. Important nonendemic families and genera of earthworms in eastern North America. Approximate numbers of species are given in parentheses. Taxonomy according to Sims and Gerard (1985).

| | | |
|---|---|---|
| LUMBRICIDAE | | |
| *Allolophora* (1) | *Dendrodrilus* (1) | *Lumbricus* (4) |
| *Aporrectodea* (5) | *Eisenia* (2) | *Murchieona* (1) |
| *Dendrobaena* (1) | *Eiseniella* (1) | *Octolasion* (2) |
| MEGASCOLECIDAE | | |
| *Pheretima* sensu lato [*Amynthas, Metaphire*] (10) | | |

extirpated, and translocated within North America by humans, the conflict between the ideas of Gates and Omodeo probably cannot be resolved. Both interpretations are rational and account for all or most known facts, but neither interpretation is testable (Ball 1975). At present, many standard references on earthworms (for example, Edwards and Lofty 1972; Sims and Gerard 1985) seem implicitly to accept Gates' interpretation, although the reasons for doing so are not clearly presented.

Whether or not the species of Lumbricidae (Table 2) that presently dominate the glaciated portions of eastern North America are true exotics, these species were apparently rare south of the glacial limits as recently as 15 to 30 years ago (Stebbings 1962; Gates 1970; Reynolds 1974; Reynolds et al. 1974). These species may therefore be considered to be exotic to the southeastern United States, and their occurrence in soils under deciduous forests in this region may be considered the result of invasion from the glaciated portions of North America, if not from the Palearctic. In the following discussion of native-exotic relationships in deciduous forests south of the glacial limits, I will therefore use "exotic" to refer to the 20–30 taxa of nonendemic earthworms (Table 2) that now dominate northern North America.

## PARADIGMS FOR EXOTIC-NATIVE EARTHWORM INTERACTIONS

About 30 years ago Stebbings (1962) observed that exotic earthworms were replacing natives in severely disturbed habitat in Missouri, an observation that was in accord with the generalization that habitat disturbance is essential for successful invasion by exotic species (Elton 1958; Orians 1986). Research on native-exotic earthworm interactions in New Zealand (Lee 1985), Australia (Abbott 1985), and eastern Kentucky (Kalisz and Dotson 1989) also supports a paradigm that envisions the following sequence of events leading to the replacement of natives by exotic earthworms: (1) severe disturbance of the habitat, (2) extirpation of native earthworm populations, and (3) occupation of vacated habitat by introduced earthworms. This paradigm was developed in regions with extensive and nonfragmented forests and has not been tested in regions where forests occur as remnant patches in a generally disturbed and altered landscape. The following discussion will compare the validity of this paradigm for the Cumberland Plateau of Kentucky, a nonfragmented forest region, and for the Bluegrass, a fragmented forest region (Fig. 1). My objective is to generalize about conditions which determine whether or not exotic species of earthworms occur in a particular forest and whether native or exotic species are dominant. My generalizations are based partly on data from studies dealing specifically with earthworms on the Cumberland Plateau (Dotson and Kalisz 1989; Kalisz and Dotson 1989), on records collected during research projects dealing with topics other than earthworms, on field notes relating to soil pits dug for teaching and demonstration, and on data from a current cooperative project with the Kentucky Chapter of The Nature Conservancy dealing with earthworms in the Bluegrass.

## STUDY AREAS AND METHODS

Both the Cumberland Plateau and the Bluegrass (Fig. 1) were originally covered by deciduous forest (Braun 1950), but the natural environments and human-use histories of the two regions are very different (Table 3). After the original forests were cut, most forests on the Cumberland Plateau were allowed to immediately redevelop, but most land in the Bluegrass was permanently converted to other uses. Forest now forms the matrix of the Cumberland Plateau; severe disturbance is restricted to spots or strips. The Bluegrass, on the other hand, is a perfect example of a fragmented forest: fields, pastures, and urbanized areas form the matrix while forests occur in remnant patches and strips.

Data were collected from a variety of severely and slightly disturbed forests; slight disturbance was defined to include logging but not forest clearance and conversion to agricultural fields, pastures, urbanized areas, or other types of intensive human use. Data from the Cumberland Plateau were collected primarily on the extensive forest lands of the University of Kentucky's Robinson Forest and the Daniel Boone National Forest (centrally located at 37° 30′N, 83° 30′W). In the Bluegrass, collections were made at 12 locations in seven counties (centrally located at 38° N, 84° 30′ W); the sample locations were forested and had never been cleared and converted to other uses, although cutting and grazing may have occurred in the past. The smallest Bluegrass forest was 6 ha, and the largest occurred as part of a continuous forested strip on the cliffs (Palisades) bordering the Kentucky River. In all Bluegrass forests, samples were collected away from forest edges and in locations that seemed to be minimally disturbed. Volumetric soil samples were collected from 0.10 or 0.25 $m^2$ areas to a depth of 10, 30, or 50 cm, and earthworms were hand-sorted in the field and identified or transported to the laboratory in cartons of moist soil for identification. Additional methodological details are given in Dotson and Kalisz (1989) and Kalisz and Dotson (1989).

TABLE 3. Summary of the characteristics of the Cumberland Plateau and Bluegrass Physiographic Regions of Kentucky.

| CUMBERLAND PLATEAU |
| --- |
| Mountains (<10% of land area is flat) |
| Acid soils from sandstone, siltstone, and shale (Pennsylvanian) |
| Settled for 100 to 150 years |
| >80% forested |
| BLUEGRASS |
| Rolling karst plain |
| Calcareous soils from limestone and shale (Ordovician) |
| Settled for 200 to 250 years |
| >30% forested |

## RESULTS

In general, soils of undisturbed or slightly disturbed forests on the Cumberland Plateau were occupied only by native earthworm species; exotic species occurred in areas that had been severely disturbed (Table 4). As documented in a previous study (Kalisz and Dotson 1989), native earthworm taxa showed variable sensitivity to disturbance: the genus *Diplocardia* was relatively insensitive and persisted even on some cleared and cultivated sites, whereas *Komarekiona eatoni* was very sensitive and occurred only on minimally disturbed sites. Seven native and six exotic taxa were recorded on 57 sites on the Cumberland Plateau; typically two or three species occurred on a site. An all-exotic earthworm fauna was found on only a single Cumberland Plateau sample site (Table 4); this was a house site with seasonally wet and sandy soils and was occupied by a sparse population of a single species of Asian origin, *Amynthas hupeiensis.*

TABLE 4. Occurrence of native and exotic earthworms taxa on slightly and severely disturbed sites in the Cumberland Plateau and Bluegrass Physiographic Regions of Kentucky.

| Earthworm Taxa | Number of Plots | |
|---|---|---|
| | Slightly Disturbed | Severely Disturbed |
| *Cumberland Plateau* | | |
| All Native | 37 | 12 |
| All Exotic | 0 | 1 |
| Mixed | 0 | 7 |
| *Bluegrass* | | |
| All Native | 3 | 0 |
| All Exotic | 15 | 4 |
| Mixed | 3 | 0 |

The relationship between the occurrence of exotic earthworm species and disturbance was different in the Bluegrass: exotic species dominated the landscape regardless of the intensity of disturbance. Exotic earthworms occurred on all severely disturbed sites and in 18 of the 21 slightly disturbed forests (Table 4). Nine exotic and five native taxa were recorded. *Diplocardia* was the most common native taxon. This endemic genus was recorded from 5 of the 21 slightly disturbed forests, from 2 of the 3 sites that had all native earthworms, and from all 3 of the sites that had both native and exotic earthworms.

## DISCUSSION

> So far the brunt of these invasions has been borne by the communities much changed and simplified by man. But some invaders are also penetrating the more stable and mature communities of ocean and natural forest.
>
> C. S. Elton, 1958:154

My interpretation is that the conditions under which exotic earthworms replace native earthworms depend on both the disturbance history of the sample site and on the state of naturalness of the landscape containing the sample site. In a non-fragmented forest landscape, such as the Cumberland Plateau, Elton's observation (above) seems to hold, and replacement of native earthworm assemblages seems only to follow severe habitat disturbance. When dealing with forest remnants in disturbed landscapes such as the Bluegrass, however, replacement of native earthworms seems to occur even when the above- and below-ground habitat appears undisturbed and large enough to support viable native populations. This discrepancy is not surprising: although soil volumes and above-ground areas larger than the effective universe of individual earthworms remain intact, forest remnants differ from non-fragmented forests in many ways including plant and animal species diversity and abundance, microclimate, and inputs of mass and energy (Saunders et al. 1991), possibly including human inputs of toxic chemicals. Although many of these effects of fragmentation are invisible, they apparently reduce the ability of native earthworms to survive or to compete with introduced species.

My conclusion is that native earthworms will likely disappear from fragmented landscapes, even from relatively undisturbed forest remnants, whereas they will disappear only from the immediate areas of severe disturbance in regions with non-fragmented habitat. In the latter case the sequence of events is likely as documented in the literature (Abbott 1985; Lee 1985; Kalisz and Dotson 1989): habitat destruction leads to extirpation or declines in populations of native earthworms, then the vacated habitat is occupied by introduced species. Based on this paradigm, direct competition between native and exotic species is not required in the replacement process. In the fragmented forest, on the other hand, visible habitat disturbance does not seem to be an essential step in the replacement process. This means (1) exotic earthworms directly out-compete and displace native species in relatively undisturbed habitat, or (2) invisible but ecologically significant changes in earthworm habitat result from forest fragmentation and lead to replacement of native species by exotics. The first of these two alternatives has not been documented, but could conceivably occur in forest remnants due to the large population of exotic earthworms in the encompassing disturbed landscape and to the greatly increased perimeter across which exotic species could invade the forest remnant. The second alternative represents a special case of the paradigm used for non-fragmented forests; habitat disturbance results in attainment of dominance by exotic species, but in this special case disturbance involves "invisible" effects rather than alteration in the appearance or physical structure of the habitat.

Regardless of the exact nature of native-exotic earthworm interactions, I do not believe that native earthworms will survive for long periods of time in forest remnants. Exceptions may be members of the endemic genus *Diplocardia.* Since members of this genus typically live in the subsoil and consume soil humus (Dotson and Kalisz 1989), they may be protected from many types of disturbance that adversely affect species that live nearer the surface and may be much less sensitive to changes in vegetation than species that consume fresh plant litter. Although exotic earthworms will ultimately dominate fragmented forest regions, endemic species may persist longest in forest remnants on steep or otherwise inaccessible areas along rivers (for example, the Kentucky River Palisades of the Bluegrass) due both to the length and continuity of the habitat and to the relatively low frequency and intensity of human disturbance and import of exotic species.

## ACKNOWLEDGMENTS

This work was supported by McIntire-Stennis funds and by a grant from The Nature Conservancy. I thank Barbara Fischer and Glen Kalisz for help in the field. This is a contribution of the Kentucky Agricultural Experiment Station, paper no. 91-8-227.

## REFERENCES

Abbott, I. 1985. Distribution of introduced earthworms in the northern jarrah forest of western Australia. Australian J. Soil Res. 23:263–70.

Ball, I. R. 1975. Nature and formulation of biogeographical hypotheses. Syst. Zool. 24:407–30.

Braun, E. L. 1950. Deciduous forests of eastern North America. Hafner Publishing Company, New York, N.Y.

Dotson, D. B. and P. J. Kalisz. 1989. Characteristics and ecological relationships of earthworm assemblages in undisturbed forest soils in the southern Appalachians of Kentucky, USA. Pedobiologia 33:211–20.

Edwards, C. A. and J. R. Lofty. 1972. Biology of earthworms. Chapman and Hall, London.

Elton, C. S. 1958. The ecology of invasions by animals and plants. Chapman and Hall, London.

Gates, G. E. 1963. Miscellanea megadrilogica. VII. Proc. Bio. Soc. Washington 76:9–18.

———. 1966. Requiem—for megadrile utopias. A contribution toward the understanding of the earthworm fauna of North America. Proc. Biol. Soc. Washington 79:239–54.

———. 1970. Miscellanea megadrilogica. VIII. Megadrilogica 1(2):1–14.

———. 1976. More on oligochaete distribution in North America. Megadrilogica 2(11):1–6.

James, S. W. 1990. Oligochaeta: Megascolecidae and other earthworms from southern and midwestern North America. Pages 379–86 *in* Soil biology guide. D. L. Dindal (ed.). John Wiley & Sons, New York, N.Y.

Kalisz, P. J. and D. B. Dotson. 1989. Land-use history and the occurrence of exotic earthworms in the mountains of eastern Kentucky. Amer. Midl. Nat. 122:288–97.

Lee, K. E. 1985. Earthworms. Academic Press, New York, N.Y.

Omodeo, P. 1963. Distribution of the terricolous oligochaetes on the two shores of the Atlantic. Pages 127–51 *in* North Atlantic biota and their history. A. Love and D. Love (eds.). Pergamon Press, London.

Orians, G. H. 1986. Site characteristics favoring invasions. Pages 133–48 *in* Ecology of biological invasions of North America and Hawaii. H. A. Mooney and J. A. Drake (eds.). Springer-Verlag, New York, N.Y.

Pielou, E. C. 1991. After the ice age. University of Chicago Press, Chicago, Ill.

Reynolds, J. W. 1974. The earthworms of Maryland (Oligochaeta: Acanthrodrilidae, Lumbricidae, Megascolecidae and Sparganophilidae). Megadrilogica 1(11):1–12.

Reynolds, J. W., E. E. C. Clebsch and W. M. Reynolds. 1974. Contributions to North American earthworms. The earthworms of Tennessee (Oligochaeta). I. Lumbricidae. Bull. Tall Timbers Res. Sta. 17:1–133.

Saunders, D. A., R. J. Hobbs and C. R. Margules. 1991. Biological consequences of ecosystem fragmentation: a review. Conserv. Biol. 5:18–32.

Schwert, D. P. 1979. Description and significance of a fossil earthworm (Oligochaeta: Lumbricidae) cocoon from postglacial sediments in southern Ontario. Can. J. Zool. 57:1402–05.

———. 1990. Oligochaeta: Lumbricidae. Pages 341–56 *in* Soil biology guide. D. L. Dindal (ed.). John Wiley & Sons, New York, N.Y.

Sims, R. W. and B. M. Gerard. 1985. Earthworms. E. J. Brill, London.

Smith, F. 1925. Certain differences between text-book earthworms and real earthworms. Trans. Illinois State Acad. Sci. 17:78–83.

Stebbings, J. H. 1962. Endemic-exotic earthworm competition in the American Midwest. Nature 196:905–06.

# The Diaspora of the Asian Tiger Mosquito

George B. Craig, Jr.[1]

On August 2, 1985, a mosquito new to the Americas was discovered in Houston, Texas. This invader (Fig. 1), popularly known as the Asian tiger mosquito, *Aedes* (*Stegomyia*) *albopictus* (Skuse 1894), was already widespread and abundant in much of Houston. During the next year, teams from the Centers for Disease Control (CDC) found this mosquito to be widely dispersed in the eastern United States, not only in most larger southern cities but as far north as Baltimore and Chicago. Subsequently, local populations have enlarged and spread to the point where this mosquito has become a major biting pest in southern cities such as Houston, Jacksonville, and New Orleans.

This mosquito breeds in small containers holding little water. It came to the U.S. in scrap tires from Japan. The used tire trade has spread it through much of the eastern United States. However, it breeds readily in discarded beverage cans, plastic bags, candy bar wrappers, anything that will hold a quarter inch of water. Blocked rain gutters, bird baths, ornamental ponds, domestic clutter all are productive. Vases and urns in cemeteries are major producers in Florida. However it also breeds in treeholes and other plant containers. In Florida, bromeliads that hold water may be found in nearly every yard in some towns. The adult mosquito does not travel far from its source; perhaps 500 yards may be its limit.

The public health community has been deeply concerned. In Asia, this highly domesticated mosquito is a major vector of the virus of dengue or break-bone fever. There is no protective vaccine for this disease. In the 1980s, dengue has been epidemic in many of the islands of the Caribbean. Every year, vacationers returning from Jamaica, Puerto Rico, and the Virgin Islands, bring the virus back with them in their bloodstream. *Aedes albopictus* breeding around homes in suburban environments makes the potential for explosive epidemics obvious.

In the past eight years, a considerable amount of research has been done on *A. albopictus* in the United States. The purpose of this paper is to summarize what we now know about its field biology and factors influencing its spread, both in the U.S. and other recent extensions from its original home in the Oriental region. A comprehensive monograph on its biology in Asia has been presented by Hawley (1988). The most recent review on *A. albopictus* in the Americas is by Rai (1991). We do not consider here the extensive literature on this mosquito as a subject for laboratory investigation. Omitted are studies on molecular biology, genetics, systematics, tissue culture, physiology, behavior, disease relationships, and control. Cell lines from this species are used in hundreds of laboratories, thus swelling the literature. The latest review on vector competence for arboviruses is by Mitchell (1991).

## CURRENT DISTRIBUTION IN THE UNITED STATES

As of 1992, *A. albopictus* has been reported from 351 counties in 25 states in the continental United States. These figures come from the *A. albopictus* Task Force of the CDC under the direction of Dr. Chester Moore (Field Services Sec-

1. Vector Biology Laboratory, University of Notre Dame, Notre Dame, IN 46556

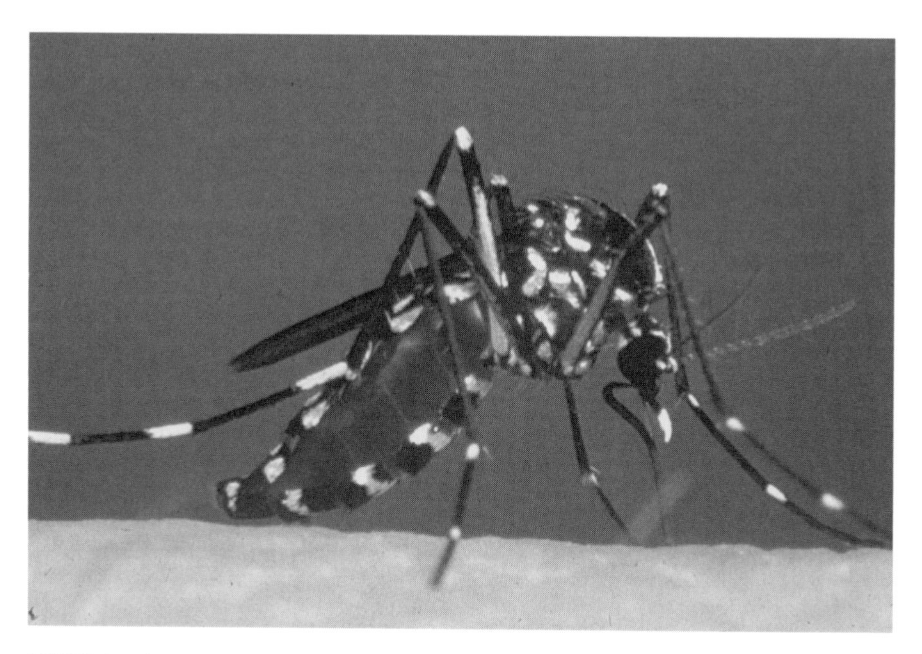

FIGURE 1.—*Aedes albopictus,* female feeding (photo by Leonard Munsterman, Univ. Notre Dame).

tion, Medical Entomology-Ecology Branch). Figure 2 shows current distribution, with 347 positive counties in 23 states. Of course, these are minimum figures. As with bird watcher statistics, positives only occur where people are looking. Most counties in the U.S. do not have the capacity to identify mosquitoes.

In my opinion, the mosquito probably occurs in every county east of the Mississippi and south of the Mason-Dixon Line (Appalachian Mountains excepted). The experience of Dr. Michael Womack (Biology Department, Macon College, Macon, Georgia) is instructive. In 1989, Georgia reported 3 positive counties. In two years of looking, Dr. Womack brought the number to 57 counties. He went on two-week military reserve duty in San Antonio and added 35 Texas counties. We need more mosquito watchers.

Some local infestations have been wiped out by diligent efforts of public health authorities. This seems to be the case in Minneapolis, Minnesota, Oakland, California, and an Air Force base in New Mexico. In Indiana, Kentucky, and Ohio, some county infestations have been eradicated, perhaps aided by winter weather. In 1988, Indianapolis reported eradication; two years latter, a new infestation from another source appeared in another part of the city. In Ohio, public health workers knocked out infestations in northern and central counties but were unable to do so at a county in southern Ohio.

In the southern states, *A. albopictus* is ubiquitous along the Gulf coast, with local populations getting larger each year. It was slow to spread in Florida for reasons to be discussed below. Once started, it spread fast. It now occurs in 64 of the 67 Florida counties, being absent only in Fort Lauderdale and Miami (Fig. 3). Those communities will undoubtedly become infested in 1993. The Florida distribution statistics are among the best in the United States. That state has organized mosquito control in nearly every county. Moreover, Dr. George O'Meara of the University of Florida has closely followed the spread, county by county. Following arrival at a point source in a county, the species gradually spreads all over. In Polk County, Florida, *A. albopictus* first appeared at a large tire dump in an

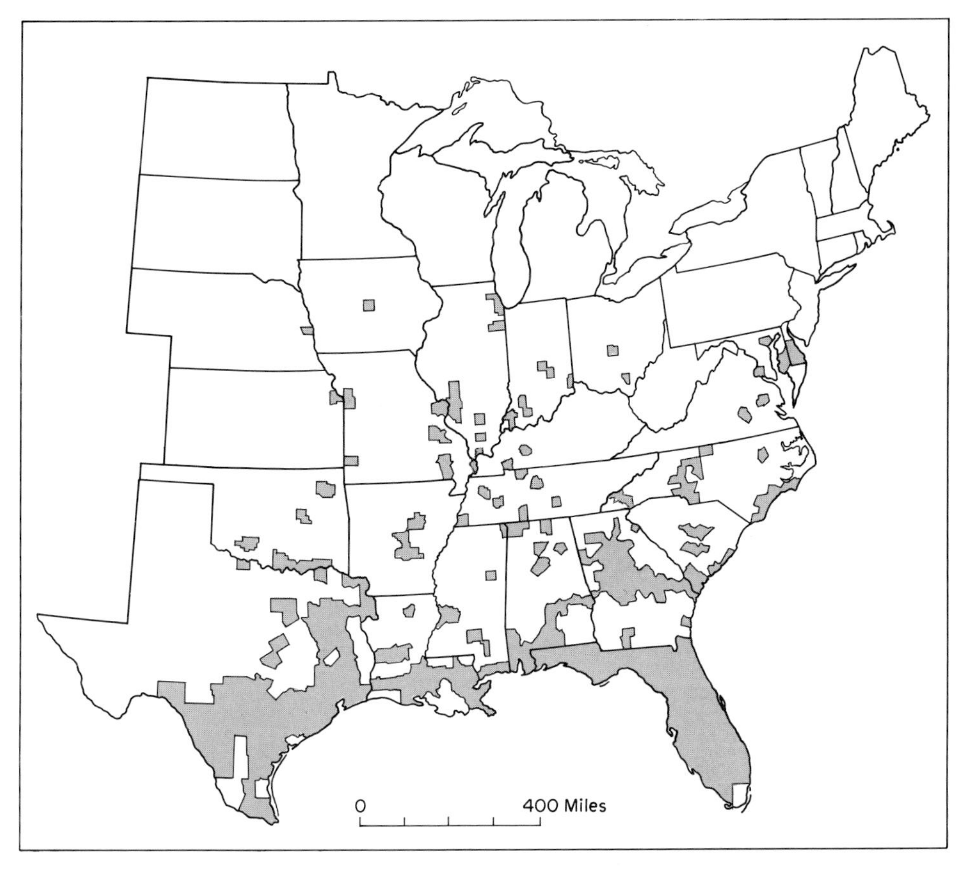


FIGURE 2.—Reported distribution of *Aedes albopictus* in the continental United States, 1992. (modified from Field Services Section, Medical Entomology-Ecology Branch, Centers for Disease Control, Fort Collins, Colorado)

isolated part of the county in 1989 (Fig. 4). By 1991, it was taken from 69 light traps scattered all over the county.

For decades *A. albopictus* has been established in Hawaii, where it is the most common biting mosquito. This species probably played a role in a major epidemic of dengue in Oahu in 1945. Intensive DDT spraying eradicated the urban vector, *A. aegypti*, but left the suburban and rural-breeding *A. albopictus* in place.

## CURRENT DISTRIBUTION—WORLD

*Aedes albopictus* originated in the Oriental Region, probably in the Indomalayan Peninsula. It is a member of Subgenus *Stegomyia*, Group C (about 50 species in the Oriental Region), Albopictus Subgroup. Among the 30+ related species in the Subgroup are such close relatives as *A. pseudalbopictus*, *A. novalbopictus*, and *A. subalbopictus*. In the early 1980s, before it began its hegira, *A. albopictus* was known from Madagascar and Mauritius on the west to Hawaii on the east, with very abundant populations in China, India, Indochina, Indonesia, Malaysia, New Guinea, and the Philippines. It did not reach Australia or the islands of the South Pacific. For Americans, the most interesting part of its distribution is the extension to the temperate zone in China (north of Bejing), Korea,

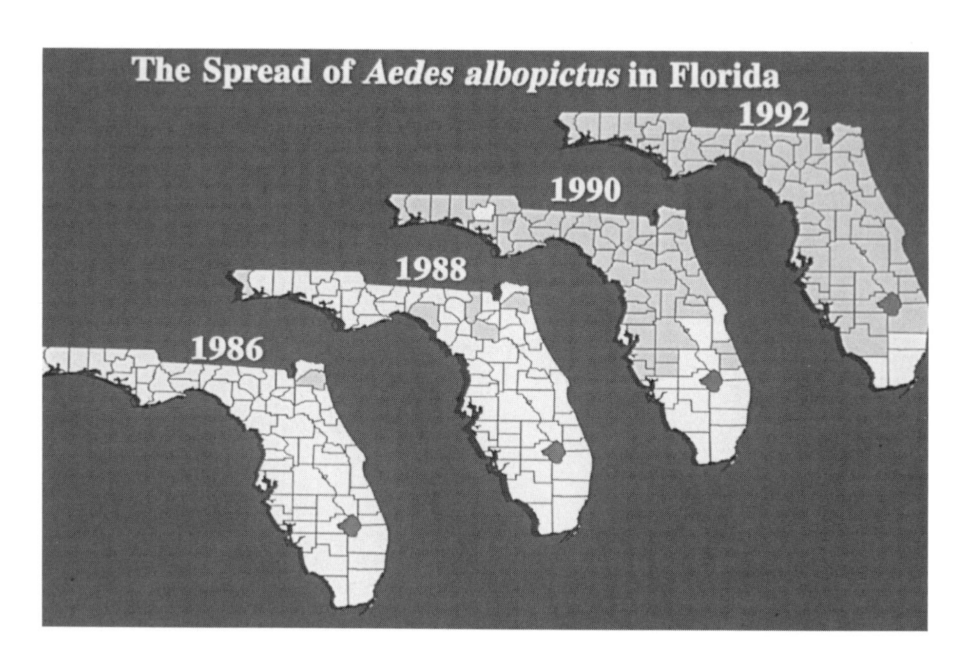


FIGURE 3.—Map of Florida showing spread of *Aedes albopictus* from a single county in 1986 to 64 of 67 counties in 1992. (from George F. O'Meara 1993)

and Japan (Fig. 5). It is from temperate Asia, probably Japan, that the U.S. infestation arose, by means of shipments of scrap tires.

In 1986, *A. albopictus* was reported from Brazil, in conjunction with an epidemic of dengue in the environs of Rio de Janeiro. It is estimated that one million people were infected in the Rio epidemic; the role of *A. albopictus* is unknown. Today, the species is firmly ensconced in the five states around Rio and Sao Paulo. The threat of *albopictus*-borne yellow fever, as well as dengue, is definitely a reality in Brazil.

The Pan American Health Organization has made major efforts to alert public health authorities in Latin America to the danger of this distinctive and easily identified mosquito. Yet, reports are limited to the U.S. and Brazil. Mexico remains a particular puzzle. It was reported from Matamoros, across the Rio Grande from Brownsville, Texas (where, according to CDC personnel, it is common). Apparently, the infestation did not persist. Note from Fig. 1 that the species occurs for hundreds of miles along the southern Rio Grande. Many trucks carry scrap tires from Houston, Texas, all over Mexico. *IF* it is true that *A. albopictus* does not occur in Mexico, we certainly should be conducting research to find the mysterious factor(s) that are keeping it out.

Elsewhere, *A. albopictus* is spreading with commercial shipping. In the last three years, it has been found in tires in ports in Capetown, South Africa, and Brisbane, Australia. In 1993, it was reported in Auckland, New Zealand, in a load of tires just off a steamer from Japan. In Europe, recent infestations are reported from Genoa and surrounding communities in Italy and in Tirane, Albania. Since the primary trading partner of Albania in recent years is the Peoples Republic of China, it is interesting to speculate that the Albanian infestation came from a ship from China that passed through the Suez Canal. Certainly, the most surprising infestation is that in Nigeria, reported by CDC personnel in connection with a yellow fever outbreak. In September of 1991, *A. albopictus* eggs

FIGURE 4.—Ground view in Polk Co. (Florida) tire dump before shredding; biting rate 500+ *Aedes albopictus* per hour. (photo by Sally Paulson, Florida Atlantic Univ.)

were collected in three separate locations in Delta State; this was not a port infestation. It was far inland, associated with rural and forested sites.

Populations of *A. albopictus* have been found on all continents except Antarctica. Why is this species on the move at the end of the 20th century? One is reminded of another container-breeder in Subgenus *Stegomyia*, *A. aegypti*, the yellow fever mosquito. This species evolved in the Subsaharan Ethiopian Region. All of its 50 species of close relatives in Group A of *Stegomyia* are confined to Africa. In that continent there is one subspecies of *A. aegypti* that is limited to forests, confined to treehole breeding and does not feed on man. There is also a domesticated subspecies that lives in villages, breeds in domestic containers, and feeds largely or entirely on man. Morphological differences in the subspecies are conspicuous. It is evident that the domestic form colonized ports, got into water barrels on ships, and spread all over the tropical and subtropical world in the age of sail.

When man began to study mosquitoes, *A. aegypti* was already the most common urban mosquito in the warmer parts of the Americas, Asia, and Europe. In colonial times in the U.S., there were yearly epidemics of yellow fever in the summer, not only in southern ports but in Boston, New York and Philadelphia. Yankee sailing captains would pick up a load of slaves (plus *A. aegypti* plus virus) in West Africa, dump the slaves in Cuba and Jamaica, then sail back to Boston with rum and sugar and the makings for yellow fever. However, eggs of *A. aegypti* are not cold-hardy; they are killed immediately on freezing. Thus, overwintering in the U.S. was confined to freeze-free areas of the Gulf Coast and Florida. *Aedes albopictus*, on the other hand, has temperate strains with freeze-tolerant eggs. For the past seven years, it has overwintered very nicely in Chicago.

The secret to international dispersal for *A. albopictus* would seem to be containerized shipping. Scrap tires provide rainwater-holding capacity, serving as "artificial treeholes." The safe, protected environment of the sealed container

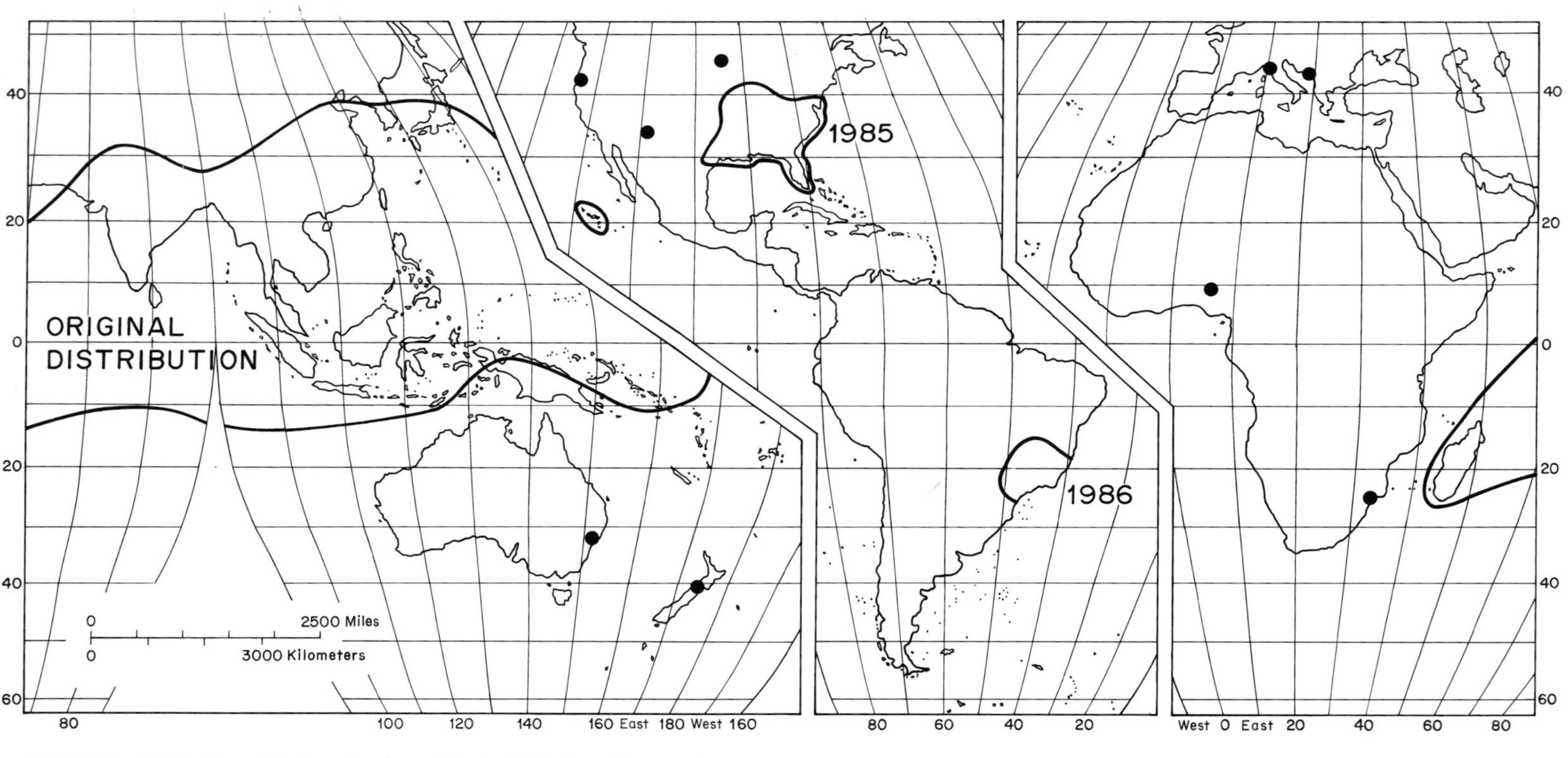


FIGURE 5.—Distribution of *Aedes albopictus*, March, 1993. (● = discovered in past three years; all ● populations in the U.S. have been eradicated)

provides a splendid mobile pram between continents. These containers are lifted directly onto trucks and trains in the harbor, often without being opened. Larval mosquitoes then are dispersed from their new home, where they move from tires to all sorts of natural and artificial water holding containers. Ports are more readily colonized by the inner-city-breeding *A. aegypti*. But rural and suburban Asian tiger mosquitoes are more adaptable to a variety of breeding sites and are more cold-tolerant.

## *AEDES ALBOPICTUS*—WHENCE?

When *A. albopictus* was first found in the U.S., public health workers wanted to know its origin, perhaps to stop further introductions. The logical inference was Honolulu or Manila or Singapore or some other tropical city. In the early 1970s, the mosquito had been found in the Los Angeles harbor in ships returning war material from Vietnam. In 1983, CDC found two specimens in light-trap collections from Memphis, Tennessee; intensive search failed to uncover a breeding population.

In 1987, workers from Notre Dame and the CDC published a paper claiming that the U.S. infestation came from north Asia, probably Japan (Hawley et al. 1987). Their evidence has held up in subsequent years. They observed:

1. In the early 1980s, the U.S. started importing millions of scrap tires from Japan for use in making spare parts such as washers and gaskets for the automobile industry. (American scrap tires are usually made of synthetic rubber and are steel-belted; Japanese tires are natural rubber and often lack the steel belt.)
2. Eggs from tropical strains were killed immediately by freezing; eggs from Japan and U.S. strains survived.
3. Females from temperate areas were photoperiod-sensitive; when reared on short-day length, resulting eggs would not hatch until diapause was broken. This prevented premature hatch in the fall. Females from the tropics were not photosensitive; they laid hatching eggs, regardless of light regime. Japanese and U.S. strains were 100 percent diapausing.

Later, electrophoretic analysis of isozymes by several workers showed close affinity of U.S. and Japanese populations but little similarity to tropical populations.

The final, "smoking gun" evidence was provided by CDC workers. They went to the port of Seattle and inspected tires coming off freighters from Japan. It was not a major surprise when they found living larvae of *A. albopictus* and other mosquito species in these tires.

This kind of detective work has been applied to *A. albopictus* introductions elsewhere. Brazilian *A. albopictus* are completely non-diapausing, suggesting either (a) a tropical origin or (b) rapid selection of a temperate strain for the tropical response. In the laboratory, we have changed a diapausing to a non-diapausing strain in five generations of selection. Selection in the reverse direction does not work at all. The Brazilian results were duplicated in a strain from Nigeria. This led to the conclusion that the lack of photosensitivity was evidence of temperate origin. On the other hand, a strain from Albania was highly photosensitive, possibly originating from north China. Further refinement of these techniques would certainly be of value.

## *AEDES ALBOPICTUS*—WHITHER?

Once establishment in the U.S. was well-recognized in 1986, the next question was "How far is it going to go?" The initial supposition was that *A. albopic-*

*tus* is the Asian counterpart of *A. aegypti* and, hence, would be limited to the warm, moist regions along the Gulf coast. That illusion vanished quickly after seeing *A. albopictus* overwinter in Chicago.

The habitat and behavioral distinctions shown in Table 1 are not absolute, as *A. albopictus* goes with tires. Close to downtown Chicago is a major site at a tire rendering plant. Near this site, it was breeding in water, in discarded machinery. Most remarkable was a population on a railroad siding, where a gondola car held old machinery covered by a canvas tarp. The tarp had 27 dents holding rain water; all 27 were swarming with larvae of *A. albopictus*, ready to take to the rails.

TABLE 1. Some differences between the two *Stegomyia* species in the United States.

| CHARACTERISTIC | AEGYPTI | ALBOPICTUS |
|---|---|---|
| 1. Eggs withstand freezing | No | Yes |
| 2. Photoperiod-induced egg diapause | No | Yes |
| 3. Breed in plant containers | No | Yes |
| 4. Habitat type | Urban<br>Inner city<br>Inside houses | Suburban, rural<br>Forest edge,<br>likes vegetation |

The northern distribution in the U.S. is certainly limited by cold tolerance. In 1987, Nawrocki and Hawley studied distribution of the species in north Asia and concluded that the −5°C January isotherm is a good predictor of distribution. Their map (Fig. 6) of the eastern U.S. predicts *A. albopictus* will overwinter in Chicago but not Minneapolis. The map predicts the species will eventually extend to southern Michigan and northern Ohio; Detroit, and Cleveland would be likely centers. On the east coast, one would predict that it would extend beyond the current Delaware-Maryland sites to Philadelphia and southern New Jersey.

The western distribution is probably controlled by lack of rainfall and low humidity. Distribution stops at the edge of the Great Plains. In 1992, Texas distribution was extended as far west as Del Rio in Val Verde County, the western limit to date. Also in 1992, sites added include Tulsa and Oklahoma City (Oklahoma), Omaha (Nebraska), and Ames (Iowa). The latter was probably added instead of Des Moines (Iowa) because it is the home of Iowa State University and expert mosquito watcher Wayne Rowley. I would not expect it to colonize coastal Oregon and Washington because their wet season, while abundant, occurs in the winter when eggs are in diapause. California may be too dry. In the days of sailing ships, *A. aegypti* never became established, although there probably was ample opportunity. Mosquitoes breeding in water barrels must have frequently been brought into Los Angeles and San Francisco by clipper ships out of Panama or the Orient. California infestations should be spotted early; most of the state is divided into mosquito control districts and there is a plethora of trained mosquito watchers. Recently, California researchers at Berkeley have shown experimentally that *A. albopictus* could thrive in some California environments.

Factors limiting southern distribution pose interesting zoogeographical problems. After the Asian tiger mosquito became established in Houston, it spread rapidly to the north and east through the used tire trade. Few Ameri-

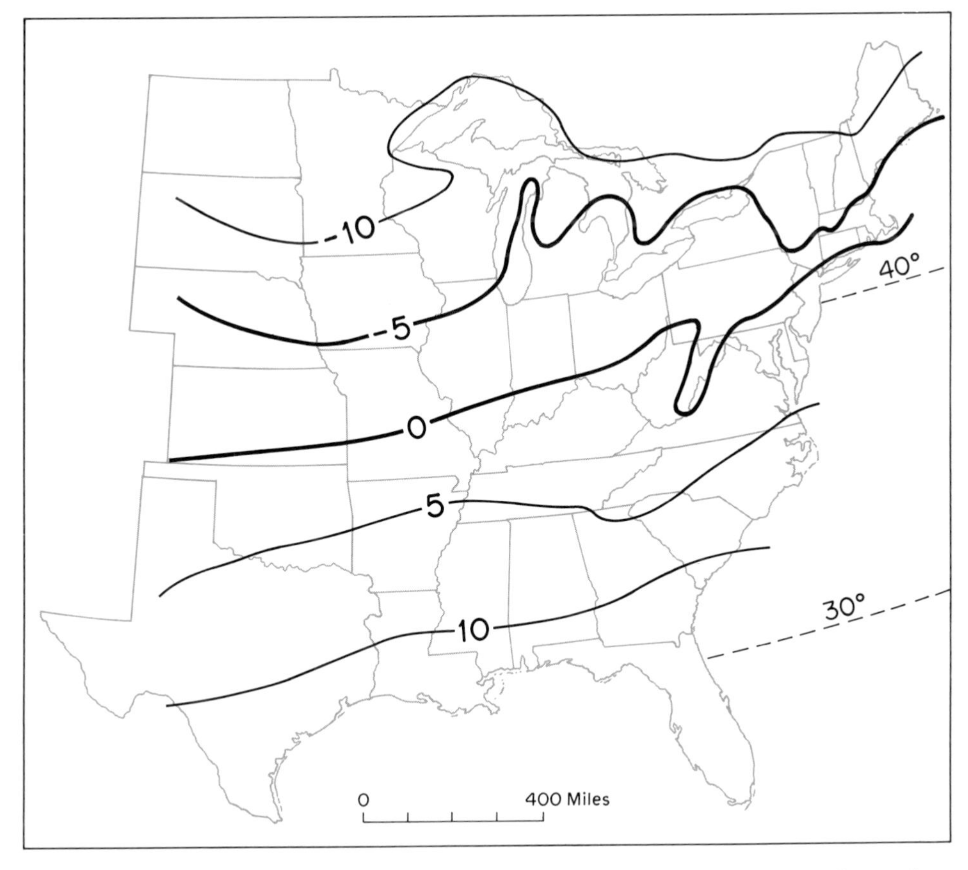


FIGURE 6.—Daily mean January (°C) isotherms for the eastern United States and southern Canada (from Nawrocki & Hawley, 1987)

cans appreciate the size of this trade. The Environmental Protection Agency estimates there are three billion used tires sitting around the American landscape and piles are growing at 250,000,000 tires per year (Fig. 7). At any rate, it is easy to understand why the new mosquito did so well in Indianapolis; this city has the same latitude as Tokyo and the mosquitoes were already adapted to living in this climate when they arrived. They were and are temperate climate mosquitoes.

Spread in south Texas and Florida was much slower. Florida distribution was limited to Jacksonville in 1986–1988 and it was not until 1989 that *A. albopictus* moved significantly into peninsular Florida. In 1989–91, movement was dramatic and rapid. In 1993, it has moved down to the counties just above the southern tip; completion of Broward, Dade, and Monroe counties should occur in 1993. With the southern movement came a change in the mosquito. It lost photosensitivity and the ability to produce diapausing eggs. This occurred in both central Florida and south Texas. In 1985–87, Houston and New Orleans workers found the mosquito only from mid-March to late September. It had a winter diapause, when the eggs did not hatch. Today, in those places and in Florida, larvae of the Asian tiger mosquito are found 12 months per year. The lack of response to different day lengths exhibited by the tropical genotype has re-evolved in the temperate genotype.

FIGURE 7.—Scrap tire pile near Westley, California, containing 67,000,000 tires but no *Aedes albopictus*. (credit unknown)

## EVOLUTION OF *AEDES ALBOPICTUS* IN THE UNITED STATES

We have shown that photoperiod-induced diapause is a genetic characteristic that differs in different populations (Fig. 8). The population that originally invaded the U.S. was photoperiodic. We developed photoperiodic curves by holding newly emerged females for 10 days in photoperiod boxes varying from 12 to 14 hours of light, at 15 minute intervals. Ten females were blood-fed and eggs were collected separately from each one. Each egg batch was counted separately, subjected to a powerful hatching stimulus, bleached to determine embryonation, and the percent of viable hatched eggs was determined. A typical population from Memphis, gave a curve with no hatch at 13:00 hours of light, 50 percent at 13:30 and 100 percent at 14:00 hours. The 50 percent point is the CP or Critical Photoperiod.

In our initial tests of U.S. populations in 1987–88, we found a north-south gradient, with strains from Chicago showing a CP around 13:45 and strains from New Orleans around 13:15 hours. After only a few years in the U.S., there was already fine tuning, with adaptation to local day lengths. This is a well-known phenomenon in photoperiod biology but the speed of adaptation in this case is startling. In subsequent years, the north-south differentiation has become more extreme. Hawley went to Japan and collected 10 populations from north to south. Tests of CP gave a sharper north-south cline, indicating the far longer period of local adaptation.

During 1989–91, we studied the *A. albopictus* in a large tire dump in Polk County in central Florida. This dump had 1.3 million tires. It was surrounded by the Great Green Swamp and is about 10 miles due west of Disneyworld. In 1989, the Florida State Police closed the dump and stationed a van at the entrance to prevent additional dumping; thus, there were no new genes introduced into the dump population. We sampled the population every 3–4 months for three years. Initially, in January of 1989, the photoperiod curve was standard and the CP was

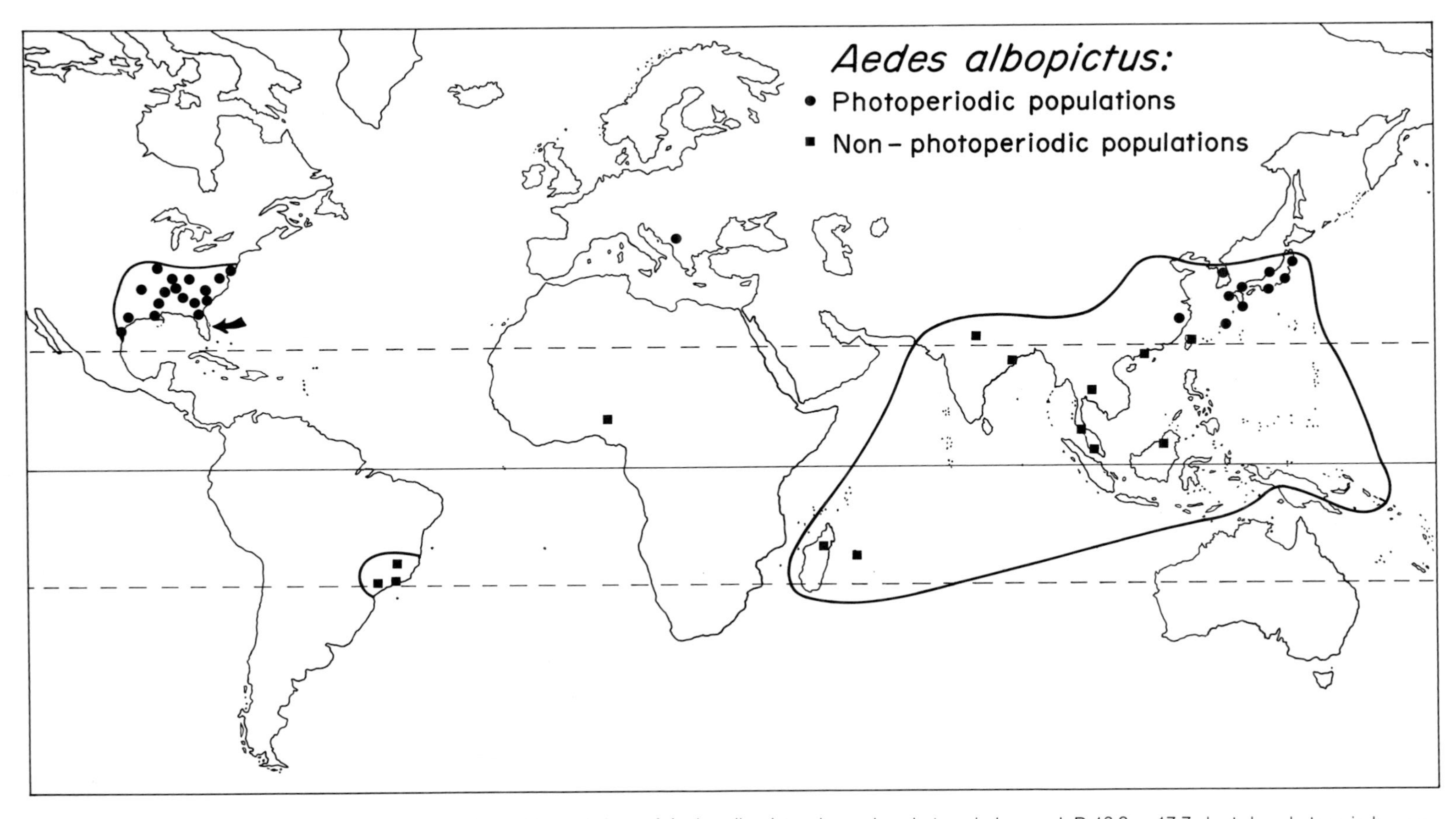


FIGURE 8.—Distribution of photoperiodic response in world populations of *Aedes albopictus*. Long day photoperiods were L:D 16:8 or 17:7 short day photoperiods were 8:16 or 9:15. All tests were done at 22°C. (modified from Hawley & Craig, 1989) note arrow indicating central Florida; changed from circle in 1989 to square in 1991.

about 13:15 hours. In March of 1989, however, the CP was lower and about 15 percent of the eggs hatched, even from mothers reared at 12:00. At each collection, the CP shifted to the left and the proportion of photo-insensitive females was higher. By May and June of 1991, the photoresponse was gone. Collections at this time gave 80–90 percent hatch from females reared at 12:00 hours of light.

We take these results as evidence that photoperiod sensitivity was progressively lost in Florida *A. albopictus* through genetic selection. Subsequent tests on 12 other south Florida populations collected in 1992 and 1993 gave confirmatory results. On the other hand, a strain from the Polk dump collected in January of 1989 and reared in the laboratory, over 13 generations without selection, still shows high photoperiod sensitivity. The selection outdoors did not apply. We have repeated these experiments at the Florida Medical Entomology laboratory at Vero Beach, using natural daylength around the 1st of January. Results were as expected. Northern *A. albopictus* diapause under the Vero Beach January light conditions. Florida populations no longer diapause; neither do strains from San Antonio, Texas. It is evident that the selection for loss of photosensitivity is a phenomenon of southern latitudes.

How does this information apply to the spread of *A. albopictus*? First, it becomes evident why it took so long for the species to invade peninsular Florida, as compared to the rapid invasion of the north and east. A genetic change had to occur, and once it *A. albopictus* flushed southward. Second, south Florida and south Texas now have populations similar to their tropical ancestors and different from their Japanese-U.S. antecedents of a short time ago. The species is NOW prepared to invade the Caribbean and Mexico. South Florida is like an arrow pointing at places like Puerto Rico, Haiti, and Jamaica. A shipload of old tires from Miami should carry this potential dengue vector to those tropical islands where dengue is endemic.

Interestingly, the United States has recently passed quarantine legislation to prevent importing of water-containing tires from Asia but we have shown no interest in protecting Latin America from tires that we export. If our newly-adapted *A. albopictus* gain a foothold in the Caribbean islands and add to the dengue burden, Americans will bear a heavy burden of guilt by negligence.

Other evidence of evolutionary change of *A. albopictus* has also been found. Ability of eggs to withstand freezing is also under genetic control. Field over-wintering experiments (Hawley et al. 1989) in three winters from 1986–89 have provided evidence of rapid geographic differentiation in U.S. populations. Eggs laid by laboratory-reared females were placed in screened outdoor containers in autumn in northern Indiana. In the following spring, the eggs were returned to the laboratory and hatched. Each year, eggs from northern sites survived at a rate higher than those from southern sites. Clearly, geographic differentiation has established a north-south cline for this character. The same kind of cline was found in eggs from Japan. Laboratory selection has also been shown to increase cold tolerance. If this is true, we may be wrong in stating that Chicago is the northern limit for the Asian tiger mosquito in the United States.

Clearly, *A. albopictus* is a highly adaptable species. Dr. Dawn Wesson in our laboratory has selected U.S. populations for resistance to organophosphate insecticides. In a few generations of laboratory selection, she raised the level of OP resistance by 30 or more times. It is difficult to be sanguine about a disease vectoring species with such evident evolutionary plasticity.

## DISEASE THREAT—VERTEBRATE PATHOGENS

Mitchell (1991) reviews vector competence of North American *A. albopictus* for North American arboviruses. Laboratory studies have shown that this species is at least as effective as the natural vector for all of the major agents causing human encephalitis: Eastern Equine Encephalitis (EEE), Western Equine Encephalitis (WEE), St. Louis Encephalitis (SLE), Venezuelan Equine Encephalitis (VEE), and La Crosse (California Serogroup) Encephalitis (LAC). It is ineffective in transmitting and other members of the California Serogroup such as Jamestown Canyon Encephalitis (JCE). Viruses not endemic to the U.S. but readily transmitted include Japanese B Encephalitis, Ross River Virus, West Nile Encephalitis and Chikungunya Virus. In other words, this species has an extremely wide range of vectorial capacity test tube for growing arboviruses; its spectrum is as wide as any mosquito.

Compounding the threat is the anthropophilic behavior of this mosquito. Unlike swamp and marsh mosquitoes, this species breeds close to man and shows the widest range of host preference of any mosquito. Savage et al. (1992) report it feeding not only on large and small mammals but on birds, turtles and snakes. It can enter into sylvan zoonoses and bring such viruses to man.

In the U.S., four arboviruses have been isolated from natural populations of *Aedes albopictus* to date. These are Potosi (POT), Tensaw (TEN), Keystone (KEY) and EEE. The first three may not be pathogenic to man but EEE causes severe and often fatal disease in humans. It also is harmful to domestic animals (equines, pheasants, emus) and has killed significant numbers of endangered species, e.g. the whooping crane. Potosi virus was first discovered from *A. albopictus* in a tire dump at Potosi, Missouri in 1989. Infection rates in the mosquito were high. This virus probably resides in deer and may increase with the exploding urban deer problem in the United States. More information on human pathology of POT is urgently needed.

The most alarming discovery is the finding of EEE in *A. albopictus* in the Polk County tire dump in Florida (Mitchell et al. 1992) (Fig. 9). This site is discussed above in connection with photoperiodism; it is located in the Great Green Swamp, where an epizootic of EEE was prevalent in aquatic birds in the summer of 1991. Our laboratory made extensive collections of mosquitoes at that time in order to study host choice. We sent the frozen mosquitoes to CDC in Fort Collins for virus assay. We were astonished to find that the *A. albopictus* were infected with EEE at a rate of 1 infected per 600 mosquitoes assayed. Comparable rate for the natural vector in New Jersey salt marshes is 1 out of a million. Active mosquito abatement and tire destruction were undertaken by the Polk County Mosquito Abatement Program. In June of 1992, personnel from CDC and Notre Dame's Vector Biology Laboratory collected in the same area; no infected *A. albopictus* were found.

The discovery of EEE virus in a natural population of *A. albopictus* stimulated a great deal of media attention. The Florida Tourist Board was less than enthusiastic, especially because the state was hit hard by an SLE epidemic in 1990 and an EEE epidemic in 1991. The Florida State Department of Environmental Regulation undertook immediate action to eliminate the tire pile. Shredding began in October of 1991 and was completed by June of 1992. They began with 1,848,867 passenger tire equivalents (PTE) and ended with 18,946 tons of shred scraps, arranged in neat piles (Fig. 10). It was impossible to count all the tires so they used PTE's which uses 20 pounds per tire as the standard. The final cost for the project was $1,282,132. This comes to about $1.00 per tire, without considering transportation and without disposing of the scraps (fire still a possibility, but not as great due to the reduction in air spaces). Many states have recently enacted taxes of about $1.00 per new tire sold in order to get funds for tire disposal. Based on Florida's experience, this amount is not nearly enough.

In my opinion, La Crosse Encephalitis poses a greater threat to the U.S. than either dengue or EEE. LAC is prevalent in the eastern half of the country, espe-

FIGURE 9.—1991 photo of 1.8 million tire dump at Great Green Swamp, Polk Co., Florida, 7 miles from Disneyworld. Eastern Equine Encephalitis found here in 1991, with infection rate of 1/600 *Aedes albopictus*. (photo courtesy of Tom LeDew, Florida Dept. Environmental Regulation)

FIGURE 10.—1992 photo of former tire dump, Great Green Swamp, Polk Co., Florida, following six months of shredder machine operation. Black piles contain 26 tones of shredded rubber. (photo courtesy of Tom LeDew, Florida Dept. Environmental Regulation)

cially in a broad crescent from Minneapolis to Albany, New York. It is a disease of children under 16, seldom killing but often causing latent brain damage. The natural vector is *Aedes triseriatus,* a mosquito that originates in treeholes but has exploded into the peridomestic environment by breeding in scrap tires. Most infected children in Minnesota and Ohio have scrap tires accumulated in the back yard. The reservoir is in chipmunks and squirrels. Here is a zoonotic virus that is already in place. *Aedes albopictus* is a splendid laboratory vector, showing high rates of both oral and transovarial transmission; venereal transmission is probable. In southern Indiana, we have found chipmunks with antibodies to LAC in a large tire pile with abundant *A. albopictus* and *A. tiseriatus.* Collection of Asian tiger mosquitoes carrying the LAC virus is more than imminent. All that remains is for a child to be bitten.

In areas such as East Saint Louis, Illinois, where *A. albopictus* and scrap tires and the LAC virus are all found, the results could be tragic (Figs. 11 & 12). As a further complication, this area has a heavy population of rats; in host choice assays, we found that more than 50 percent of the *A. albopictus* had fed on rats. The domestic rat, itself a non-native from Europe, has never been implicated in the LAC cycle but the virus grows readily in laboratory rats. It looks as though all the parts are in place for a new kind of LAC epidemic.

I will not dwell on the dangers to the U.S. from *albopictus*-transmitted dengue. In southern cities, where *A. albopictus* has become a first class biting pest, all that is needed is airport arrival of an incipient case of dengue from the Caribbean (or the Orient or Africa). Formerly, dengue was considered a very painful and debilitating disease but one of insignificant mortality. However, the picture has changed; Dengue Haemorrhagic Fever or Dengue Shock Syndrome is due to sequential infection of two of the dengue serotypes, 1–4. In the past, each island in the Caribbean had its own serotype. Modern transportation and trade is causing much more mixing of serotypes and hence, more DHF.

I believe the dengue-DHF-*A. albopictus* threat is much greater to the islands of the Caribbean than it is to the United States. In those islands, dengue is now maintained by urban *A. aegypti.* There is no rural or sylvan vector comparable to *A. albopictus.* It should be remembered that in tropical Asia, one hears more about the *aegypti*-borne epidemics of dengue in the large cities but that the majority of the people get their dengue on the farm and in the small towns where *A. albopictus* is prevalent. At the least, *A. albopictus* will add a new complexity to dengue in the Americas.

Dog heartworm, *Dirofilaria immitis*, is vectored by *A. albopictus.* In Japan, three percent of Tokyo *A. albopictus* were found to be infected. Earlier work in North Carolina found that this species is entirely refractory. However, recent work at our laboratory (Scoles and Craig 1993) found marked genetic variability in U.S. populations for this character. North Carolina populations were indeed refractory, but those from New Orleans showed 40 percent of mosquitoes developing infectious L3 larvae; other U.S. populations were intermediate. This mosquito breeds close to homes, encounters dogs and feeds readily on them (Savage et al. 1993).

## CONTROL PROBLEMS

Control of *A. albopictus* poses difficult problems. As with all mosquitoes, source reduction is to be preferred. However, the small water containers used by this species are often dispersed and cryptic. In the 1960s, the U.S. Public Health Service (USPHS) spent $100,000,000 in a futile attempt to eradicate another container breeder, *A. aegypti,* from the southeastern United States. After six

FIGURE 11.—Scrap tires full of *Aedes albopictus* in East St. Louis, Illinois. Note Gateway Arch across the Mississippi River in St. Louis. (photo by Leonard Munstermann)

years of diligent effort, eradication had not been achieved in a single county even though USPHS workers had a powerful tool, DDT. *Aedes albopictus* is far harder to eliminate because it breeds in plant cavities and treeholes as well as domestic containers. In the 1940s and 50s, vast DDT spraying campaigns were conducted on the islands of Mauritius and Oahu. In both cases, *A. aegypti* was eradicated but *A. albopictus* was left untouched; both islands have abundant *A. albopictus* today.

The bottom line on control in southern cities in the U.S. is that we do not have the technology. Most mosquito abatement districts do not even try. Alternatively, they use night time truck-mounted ULV adulticiding, a generally futile approach for *Stegomyia.*

The one positive accomplishment of the USPHS *A. aegypti* Eradication Campaign was the development of improved monitoring through the use of the Fay Ovitrap or LBJ (Little Black Jar). *Stegomyia* mosquitoes are active during the day and are not attracted to the conventional New Jersey Mosquito Light Trap. The Ovitrap is a pint black jar half-filled with water plus a flat stick upon which eggs are deposited. Sticks are collected and eggs counted once per week. It has been estimated that in terms of man-hours, this method is 100 times as effi-

FIGURE 12.—Single discarded tire along East St. Louis road. The disease threat is greater from individual tires close to people than from large tire piles. (photo by Leonard Munstermann)

cient as searching for larvae or collecting biting adults. It works equally well for both *A. aegypti* and *A. albopictus*.

## THE GOOD NEWS? *AEDES AEGYPTI*

As *A. albopictus* has waxed in the southern U.S., *A. aegypti* has waned. The traditional *A. aegypti* belt extends along the Gulf coast from Florida to Texas and north to Arkansas, Tennessee, and the Carolinas. Throughout this belt, the yellow fever mosquito is becoming an endangered species. George O'Meara (1993) in Florida has documented the process carefully. In county after county, the process occurs rapidly, often in 1–2 years. We cannot be sure that *A. albopictus* is responsible for the demise of its disease vectoring cousin but it certainly looks suspicious.

The mechanism for competitive replacement remains one of the great ecological mysteries of our times. The conventional explanation has been larval competition in the container breeding site. Unfortunately, this does not hold up to experimental analysis. Many workers have done laboratory experiments where various numbers of both species are reared together. The two species have little effect on one another. *Aedes aegypti* wins more often than *A. albopictus*. That does not explain what is happening in the field. Some possible hypotheses:

1. Nasci et al. (1989) has suggested that cross-specific mating will sterilize female *A. aegypti*, a form of competition by satyrism. In the laboratory, injection of matrone, the male accessory gland pheromone, into female *A. aegypti* renders them refractory to subsequent insemination. The reciprocal cross, female *A. albopictus* and male *A. aegypti* has no such effect. However, later work by Nasci and other workers (S. Paulson, G. Craig) has shown that cross-mating does not occur often enough.
2. Klowden and Chambers (1992) have shown greater reproductive efficiency in *A. albopictus*. As compared to *A. aegypti*, they have more egg production with small blood meals and greater resistance to starving. It is hard to see how this would account for the very rapid replacement in the field, although it might serve over the long-term.
3. An intestinal gregarine protozoan parasitizes both species. *Ascogregarina culicis* occurs in *A. aegypti* and *Ascogregarina taiwanensis* occurs in *A. albopictus*. Munstermann and Wesson (1990) point out that *Asc. taiwanensis* came to the U.S. along with its host and it is now widespread. Spores are eaten by young larvae, the sporozoites penetrate and grow in midgut cells, emerge and go into the ends of the Malpighian tubes, where pairs of gamonts fuse sexually, cysts are produced and new spores are defecated by emerged adults. Both parasites are very common, nearly ubiquitous. Each mosquito tolerates its own gregarine very well; perhaps 10–25 percent mortality occurs in heavy infestations. In cross-species infections, midgut cell invasion occurs but the sexual part of the cycle is not completed. Mortality is dose-dependent. A heavy dose of *A. taiwanensis* will annihilate *A aegypti* but mortality is much less in the reciprocal cross-infection. In the past, workers doing laboratory experiments on competition in laboratory microcosms have used parasite-free laboratory colonies. It is suggested that the parasite in *A. albopictus* gives it a major edge in field competition.

## CONCLUSION

Some medical workers have suggested the U.S. owes a debt of gratitude to *A. albopictus* for displacing the yellow fever mosquito, *A. aegypti*. This is incredibly short-sighted. These medical types fail to understand that:

1. *Aedes albopictus* has a far greater range because of its cold tolerance,
2. *Aedes albopictus* has a more inaccessible habitat because of plant breeding and therefore is far harder to control.
3. These two species have a different habitat, with *A. aegypti* in the inner city and *A. albopictus* in suburban and rural areas; the U.S. has far more of its population in the latter areas,
4. *Aedes albopictus* has already achieved far larger populations in the United States. *Aedes aegypti* was never a major biting pest in the United States. Today in New Orleans, 80 percent of the mosquito complaint telephone calls are due to *A. albopictus*, not their more traditional mosquito problems.
5. *Aedes albopictus* grows the same range of arboviruses; it is a more efficient

vector for some of them. Nobody ever heard of *A. aegypti* entering the cycle of EEE.

6. *Aedes albopictus* may be on the verge of entering the cycle of La Crosse Encephalitis. *Aedes aegypti* has no chance of doing so because it is wiped out by winter cold.
7. *Aedes albopictus* has the widest host range of all mosquitoes. *Aedes aegypti* outside Africa is too domesticated to get into feral zoonoses. Take yellow fever in South America. *Aedes albopictus* could pick up the virus in the sylvan part of the cycle AND carry it to man in the domestic part.

What do we see for the future in the Americas? (1) Continued enlargement of foci in Brazil and the U.S., (2) Spread to the Caribbean and Central America, (3) Serious involvement in the cycle of dengue, and (4) Serious involvement with zoonotic arboviruses such as La Crosse Encephalitis. At some time, the U.S. will have to get serious about developing adequate control methodology. For the rest of the world, new infestations will continue to break out through metastasis by containerized shipping. The role of this species in Africa remains an elusive but potentially disastrous mystery.

## ACKNOWLEDGMENT

Some of the work reported herein was supported in part by NIH Research Grant AI-02753. Thanks are due to Dr. George O'Meara for much assistance with the photoperiod change research in Florida.

## REFERENCES

BEAMAN, J. R. and M. J. TURELL. 1991. Transmission of Venezuelan Equine Encephalomyelitis by strains of *Aedes albopictus* collected in North and South America. J. Med. Entomol. 28:161–64.

BLACK, W. C., J. A. FERRARI, K. S. RAI, et al. 1988. Breeding structure of a colonizing species, *Aedes albopictus* Skuse in the U.S.A. Heredity 60:173–82.

CORNELL, A. J. and R. H. HUNT. 1991. *Aedes albopictus* in Africa. First records of live specimens in imported tires in Capetown, South Africa. J. Amer. Mosq. Cont. Assn. 7(1):107–108.

CRAIG, G. B., JR. and W. A. HAWLEY. 1981. The Asian Tiger Mosquito, *Aedes albopictus*: Whither, whence and why not in Virginia? Virginia. Ag. Exp. State. Info. Serv. 91–2:1–10.

CRAVEN, R. B., D. A. ELIASON, D. B. FRANCY, et al. 1988. Importation of *Aedes albopictus* into the U.S.A. in use tires from Asia. J. Amer. Mosq. Cont. Assn. 4(2):138–42.

CULLY, J. F., P. HEARD, D. M. WESSON, et al. 1991. Antibodies to La Crosse virus in eastern chipmunks in Indiana near an *Aedes albopictus* population. J. Amer. Mosq. Cont. Assn. 7(4):651–53.

CULLY, J. F., T. G. STREIT and P. B. HEARS. 1992. Transmission of La Crosse virus by four strains of *Aedes albopictus* to and from the Eastern Chipmunk (*Tamias striatus*). J. Amer. Mosq. Cont. Assn. 8(3):237–40.

FORATTINI, O. P. 1986. *Aedes albopictus*: identification in Brazil. Rev. Saude Publica 20(3):244–45.

FRANCY, D. B., N. KARABATSOS, D. M. WESSON, ET AL. 1990. A new arbovirus from *Aedes albopictus* , an Asian mosquito established in the U.S.A. Science 250:1738–40.

FRANCY, D. B., C. G. MOORE and D. A. ELIASON. 1990. Past, present and future of *Aedes albopictus* in the U.S.A. J. Amer. Mosq. Cont. Assn. 6(1)127–32.

GRIMSTAD, P. R., J. F. KOBAYASHI, M. ZHANG, et al. 1989. Recently introduced *Aedes albopictus* in the U.S.A.: Potential vector of La Crosse virus (Bunyaviridae: California Serogroup). J. Amer. Mosq. Cont. Assn. 5(3):422–27.

GRIMSTAD, P. R. and T. G. STREIT. 1989. Vector competence of *Aedes albopictus* for La Crosse Encephalitis Virus. Proc. Ohio Mosq. Cont. Assn. 19:21–27.

HANSON, S. M., J. P. MUTEBI, G. B. CRAIG, JR., et al. 1993. Reducing the overwintering ability of *Aedes albopictus* by male release. J. Amer. Mosq. Cont. Assn. (1): in press.

HAWLEY, W. A. 1988. The biology of *Aedes albopictus*. J. Amer. Mosq. Cont. Assn. Supplement #1, 4:1–40.

HAWLEY, W. A. 1991. Adaptable immigrant. The Asian Tiger Mosquito. Nat. Hist. Mag. July, 1991: p. 55–59.

HAWLEY, W. A. and G. B. CRAIG, JR. 1989. *Aedes albopictus* in the Americas. Future prospects. Proc. Arbovirus Resch. Aust. p. 202–205.

HAWLEY, W. A., C. B. PUMPUNI, R. H. BRADY, et al. 1989. Overwintering survival of *Aedes albopictus* (Diptera: Culicidae) eggs in Indiana. J. Med. Entomol. 26(2):122–29.

HAWLEY, W. A., P. REITER, R. S. COPELAND, et al. 1987. *Aedes albopictus* in North America: probable introduction in used tires from northern Asia. Science 236:114–16.

HEARD, P. B., M. L. NIEBYLSKI, D. B. FRANCY, et al. 1991. Transmission of a newly recognized virus (Bunyaviridae: Bunyavirus) isolated from *Aedes albopictus* (Diptera: Culicidae) in Potosi, Missouri, U.S.A. J. Med. Entomol. 28(5):601–605.

HOBBES, J. H., E. A. HUGHES and B. H. EICHOLD. 1991. Replacement of *Aedes aegypti* by *Aedes albopictus* in Mobile, Alabama. J. Amer. Mosq. Cont. Assn. 7(3):488–89.

KAMBHAMPATI, S., W. C. BLACK and K. S. RAI. 1991. Geographic origin of the U.S.A. and Brazilian *Aedes albopictus* inferred from isozyme analysis. Heredity 67(1):85–94.

KAY, B. H., W. A. IVES, P. I. WHELAN, et al. 1990. Is *Aedes albopictus* in Australia? Med. J. Aust. 153 (1):31–34.

KLOWDEN, M.J. and G.M. CHAMBERS. 1992. Reproductive and metabolic differences between *Aedes aegypti* and *Aedes albopictus*. J. Med. Entomol. 29:467–71.

MITCHELL, C. J. 1991. Vector competence of North and South American strains of *Aedes albopictus* for certain arboviruses. A review. J. Amer. Mosq. Cont. Assn. 7(3):446–51.

MITCHELL, C. J., M. L. NIEBYLSKI, G. C. SMITH, et al. 1992. Isolation of Eastern Equine Encephalitis virus from *Aedes albopictus* in Florida. Science. 257:526–27.

MONATH, T. P. 1986. *Aedes albopictus*. An exotic mosquito vector in the U.S.A. Ann. Intern. Med. 105(3): 449–51.

MOORE, C. G., D. B. FRANCY, D. A. ELIASON, et al. 1988. *Aedes albopictus* in the U.S.A: rapid spread of a potential disease vector. J. Amer. Mosq. Cont. Assn. 4(3): 356–61.

MUNSTERMANN, L. E. and D. M. WESSON. 1990. First record of *Ascogregarina taiwanensis* (Apicocomplexa: Leucudinidae) in North American *Aedes albopictus*. J. Amer. Mosq. Cont. Assn. 6(2):235–43.

NASCI, R. S., C. G. HARE and F. S. WILLIS. 1989. Interspecific mating between Louisiana strains of *Aedes albopictus* and *Aedes aegypti* in the field and the laboratory. J. Amer. Mosq. Cont. Assn. 5 (3):416–21.

NAWROCKI, S. J. and W. A. HAWLEY. 1987. Estimation of the northern limits of distribution of *Aedes albopictus* in North America. J. Amer. Mosq. Cont. Assn. 3:314–17.

O'MEARA, G. F. 1993. The spread of *Aedes albopictus* in Florida. Amer. Entomol. 39(2): in press.

O'MEARA, G. F., A. D. GETTMAN, L. F. EVANS, et al. 1992. Invasion of cemeteries in Florida by *Aedes albopictus*. J. Amer. Mosq. Cont. Assn. 8(1):1–10.

PUMPUNI, C. B., J. KNEPLER and G. B. CRAIG, JR. 1992. Influence of temperature and larval nutrition on the diapause-inducing photoperiod of *Aedes albopictus*. J. Amer. Mosq. Cont. Assn. 8(3):223–27.

RAI, K. S. 1991. *Aedes albopictus* in the Americas. Ann. Rev. Entomol. 36:459–84.

REITER, P. and D. SPRENGER. 1987. The used tire trade: a mechanism for the worldwide dispersal of container-breeding mosquitoes. J. Amer. Mosq. Cont. Assn. 3(3):494–501.

ROSEN, L. 1987. Sexual transmission of dengue viruses by *Aedes albopictus*. Amer. J. Trop. Med. Hyg. 37(2):398–402.

SABATINI, A., V. RAINERI, G. TROVATO, et al. 1990. *Aedes albopictus* in Italy and possible spread of the species in the Mediterranean area. Parasitologia (Rome) 32(3): 301–304.

SAVAGE, H. M., V. I. EZIKE, A. C. N. NWANKWO, et al. 1992. First record of breeding populations of *Aedes albopictus* in continental Africa: implications for arboviral transmission. J. Amer. Mosq. Cont. Assn. 8:101–103.

SAVAGE, H. M., M. L. NIEBYLSKI, G. C. SMITH, et al. 1993. Host-feeding patterns of *Aedes albopictus* (Diptera:Culicidae) at a temperate North American site. J. Med. Entomol. 30(1):27–34.

SCOLES, G. A. and G. B. CRAIG, JR. 1993. Variation in susceptibility to *Dirofilaria immitis* among U.S. strains of *Aedes albopictus*. Vect. Cont. Bull. N.C. States 2(1):98–103.

SHROYER, D. M. 1990. Vertical maintenance of Dengue-1 virus in sequential generations of *Aedes albopictus*. J. Amer. Mosq. Cont. Assn. 6(2):312–14.

SPRENGER, D. AND T. Wuithiranyagool. 1986. The discovery and distribution of *Aedes albopictus* in Harris County, Texas, U.S.A. J. Amer. Mosq. Cont. Assn. 2(2): 217–19.

STREIT, T., P. R. GRIMSTAD and G. B. CRAIG, JR. 1991. Stabilized La Crosse Virus infection in *Aedes albopictus*. Amer. J. Trop. Med. Hyg. 45(suppl.3):276.

TURELL, M. J., J. R. BERMAN AND R. F. Tammariella. 1992. Susceptibility of selected strains of *Aedes aegypti* and *Aedes albopictus* to Chikungunya Virus. J. Med. Entomol. 29(1):49–53.

WESSON, D. M. 1990. Susceptibility to organophosphate insecticides in larval *Aedes albopictus*. J. Amer. Mosq. Cont. Assn. 6(2):258–64.

WESSON, D. M., W. HAWLEY AND G. B. Craig, Jr. 1990. Status of *Aedes albopictus* in the Midwest: La Crosse belt distribution. Proc. Illinois Mosq. Vect. Cont. Assn. 1:11–15.

# The Fire Ant (*Solenopsis invicta*): Still Unvanquished*

Walter R. Tschinkel[1]

The fire ant (*Solenopsis invicta*) Buren has caused considerable commotion in the United States since its first appearance about 50 years ago (Lofgren 1986a). A native of the seasonally flooded Pantanal region of southern Brazil (Fig. 1), this ant was first noticed in the vicinity of the harbor at Mobile, Alabama around 1940. Although it probably made its way from Brazil by way of some form of shipping, its arrival has not been linked to any particular commodity. In any case, *S. invicta* joined the slower spreading *S. richteri* Forel from northern Argentina, which had appeared in Mobile in 1918. From this initial point of introduction, *S. invicta* began a rapid range expansion, partly by means of its own dispersive mating flights but mostly through the aid of man. By the mid-1950s, its range consisted of a contiguous zone around Mobile and numerous incipient populations centered on nurseries throughout the southeastern U.S. (Fig. 2). Mated queens or young colonies had apparently hitched rides on nursery stock transported throughout the Southeast. While USDA quarantine procedures reduced the rate of spread by nursery stock, this and other human modes continue to be important means of range extension for *S. invicta*. Almost all of the range expansion in the U.S. since the mid-1950s has consisted of filling in the spaces between incipient populations established during the first 10 to 15 years. In 1982 the ant appeared in Puerto Rico, and outlying populations are presently found in San Antonio and Oklahoma City (Fig. 2). *Solenopsis richteri* is now confined to northern Mississippi, having been displaced from its earlier range by *S. invicta* (Lofgren 1986a).

The prediction of *S. invicta*'s ultimate range limits is somewhat contentious and depends on certain biological assumptions. Because the ant has no true hibernation, its northward spread must ultimately be limited by winter-cold or brevity of a season warm enough for colony development and reproduction. There is evidence that the northern range limit is being approached in northern Mississippi, Alabama, Georgia, and the Carolinas. The southern latitudinal limits in South America are roughly similar to the northern ones in the U.S.—about 32°S and 34°N, respectively.

Most of the recent spread has been westward in Texas. While this brings the ant into increasingly arid zones, there is little knowledge of how or whether aridity will limit the ant's spread. The timing of rainfall may be as important as the total amount, because the mating flights and successful colony founding take place only on warm days after heavy rains. Mediterranean climates such as those of the western coastal states, with their cold winter rains and hot rainless summers, may prevent successful reproduction of fire ant colonies. Little is known about the southward spread into Mexico, though it has been speculated that the native ant fauna may become increasingly resistant to invasion as the tropics are approached.

The final range is likely to result from the combined effects of climate, physi-

*For a synopsis of a general ant life cycle, see Appendix, p. 135.

1. Department of Biological Science, Florida State University, Tallahassee, FL 32306–3050

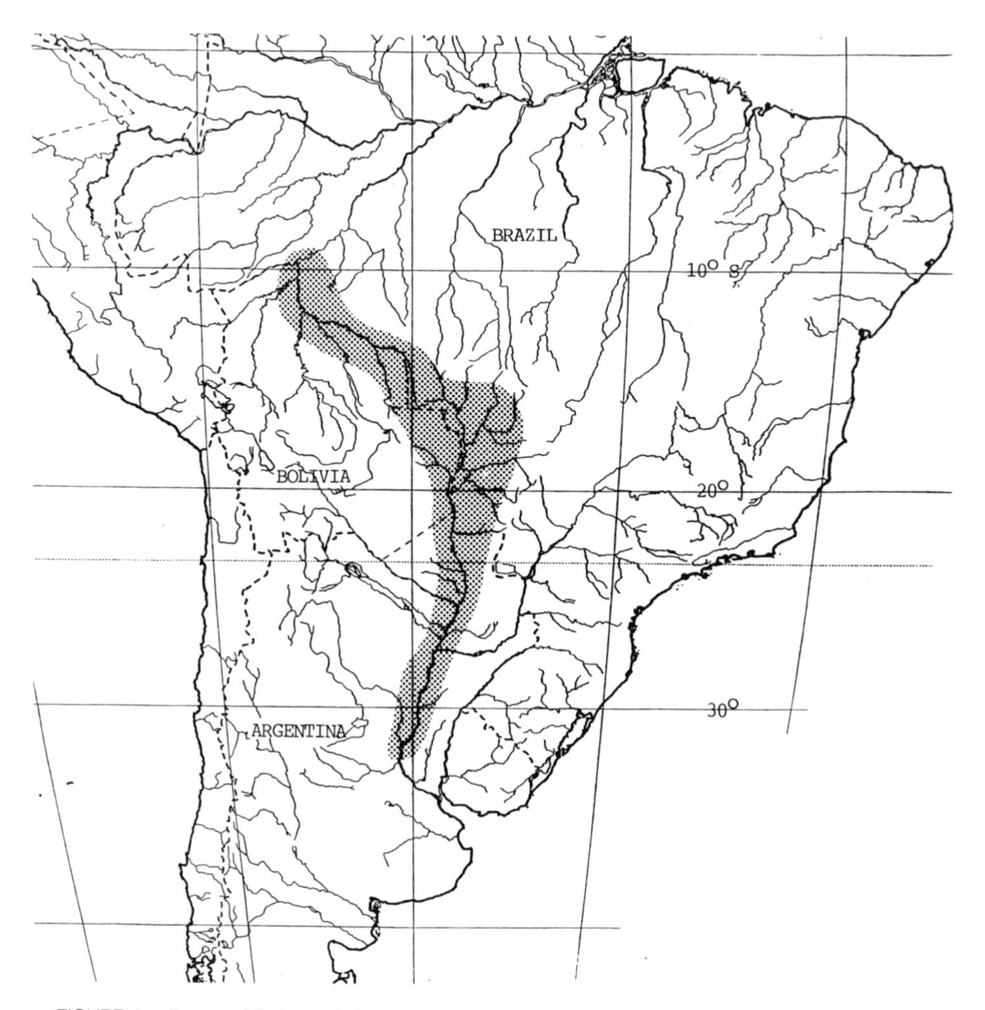


FIGURE 1.—Range of *Solenopsis invicta* in its South American homeland. The range includes the flood plains of two major river systems. Data from Buren et al. 1974 and Ross and Trager 1990.

cal, and biotic factors. In less favorable areas and at the margins of its eventual range, we can expect the fire ant to be a less dominant element of the fauna.

## IMPACT OF THE FIRE ANT IN THE U.S.

Much controversy has surrounded the claims of economic, social, and biological impacts of *S. invicta* (Lofgren 1986b; Davidson and Stone 1989). The early "parade of horrors" that formed the mythic underpinnings of such pork-barrel programs as the USDA's Mirex program has more recently given way to more careful attempts to quantify the ant's impacts (Lofgren 1986b). The human response to fire ants is often deeply emotional, as even brief conversations with many southerners will establish. This emotionality is based on the fact that fire ants readily and effectively sting people who blunder into their nests. The unaware can easily sustain dozens of stings in a single encounter. Compared to other ants, individual fire ant stings only rate a "moderate" on the "Richter pain scale," but they often make up for this through their numbers. The venom in-

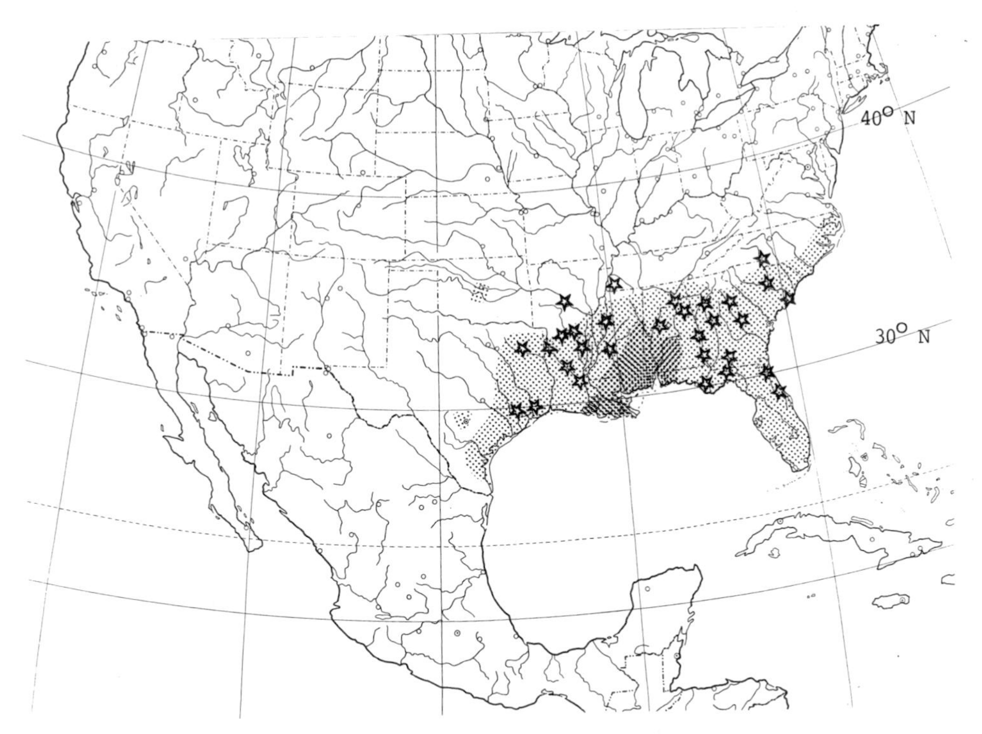


FIGURE 2.—Range of *S. invicta* in 1953 (dark shading and stars) and in 1986 (light shading). Much of the early spread was by transport of nursery stock. Most of the incipient populations in 1953 (stars) were centered on nurseries (redrawn from Lofgren 1986).

jected by the stinging worker soon causes the formation of a sterile pustule because white blood cells move in to clean up the venom-killed cells. The great majority of humans suffer only discomfort from the stings, but a small fraction are hyper-allergic to varying degrees. The response of such people may range from regional swelling to anaphylactic shock and death. Although many southerners are stung, only a small fraction need to seek medical treatment (Lofgren 1986b). In general, the stinging of people makes the fire ant a predominantly urban and suburban problem, in spite of its early attention from the agricultural sector.

While the impact of *S. invicta* directly on people is decidedly negative, its impact on agriculture is mixed (Lofgren 1986b; Davidson and Stone 1989). Claims of interference with harvesting and farm machinery have been made, though most remain sketchily documented. The ants are known to damage some crops such as eggplants, okra, and orange tree seedlings, but the quantification of proportional economic impact is difficult. On the other hand *S. invicta* has been shown to reduce hornflies and lone star ticks (pests of cattle) as well as several crop pests, including those of cotton and sugar cane. Being the general predator that it is, it is no surprise that fire ants can be beneficial in certain situations.

In the last 15 years a new phenomenon has come to light, one that may have a major biological impact. While fire ant colonies are generally monogyne (have a single egg-laying queen), there exist population enclaves throughout the Southeast in which colonies are polygyne (multiple egg-laying queens) (Glancey et al. 1987). In east Texas, fire ants exist as a mosaic of populations, about half of which are polygyne. The change in queen number brings about a major shift in colony biology—colonies are no longer territorial, so that ant density is no longer limited by territoriality; mating flights are small and most resources are invested in worker production, resulting in very high growth rates; colonies (nests) reproduce by budding or fission and disperse on foot, causing the population to spread as a contiguous front (Fig. 3) (Porter et al. 1988). Fire ant densities in polygyne populations are up to six times that in monogyne, causing a massive shift in the fauna. As the polygyne population moves in, the diversity and abundance of native arthropods and even vertebrates plummets. What food material sustains these high ant densities is not clear, but the prospect of the displacement of many native species is a serious concern (Porter et al. 1988).

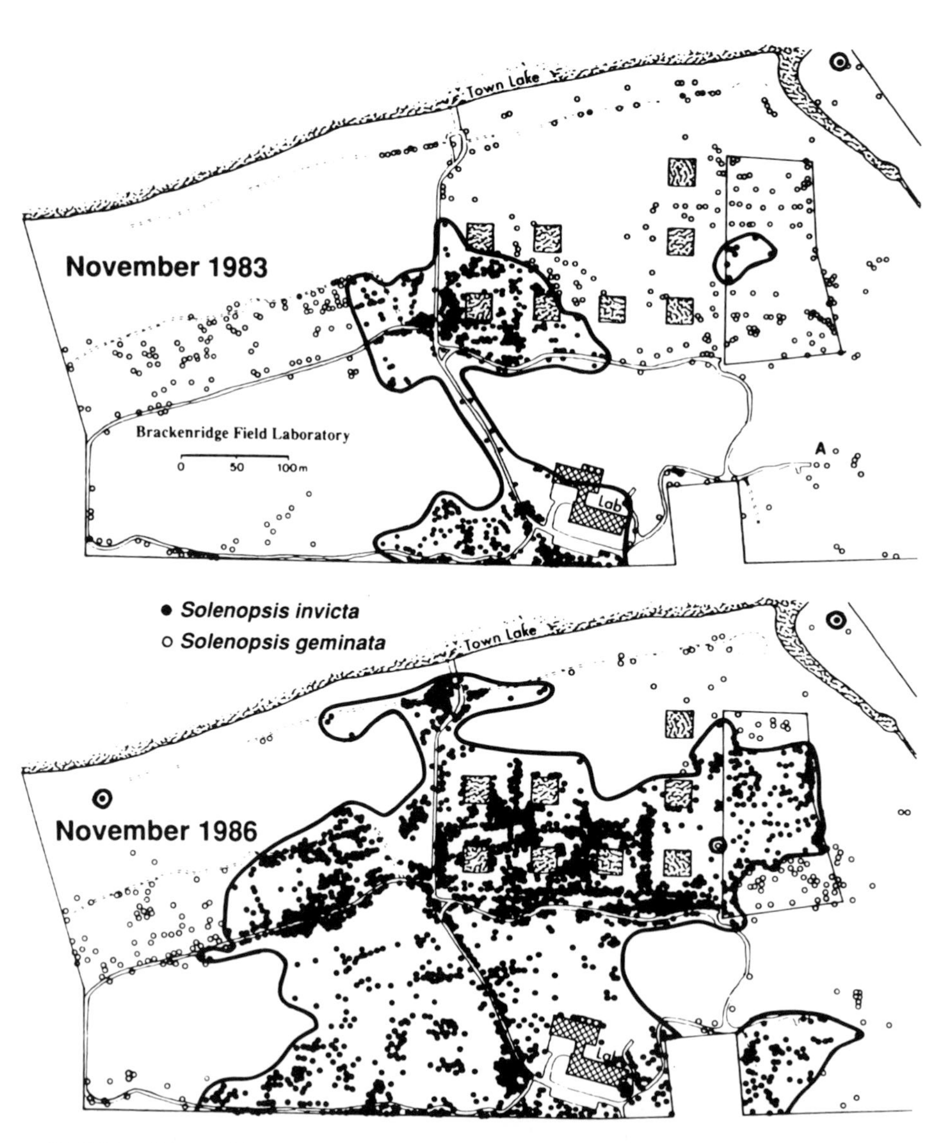


FIGURE 3.—Spread of the polygyne form of *S. invicta* at Brackenridge Field Laboratory near Austin, Texas. The polygyne form reproduces colonies by fission rather than independently through nuptial flights of queens. The population of colonies thus spreads as a contiguous front. Open circles, *S. geminata*; closed circles, *S. invicta* polygyne form (from Porter et al. 1988).

## WHY IS *S. INVICTA* INVASIVE?

The answer to this question is suggested by the types of habitats in which fire ants are abundant, both in the U.S. and in Brazil. In both cases, fire ants are found mostly in habitats that have been disturbed by man or by natural processes such as flooding, landslide, or disastrous storms. For example, in the natural coastal plain forests in northern Florida, *S. invicta* is limited to disturbed roadsides, heavily disturbed clearing sites, pond margins or areas with a high water table (Fig. 4) (Tschinkel 1988a). This distribution pattern suggests that *S. invicta* is a "weedy" species, an animal counterpart to the familiar plant weeds that invade cleared land and other disturbed sites.

Like plant weeds, some animals are adapted for opportunistic exploitation of ecologically disturbed habitat (Ito 1978). Under natural circumstances, occurrence of these habitats is unpredictable in time and space. The habitats are ephemeral, reverting to the climax vegetation through succession. In order to exploit these habitats, weeds and weed-like animals often share a number of characteristics. Habitat unpredictability selects both to invest heavily in large numbers of propagules. Spatial unpredictability is overcome by good dispersal ability. Seeds are lofted by the wind or carried by animals, while sexual ants fly long distances. Having landed in disturbed habitat, both must colonize it effectively, and because the habitat reverts to an unfavorable state, they must grow rapidly, reproduce early and continuously. Weeds and weed-like animals lead a sort of fugitive existence, from disturbed patch to patch. Success depends on their ability to arrive, as the Confederate General DeForrest said, "the firstest

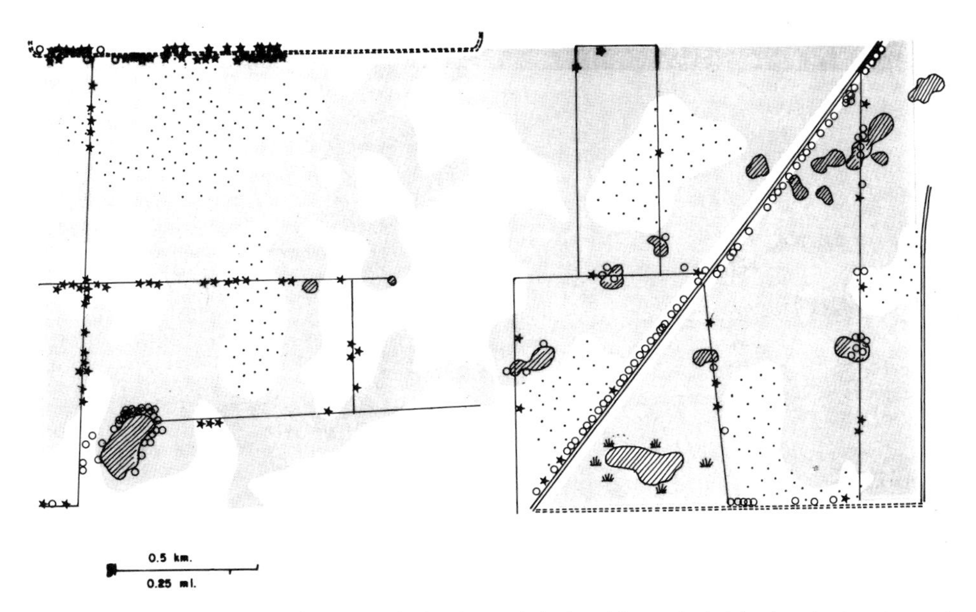


FIGURE 4.—Distribution of *Solenopsis invicta* (open circles) and *S. geminata* (stars) nests along transects in the high pinelands near Tallahassee, Florida. *S. geminata* is closely related to *S. invicta* but is probably a native species. Each symbol designates one nest within sight (2–3 m) of the transect line. *S. invicta* dominates disturbed areas and pond-margins, while *S. geminata* dominates the less disturbed upland areas. Management condition of forest is indicated by shading: shaded, mature longleaf pine forest; stippled, seed trees only; unshaded, recently clearcut, site prepared, and replanted with longleaf pine. Ponds indicated by cross-hatching. Paved roads, solid double lines; graded dirt roads, dashed double lines (from Tschinkel 1988a).

with the mostest," and to prevail in the scramble for space and resources before the habitat becomes unsuitable.

Comparing *S. invicta* with weeds gives us an immediate insight into its rapid spread in the United States. The primeval condition of the southeastern United States was unbroken broadleaf and pine forest, with small patches of early succession scattered haphazardly as windfalls, sandbars or clearings for Indian agriculture. By the early 20th century most of the forest had been cleared for living space and agriculture, creating vast new opportunities for early succession species. Availability of so much early succession habitat and an ant fauna not fully capable of exploiting it set the stage for the rapid spread of *S. invicta* (Buren 1983). The synergism between fire ants and man will continue as long as humans depend on early succession communities for their existence—man is truly the fire ant's best friend.

How well does *S. invicta* fit the stereotype of a weed? Let us survey the major biological attributes of fire ants in this light.

## DEPENDENCE UPON DISTURBANCE

As noted above, fire ants occur mostly in areas of disturbance such as lawns, pastures, roadsides, and agricultural land (Fig. 5). Less obviously, in addition to gross disturbance, *S. invicta* is favored by a high water table and seasonal flooding (Tschinkel 1988a). The Brazilian homeland, the Pantanal, is a seasonally flooded headwater of two major rivers. As a partial adaptation to this hydroregime, fire ant

FIGURE 5.—Typical aspect of fire ant colonies in a pasture. Colonies of such monogyne populations are well-spaced and defend territories against neighboring colonies. Inset is of a single nest, ca. 50 cm in diameter and 25 cm high. (photo by author)

FIGURE 6.—A floating colony of fire ants at a flooded pond-side site. The ability to float as a mat is probably an adaptation to the seasonal flooding to which *S. invicta* is sometimes subjected in South America. (photo by author)

colonies have the capacity, when the waters rise, to float as a mat of ants, surviving for weeks until the waters recede or they drift ashore (Fig. 6) (Morrill 1974). At least in the U.S., seasonal flooding eliminates most other ants. Thus in north Florida, fire ants have invaded wet-savanna even in the absence of human disturbance.

Specific disturbance of the ant community may be sufficient to favor *S. invicta.* Plots in which *S. invicta* was a minor component of the ant community were treated with poison bait, killing all ants. Upon recolonization, the community was heavily dominated by *S. invicta* (Fig. 7) (Buren et al. 1978; Summerlin et al. 1977). Apparently the colonizing ability and speed of fire ants exceeded that of our native ants. These findings make it likely that, by eliminating native ants, large-scale control programs such as the Mirex program ultimately helped the fire ant spread more rapidly and achieve higher dominance.

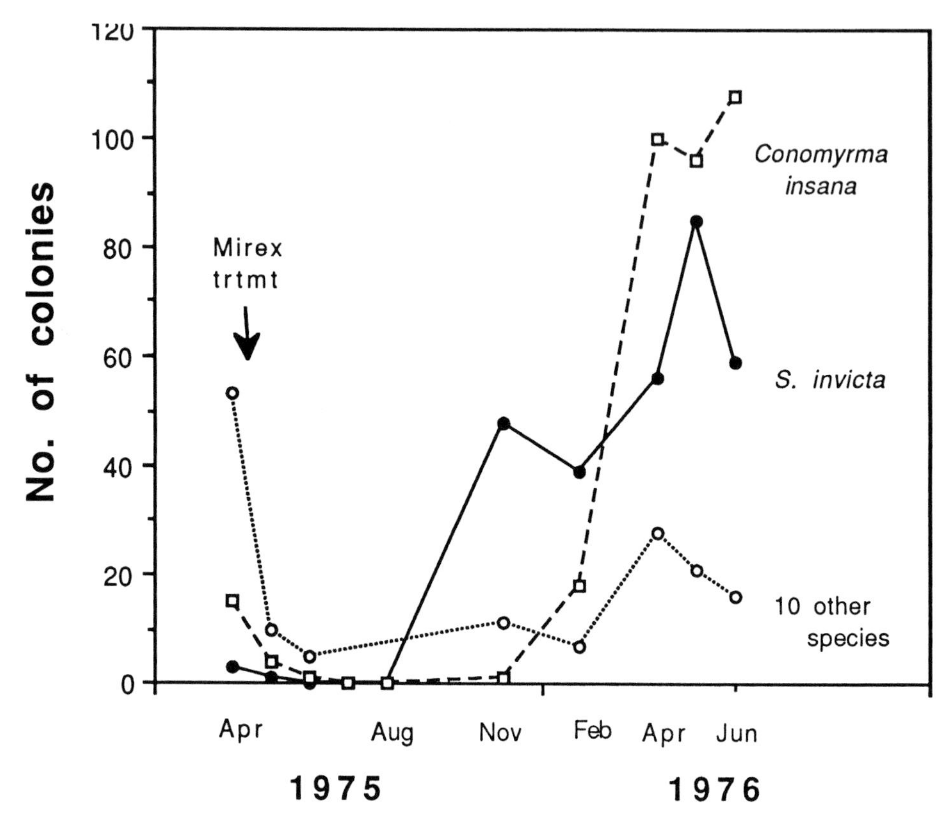


FIGURE 7.—Disturbance of the ant community and its effect on subsequent recolonization by several species. Most ants were killed with a single treatment of poison bait (Mirex). The fire ant *S. invicta* changed from a minor member of the ant community before treatment to a dominant member after. *Conomyrma insana* colonies, while more numerous than *S. invicta* colonies, are only about one-twentieth the mature size. Data from Summerlin et al. (1977).

## INVESTMENT IN REPRODUCTION

The fire ant invests heavily in propagules in the form of sexual males and females. Having entered the reproductive phase, fire ant colonies invest about one-third of their annual production in sexuals. This translates into about 4,000 to 6,000 sexuals per year, very high in comparison to the small number of other ant species for which data are available, and in the range of plant weeds (Tschinkel 1992b).

## DISPERSAL

Like the winged seeds of many plant weeds, fire ants disperse by means of winged sexual females. Mating flights take place on warm days after heavy rains. As late spring weather fronts sweep across the southeastern United States, they are followed by enormous synchronized mating flights of fire ants. Millions of winged males and females leave their natal nests, fly up into the sky, and mate 300 to 800 feet above the ground (Lofgren et al. 1975). The female stores a lifetime supply of sperm, about 7 million, in her spermatheca (Tschinkel 1987), and after a dispersal flight varying from a few dozen meters to a kilometer or more (Lofgren et al. 1975), she descends to the ground.

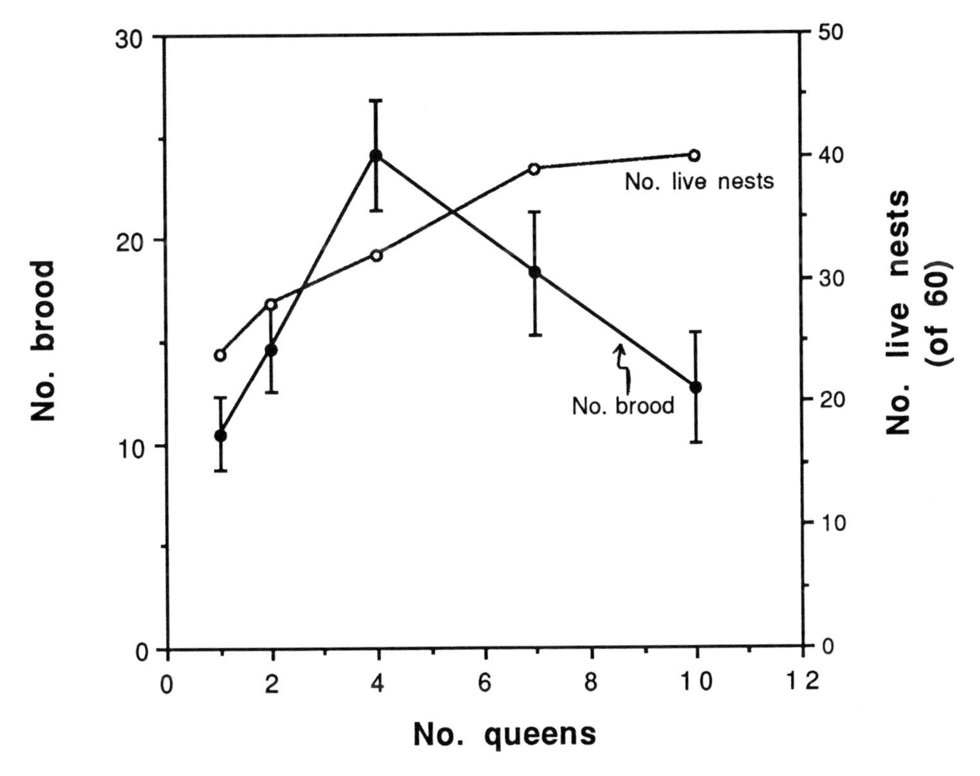


FIGURE 8.—The effect of cooperative colony founding and queen number on brood production and colony survival. Progeny production and the likelihood of founding success are maximum at intermediate group sizers (Tschinkel, unpubl. data).

## EFFECTIVE COLONIZATION

Unlike plant weeds, fire ant females choose their landing site while still airborne. Sites that are *partially* vegetated as a result of recent disturbance are especially attractive to them, and they often colonize these in large numbers. On one 1200 $m^2$ study plot in Tallahassee, I estimated that between 10,000 and 20,000 mated queens attempted colonization in the course of one summer (Tschinkel 1992a).

Having landed, the newly mated queen breaks off her wings, and usually within a radius of 4 to 5 m of her landing spot, she digs a 5 to 15 cm deep tunnel at the end of which she forms a chamber. She seals herself into this nest for the duration of the founding period and rears the first brood of 5 to 35 minim workers from reserves stored in her body, losing 70 percent of her energy content in the process (Tschinkel, unpubl.).

Fire ants have evolved two ways of improving the success of this population founding phase. Queens often join one another to rear the first worker brood, bringing about improved survival and higher worker production (Fig. 8) (Tschinkel and Howard 1983). Those nests that succeed in producing minim workers engage in brood-raiding as soon as the nests are opened to the surface (Tschinkel 1992a). In brood-raiding, workers from incipient nests seek out neighboring incipient nests in order to steal their brood, and because the host workers often steal them back, a reciprocal raid develops. These can sometimes involve many nests (Fig. 9), although most raids are small. The winners are

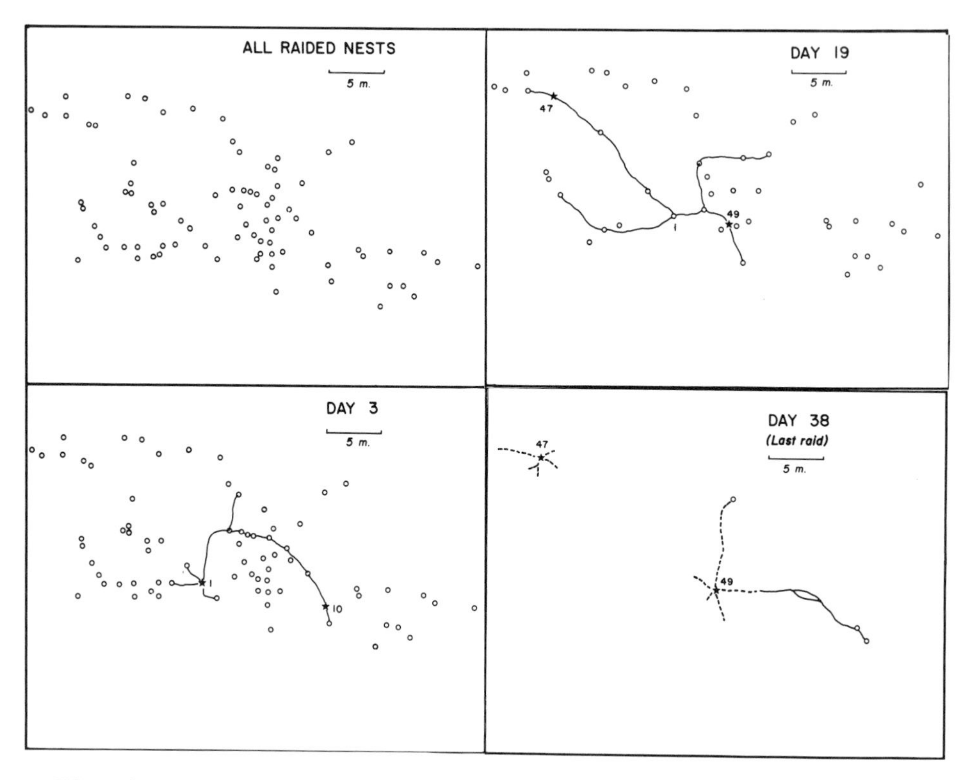


FIGURE 9.—The course of a very large brood raid. This raid lasted for 38 days and included hundreds of meters of raiding trails. By the end, over 80 incipient colonies had coalesced into two colonies (stars, 47 and 49) (from Tschinkel 1992a).

usually the nests that had the most workers initially, i.e., those founded by groups of queens, underscoring the importance of cooperative founding. The losing workers generally join the winners, and the losing queens can often be seen wandering on the surface, sometimes trying to enter incipient nests. The outcome of raiding is the coalescence of many small nests into a small number of much larger ones. Workers soon kill all but one queen. The lone survivor and new queen mother is not necessarily one that founded that nest. The nest now stands on the threshold to growth.

## RAPID COLONY GROWTH

The importance of cooperative founding and brood-raiding becomes clear in the next weedy property—rapid growth. The ephemeral habitat requires that growth be rapid and reproduction early. Early growth in fire ant colonies is exponential, i.e., multiplicative (Tschinkel and Howard 1983; Tschinkel 1988b). A colony beginning growth with 50 minims will be 10 times as large as one beginning with 5 minims as long as exponential growth continues. This brings the importance of initial brood number into sharp focus because the winner of the scramble for space will be the colony which achieves the largest size at the earliest date.

Like other weeds, fire ant colonies are capable of rapid growth. Lab and field studies suggest that colony growth rates are strongly food-limited (Tschinkel 1988b). In a few cases, lab-reared newly mated queens produced 500,000 work-

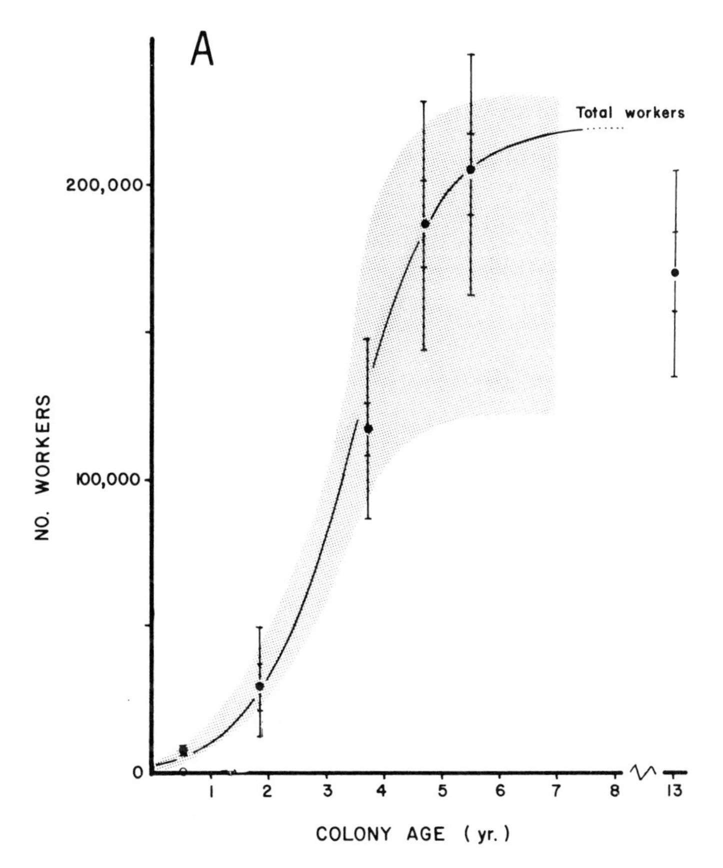


FIGURE 10.—Mean growth curve of fire ant colonies in moderately productive habitat. Growth is logistic, and colonies reach full size in 5 to 7 years. Shading shows the limits of seasonal variation in colony size (modified from Tschinkel 1988b).

ers in their first year of life (Tschinkel and Porter 1988). In average habitat, fire ant colonies grow to sizes of about 200,000 workers in 5 to 6 years (Fig. 10) (Tschinkel 1988b). Growth is forced by a feedback loop in which the older larvae produce a substance which stimulates the egg-laying rate of the queen (Tschinkel 1988c). In mature colonies, queens may lay up to their own body weight in eggs daily, about 3,000 to 5,000 eggs per day. Such queens are 75 percent ovary, by weight, and are egg-laying machines (Fig. 11).

## EARLY AND CONTINUING REPRODUCTION

The ephemeral nature of the habitat also selects for early and continuing reproduction in weeds. When fire ant colonies are only 10 percent of their final size, they begin to invest 30 percent of their annual production into sexuals (Tschinkel 1992b). Sexual production therefore increases in direct relation to colony size.

Moreover, sexual production and release is less seasonal than in many other ants. *Solenopsis invicta* produces sexuals during any part of the winter, spring, and early summer warm enough to allow it. They release these on mating flights over several months, beginning with the first warm spring rains and ending with exhaustion of the supply of sexuals (Fig. 12) (Lofgren et al. 1975). In the Tallahassee area, flights begin in early May and taper off through July.

FIGURE 11.—A fire ant queen and her retinue. Workers in the retinue groom the queen, feed her a special liquid diet, and take away the eggs she lays. This queen was laying about 70 eggs per hour. Ovaries make up most of the abdominal volume. (photo by author)

## CAPACITY FOR VEGETATIVE REPRODUCTION

Many weeds, having established in a suitable patch of habitat, increase their dominance in it by vegetative reproduction—runners, asexual seed production, root shoots, and so on. In an analogous sense, fire ants have undergone a similar change of biology. Throughout the range of the fire ant, both in Brazil and the U.S., are found population enclaves of polygyne colonies. As noted above, these colonies invest little in sexual production. Rather, most resources are invested in colony growth (worker production), which results in asexual-like colony reproduction by budding or fission. Like many asexually reproducing plant weeds, these polygyne populations achieve very high dominance of their habitat.

## NON-WEEDY TRAITS

Of course, no category is perfect. Several aspects of fire ant biology do not fit the weedy syndrome. Large size is not often typical of weeds, and fire ant colonies rank among the largest ant colonies, often containing up to 250,000 workers.

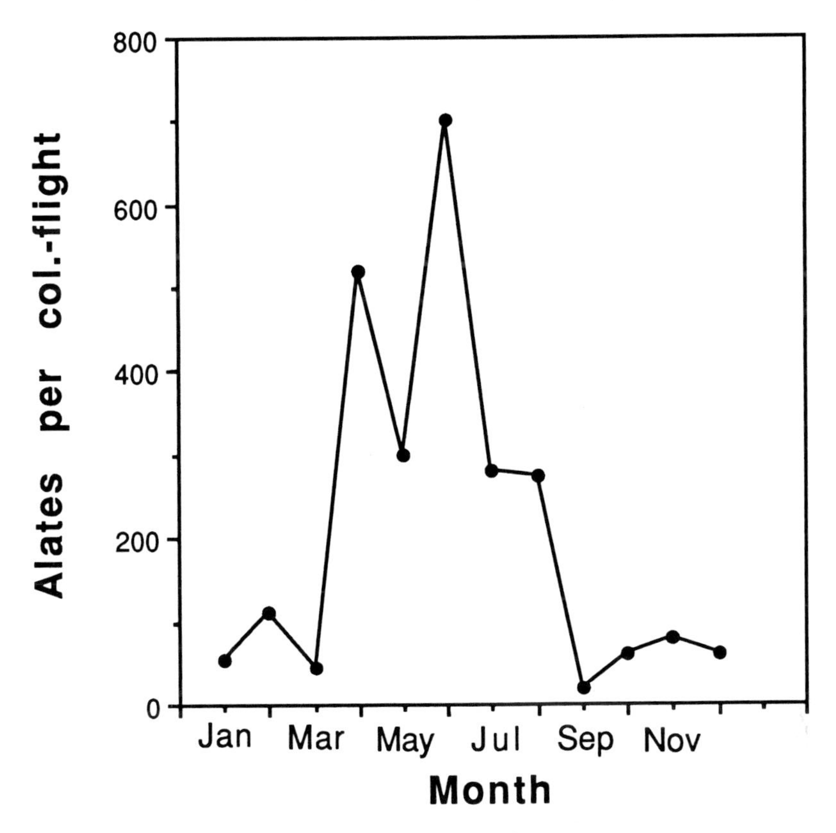


FIGURE 12.—The release of winged sexuals (alates) on nuptial flights throughout one year in north Florida. Large mating flights can occur during most of the late spring and early summer (data from Morrill 1974).

Across species, ant colonies probably average a few hundred workers. Competitiveness of weeds is thought to rely heavily on speed of growth and less on direct interactions among individuals. However, soon after the brood-raiding phase, monogyne fire ant colonies become territorial, excluding members of neighboring colonies from their foraging territory by means of aggression and warfare (Fig. 13) (Wilson et al. 1971). Long life is another non-weedy property. While an individual queen runs out of sperm (and therefore worker-producing capacity) after 5 to 7 years (Tschinkel 1987), she is often replaced with a new one (Tschinkel and Howard 1978), making colonies potentially immortal, at least until the habitat becomes unfavorable. Thus fire ants act like weeds early in the population cycle, and competitors later.

## THE FUTURE

The chimera of fire ant eradication (Lofgren and Weidhaas 1972) that floated over the decades of the 1960s and 1970s seems to have faded. The fire ant is here to stay. With respect to the monogyne form, public demand for relief has softened. Most households accomplish their own fire ant control with the many baits and mound drenches available for this purpose. In agriculture, the marketplace has determined, on the one hand, just how much of a problem the fire ant really is and, on the other, ants may also have value. These issues will continue to draw refining discussion and new data.

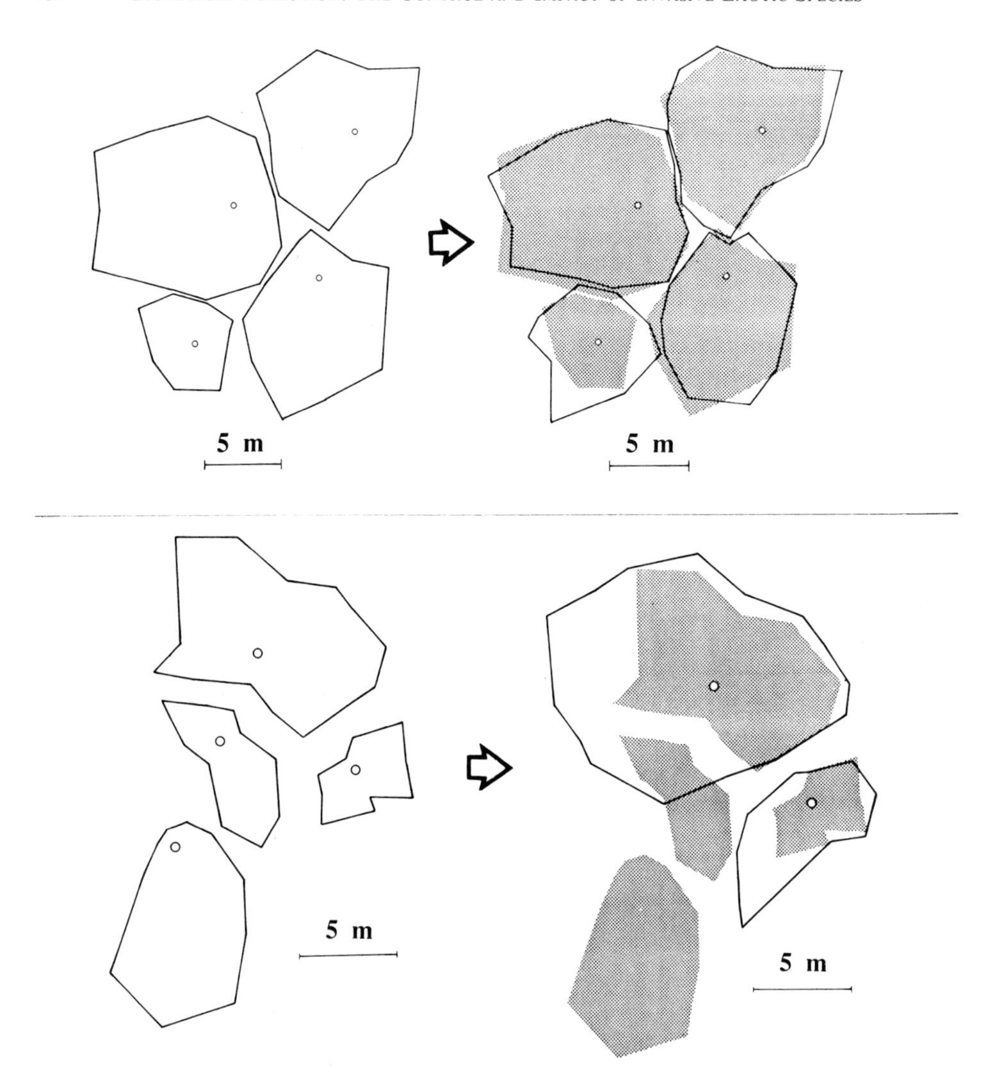


FIGURE 13.—Map of two groups of adjacent territories of fire ant colonies near Tallahassee, Florida. The mound position is shown by the circles. In the lower group, two colonies were removed, causing the remaining two to expand their territories (lower right) and demonstrating that territory size results from active defense. Shaded areas show original territory size. The controls showed little net change in territory area over the same period (upper right) (Adams, Macom and Tschinkel, unpubl.).

The polygyne form may yet be seen as causing unacceptable dislocations in our native ecosystems, but the full-scale and nature of its impact is not yet known. Presently the only known biological deterrent to its spread is the monogyne, territorial form of *S. invicta*, leading to the ironic situation in which the Texas Department of Agriculture recommends landowners to husband their monogyne colonies.

Because chemical control is unspecific and *S. invicta* is such an effective recolonizer, the only long-term hope of reducing fire ant dominance is through biocontrol. While a number of potential biocontrol agents are in hand, no releases have been attempted (Jouvenez 1990). As is so often the case, it will probably require release of multiple agents to achieve substantial reduction of fire ant populations. The goal should be to reduce the fire ant's competitive advantage to a level comparable with native species. If this can be achieved, the fire ant will probably appear more like a normal component of our ant fauna rather than an overdominant alien (Buren et al. 1978; Buren 1983). It should be added that no biocontrol against a social insect has ever been attempted.

Finally, there is one area in which the fire ant has repaid us richly. Its abundance, ease of experimental manipulation, and high profile have enabled researchers to make it the best understood ant. Fire ant research will continue to contribute important insights into the biology of social insects, evolution, population genetics, and perhaps as yet undiscovered realms. As far as biologists are concerned, fire ants have more than compensated for the aggravation they have caused.

## APPENDIX: A PRIMER OF ANT LIFE CYCLES

All ants are social and live in colonies in which some individuals called queens (or a single one) are fertile and lay most of the eggs, while others (the workers) are sterile, raise the brood, and do all other colony work, including foraging (Hölldobler and Wilson 1990). All socially functional individuals are female. The differences between queens and workers are the result of developmental not genetic differences. Males develop from unfertilized eggs and are produced only for colony reproduction. The most typical form of colony reproduction in ants occurs when winged sexual males and females produced at the proper season (usually spring or summer) leave the nest on a nuptial flight. Mating takes place on the ground, on vegetation, or in the air, and the male dies within a short time. The female alights on the ground after a dispersive flight of variable length, breaks off her wings, and digs or seeks a chamber in which she seals herself during the founding period (claustral period). From material derived from stored nutrients and degenerating wing muscles, she lays eggs and feeds the resulting larvae until they become a brood of a few to a few dozen tiny workers (minim workers). These minims then open the nest, begin foraging, and take over all the work duties such as brood-rearing and nest construction. The queen becomes an egg-laying machine. Colony growth ensues through worker production until the colony contains enough workers to produce winged sexual females and males, closing the cycle.

Alternately, in some species, colonies may be polygyne, containing more than one egg-laying queen. Typically these queens are adopted as newly mated queens after mating flights. Colonies then reproduce by budding or fission of a portion of the workers and queens in each fragment.

## REFERENCES

Buren, W. F. 1983. Artificial faunal replacement for imported fire ant control. Florida Entomol. 66:93–100.

Buren, W. F., G. E. Allen and R. N. Williams. 1978. Approaches toward possible pest management of the imported fire ants. ESA Bulletin, 24:418–21.

Davidson, N. A. and N. D. Stone. 1989. Imported fire ants. Pages 196–217 *in* Eradication of exotic pests: analysis with case histories. D. L. Dahlsten and R. Garcia (eds.). Yale Univ. Press, New Haven, Conn.

Glancey, B. M., J. C. E. Nickerson, D. Wojcik, et al. 1987. The increasing incidence of the polygyne form of the red imported fire ant, *Solenopsis invicta*, in Florida. Florida Entomol. 70:400–02.

Hölldobler, B. and E. O. Wilson. 1990. The ants. Belknap/Harvard Press, Cambridge, Mass.

Ito, Y. 1978. Comparative ecology. Cambridge Univ. Press, N.Y.

Jouvenez, D. P. 1990. Approaches to biological control of fire ants in the United States. Pages 620–17 *in* Applied Myrmecology: a world perspective. R. K. Vander Meer, K. Jaffe and A. Cedeno (eds.). Westview Press, Boulder, Colo.

Lofgren, C. S. 1986a. History of imported fire ants in the United States. Pages 36–47 *in* Fire ants

and leafcutting ants, biology and management. C. S. Lofgren and R. K. Vander Meer (eds.). Westview Press, Boulder, Colo.
———. The economic importance and control of imported fire ants in the United States. Pages 227–55 *in* Economic impact and control of social insects. S. B. Vinson, ed.
LOFGREN, C. S., W. A. BANKS and B. M. GLANCEY. 1975. Biology and control of imported fire ants. Ann. Rev. Entomol. 29:1–30.
LOFGREN, C. S. and D. E. WEIDHAAS. 1972. On the eradication of imported fire ants: a theoretical appraisal. Bull. Entomol. Soc. Amer., 18:17–20.
MORRILL, W. L. 1974. Dispersal of red imported fire ants by water. Florida Entomol. 57:39–42.
PORTER, S. D., B. VAN EIMEREN and L. E. GILBERT. 1988. Invasion of red imported fire ants (Hymenoptera: Formicidae): microgeography of competitive replacement. Ann. Entomol. Soc. Amer. 81:913–18.
ROSS, K. G. and J. C. TRAGER. 1990. Systematics and population genetics of fire ants (*Solenopsis saevissina* complex) from Argentina. Evolution 44:2113–34.
SUMMERLIN, J. W., A. C. F. HUNG and S. B. VINSON. 1977. Residues in nontarget ants, species simplification and recovery of populations following aerial applications of Mirex. Environ. Entomol. 6:193–97.
TSCHINKEL, W. R. 1987. Fire ant queen longevity and age: estimation by sperm depletion. Ann. Entomol. Soc. Amer. 80:263–66.
———. 1988a. Distribution of the fire ants *Solenopsis invicta* and *S. geminata* in north Florida in relation to habitat and disturbance. Ann. Entomol. Soc. Amer. 81:76–81.
———. 1988b. Colony growth and the ontogeny of worker polymorphism in the fire ant, *Solenopsis invicta* Buren. Behav. Ecol. Sociobiol. 22:103–15.
———. 1988c. Social regulation of egg laying rate in queens of the fire ant, *Solenopsis invicta*. Physiol. Entomol. 13:327–50.
———. 1992a. Brood-raiding and the population dynamics of founding in the fire ant, *Solenopsis invicta*. Ecol. Entomol. In press.
———. 1992b. Sociometry and sociogenesis in colonies of the fire ant, *Solenopsis invicta*. Ecol. Monogr. In review.
TSCHINKEL, W. R. and D. F. HOWARD. 1978. Queen replacement in orphaned colonies of the fire ant, *Solenopsis invicta*. Behav. Ecol. Sociobiol. 3:297–310.
———. 1983. Pleometrotic colony foundation in the fire ant, *Solenopsis invicta*. Behav. Ecol. Sociobiol. 12:103–13.
TSCHINKEL, W. R. and S. D. PORTER. 1988. The efficiency of sperm use in queens of the fire ant, *Solenopsis invicta* Buren. Ann. Entomol. Soc. Amer. 81:777–81.
WILSON, N. L., J. H. DILLIER and G. P. MARKIN. 1971. Foraging territories of imported fire ants. Ann. Entomol. Soc. Amer. 64:660.

# Distribution and Spread of the Invasive Biennial *Alliaria petiolata* (Garlic Mustard) in North America

Victoria Nuzzo[1]

## INTRODUCTION

*Alliaria petiolata* (Bieb. [Cavara & Grande]) is a cool-season, shade-tolerant, obligate biennial herb that invades forested natural communities in the Midwestern and northeastern United States and adjacent Canada (Fig. 1). *Alliaria petiolata* is often referred to as *Alliaria officinalis* Andrz; earlier names included *Alliaria alliaria L. (Britton), Sisymbrium alliaria Scop., Sisymbrium officinalis DC* and *Erysimum alliaria L. Alliaria petiolata* (hereafter referred to as *Alliaria*) is commonly known as garlic mustard, in reference to the strong garlic fragrance produced when the plant is crushed. Other common names include hedge garlic, jack-by-the-hedge, and sauce-alone (Georgia 1920). The plant is native to northern Europe, south of 68°N (Tutin et al. 1964), ranging from England to Sweden to the western region of the former USSR, and south to Italy. From this native range, *Alliaria* has spread to North Africa, India, Sri Lanka (Rai et al. 1972 and Fernaldo 1971, *in* Cavers et al. 1979) as well as Canada (Cavers et al. 1979) and the United States (Gray et al. 1889; Fernald 1970). The North American range of *Alliaria* extends from British Columbia (Cavers et al. 1979) to New England (Gleason and Cronquist 1963) and south to Missouri (Yatskievych and Turner 1990). Regional distribution has been mapped for Wisconsin (Patman and Iltis 1961), North Carolina (Radford et al. 1965), Kansas, Minnesota, North Dakota (Barkley 1977), Canada (Cavers et al. 1979), southern New York (Brooks 1983), Michigan (Voss 1985), Illinois (Nuzzo 1991c), and Ohio (Furlow 1991).

Life history and biology of this European mustard have been investigated by Murley (1951), Trimbur (1973), Lhostka (1975), Cavers et al. (1979), Roberts and Boddrell (1983), Byers and Quinn (1987, 1988), Babonjo et al. (1990) and Kelley et al. (1991) and summarized in Cavers et al. (1979), and Nuzzo (1991a). *Alliaria* germinates in early spring, overwinters as a basal rosette, flowers the following spring, and produces seed in summer, some 15 months after germinating. Plants produce an average of 350 seeds, and individual plants may produce up to 7,900 seeds (Nuzzo, unpubl.). Seeds have a variable dormancy period, germinating in 8 months in Kentucky (Baskin and Baskin 1992) and in 20 months in Ontario (Cavers et al. 1979).

*Alliaria* invades shaded communities, habitat essentially unoccupied by other invasive alien herbs. The plant most frequently invades wet to dry-mesic deciduous forests but also occurs in the partial shade characteristic of oak savannas, forest edges, shaded roadsides, in urban areas, and occasionally in full sun, particularly in areas exposed to periodic disturbance.

*Alliaria* threatens the floristic structure of invaded natural communities (Schwegman 1989), particularly the herbaceous layer. First recognized as a prob-

1. Native Landscapes, 1947 Madron Dr., Rockford, IL 61107

FIGURE 1.—Garlic mustard in Cook County Forest Preserve District, Illinois. (photo by Bill Glass)

lem species in Ontario (Cavers et al. 1979), this European mustard was not considered a threat in the United States until the late 1980s. In a 1986 survey of 25 natural areas botanists throughout Illinois, which requested a rank-ordered list of the 10 most problematic exotic plants in the state, *Alliaria* was not listed by any of those surveyed (B. N. McKnight pers. comm.). By 1991 many Midwest states had identified *Alliaria* as a species of concern, and methods to control the plant in natural communities were under study (Nuzzo 1991a, 1991b).

## DISTRIBUTION AND SPREAD

Entry, spread, and distribution of *Alliaria* were determined using >1,150 collection records obtained from 77 North American herbaria, supplemented by sight observations made between 1989 and 1991 by field biologists, site stewards, and the author. The resulting distribution map (Fig. 2) depicts within 20-year periods the first record or observation of *Alliaria* within 15-minute topographic quadrangles (approximately 590 $km^2$ in northern Illinois) in the northcentral

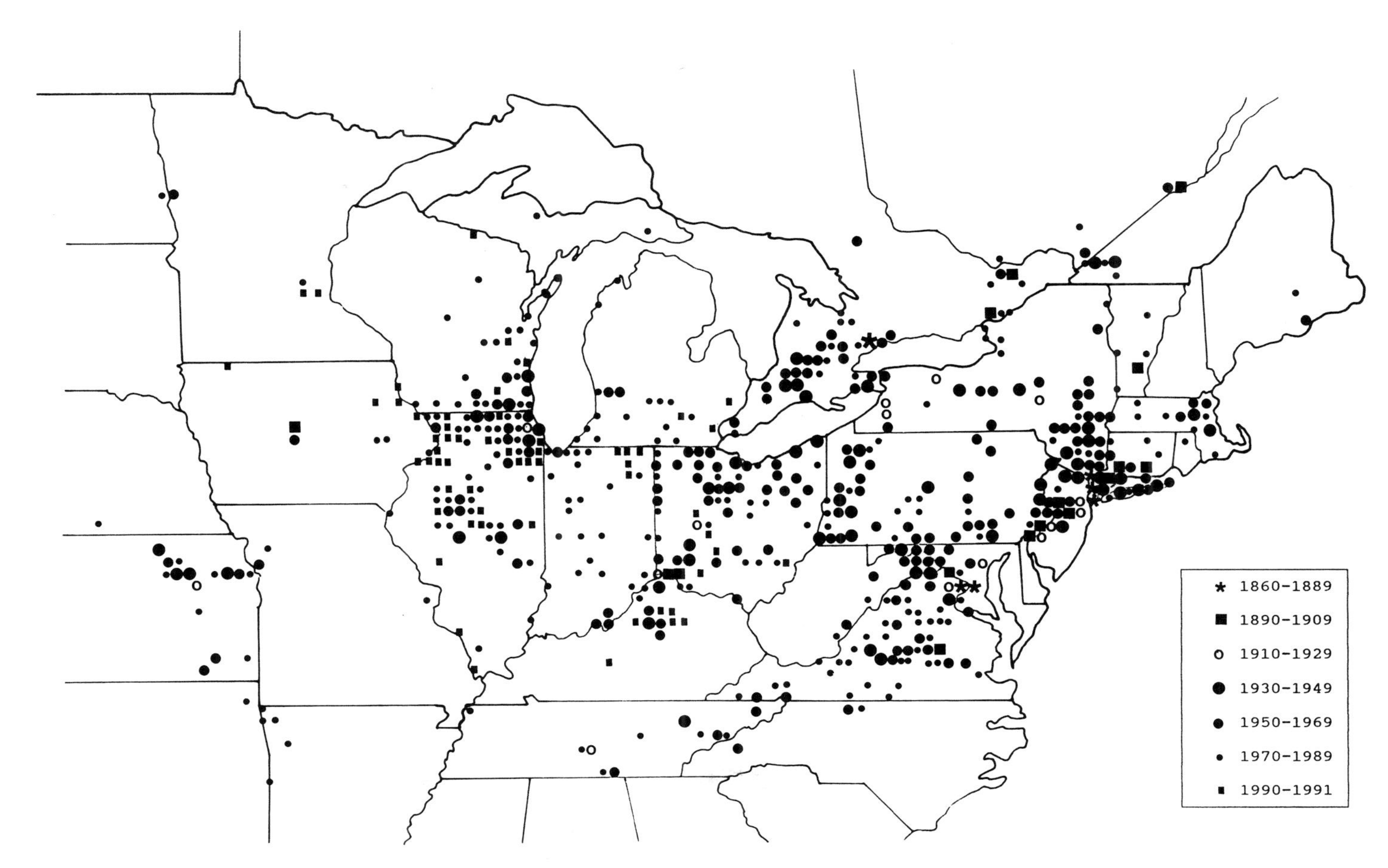


FIGURE 2.—*Alliaria petiolata* distribution in eastern North America 1868–1991, mapped by first record of occurrence within 15′ topographic quadrangles.

U.S. and adjacent Canada. Infrequent collections from western regions were not mapped.

This map is subject to certain limitations. First, distribution maps based on herbarium records represent only the minimum presence of a species and presumably underestimate actual rate of spread and distribution. Second, mapping the first record of an invasive species within equi-sized blocks provides a reliable measure of the rate of invasion through space but not of the extent of invasion within a block (Mack 1985).

*Alliaria* was first recorded in North America in 1868 on Long Island, New York (Leggett *s.n.* NYS). Spread was initially gradual, with presence reported in just six quadrangles in the next 21 years (Table 1). Multiple collections made within these quadrangles indicate that the species was locally abundant. Lack of collection outside the six quadrangles implies that the species had not spread far from the initial invasion site and/or that collectors worked only within a limited area.

TABLE 1. First record of *Alliaria petiolata* occurrence within 15-minute topographic quadrangles (approximately 590 km$^2$) in the United States and Canada.

| | 1868–1889 | 1890–1909 | 1910–1929 | 1930–1949 | 1950–1969 | 1970–1989 | 1990–1991 | TOTAL |
|---|---|---|---|---|---|---|---|---|
| United States | 5 | 16 | 13 | 58 | 158 | 233 | 57 | 540 |
| Canada | 1 | 3 | 0 | 8 | 21 | 21 | 0 | 54 |
| Total | 6 | 19 | 13 | 66 | 179 | 254 | 57 | 594 |
| Cumulative | 6 | 25 | 38 | 104 | 283 | 537 | 594 | |

In the next 20 years (1890–1909) *Alliaria* was collected from an additional 19 locations, including the St. Lawrence Valley in Canada as well as Idaho, Iowa, and Ohio. By 1929, 61 years after the first collection, *Alliaria* had been recorded from a total of 38 quadrangles in 11 states and 2 provinces. The rate of spread, or of collection, increased in the following decades. By 1989 Alliaria was recorded from 537 topographic blocks. An additional 57 blocks were recorded in 1990–1991, all but two based on sight observations. The paucity of herbarium collections since 1989 may reflect the tendency for an invasive species to be well-collected in the early stages of invasion and under-collected in the later stages, once it is established in a locale and considered to be an 'undesirable' or 'weedy' species.

By the end of 1991 *Alliaria* occurred in a minimum of 594 topographic quadrangles in 30 states and 3 provinces. What is not apparent is the extent of infestation within each quadrangle. In Illinois virtually all available habitat in the Chicago region supports *Alliaria,* while southern counties have very localized infestations. Throughout the state 30 percent of state parks and 31 percent of dedicated nature preserves support populations of this species (Nuzzo 1991c).

Graphing the quadrangle occurrences cumulatively by 30-year periods produces a distinct j curve (Fig. 3), reflecting the exponential spread of this plant. *Alliaria* spread gradually during the first 60 years (1868–1929), at a rate of approximately 0.6 quadrangle or 366 km$^2$/year. The rate of spread increased substantially beginning in 1930 (3.3 quadrangles or 1,950 km$^2$/year between 1930 and 1949) and jumped dramatically again after 1950 (8.9 quadrangles or 5,280 km$^2$/year). As of 1991 there was no indication of leveling off. On the basis of the 433 quadrangle occurrences between 1950 and 1989, *Alliaria* is spreading, or being recorded, at the rate of some 10.8 quadrangles or 6,400 km$^2$/year. It must

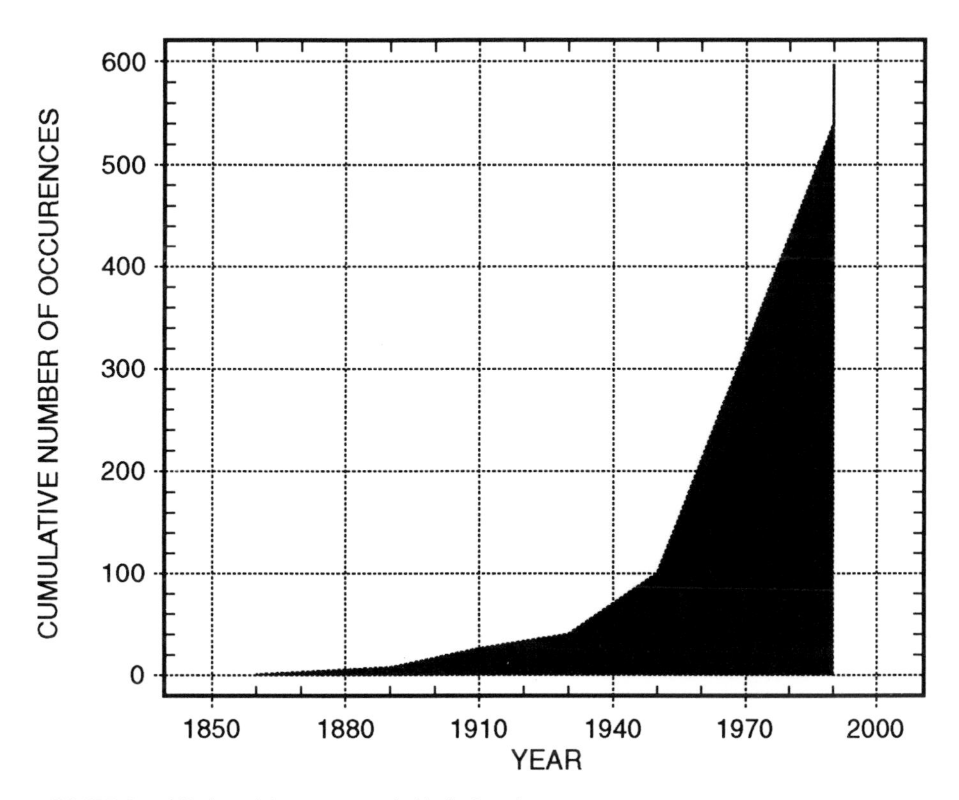


FIGURE 3.—*Alliaria petiolata* presence in North America.

be emphasized that this figure reflects only presence, and not abundance, of plants within each topographic quadrangle.

The rate of spread of *Alliaria* is greater than that recorded for purple loosestrife (*Lythrum salicaria*; 645 km$^2$/year since 1940 [Thompson 1991]) but considerably less than documented for cheat grass (*Bromus tectorum*; all available habitat in 30 years [Mack 1981]). The rapid increase in *Alliaria* spread is consistent with increases recorded for other alien species, beginning approximately 40 to 50 years after initial invasion (Lacey 1957; Thompson et al. 1987).

Spread progressed from the northeastern seaboard westward, a pattern typical of many invasive European species. Regional spread of *Alliaria* has two components; establishment of multiple satellite populations (*sensu* Auld et al. 1978; Auld and Coote 1980) often separated by great distances, and spread as an advancing front from population centers. Eventually the satellite populations coalesce. Both modes are noticeable in Fig. 2, which depicts the minimum presence of *Alliaria* as of 1991. Infrequent collections from western North America imply that the species may be a sporadic rather than established component of the regional flora.

Within individual communities *Alliaria* population size may fluctuate widely from year to year, reflecting both the biennial nature of this plant and the 20-month seed dormancy of northern plants. Across a region *Alliaria* presence consistently increases through time. In seven northern Illinois forests *Alliaria* occurred at an average frequency of 24 percent in 1989, 34 percent in 1990, and 46 percent in 1991 (Nuzzo, unpubl.). *Alliaria* abundance in these same communities, as measured by percent cover, showed a modest but non-significant increase during the same time period. This implies an invasion strategy whereby *Alliaria* initially spreads relatively rapidly through a site at low density and subsequently establishes higher density populations. On a generational basis *Alliaria's* presence within a community also consistently increases, by an average of 250 percent (Nuzzo, unpubl.). The increase is considerably greater at sites subjected to disturbance, where presence of *Alliaria* increased more than 14-fold between generations.

*Alliaria,* like many invasive alien herbs, is disturbance-adapted, and both natural and anthropogenic disturbance factors are associated with the rapid invasion rate in natural communities (Nuzzo 1991c). Naturally disturbed habitats such as floodplains and riverbanks and anthropogenically disturbed habitats such as roadsides, heavily used preserves, and urban areas are the primary dispersal corridors for *Alliaria.* Disturbances create habitat suitable for initial entry, and continued disturbances maintain habitat suitable for expansion. However, once established in a locale, additional disturbance may not be necessary for continued spread of *Alliaria* (*sensu* Mack 1985). Exponential expansion of other weed species has been correlated with various anthropogenic disturbances (Lacey 1957; Mack 1981; Forcella and Harvey 1983; Thompson et al. 1987).

## HABITAT

Habitat data were compiled from the 705 U.S. collection records that indicated habitat. *Alliaria* was most frequently collected in forests, along roads, and near rivers; > 75 percent of collections were made near one of these habitats. Eight percent were made in urban areas, 4 percent in arboreta or on campuses, and 3 percent along railroads.

Roads (21%) and rivers (23%) were the primary collection locations indicated on 296 records from northeastern states (Conn., Del., Mass., Md., Maine, N.H., N.J., N.Y., Pa., Vt., and the District of Columbia), followed by forested riverbanks (14%), other forested areas (13%), and urban areas (9%) (Fig. 4). In 146 records from southeastern states (Ky., N.C., Va., and W.Va.) slightly more collections were made on forested riverbanks (21%) than along roads (18%), rivers (16%), or forests (16%). In contrast in the Midwestern states (Iowa, Ill., Ind., Kans., Mich., Minn., Mo., Ohio, and Wis.), 30 percent of the 260 collections were made in forests, 14 percent along roads, and 10 percent along rivers or other wet areas. Interestingly, 5 percent of the Midwestern collections were made along railroads.

Regional differences in collection locations have many causes, and cannot be attributed solely to habitat preference of the collected species. However, the decline in river-associated habitat from Northeast to Southeast to Midwest and concomitant increase in non-riverine forest habitat may indicate that *Alliaria* preferentially invades drier forest communities in the Midwest than in the Northeast. This is supported by the higher presence along railroads, which are generally indicative of drier habitats.

## CONTROL

Active management to eliminate *Alliaria* from natural areas and to limit invasion into new locales may slow the spread of this plant. Generally effective methods include removal of the flowerstalk prior to seed production, dormant-season prescribed fire, and dormant-season herbicide application (Nuzzo 1991a, 1991b). Such management is feasible in isolated forested communities and in regions where *Alliaria* has low presence. *Alliaria* is self-compatible (Cavers et al. 1979; Babonjo et al. 1990), and a single plant is sufficient to populate or repopulate a site.

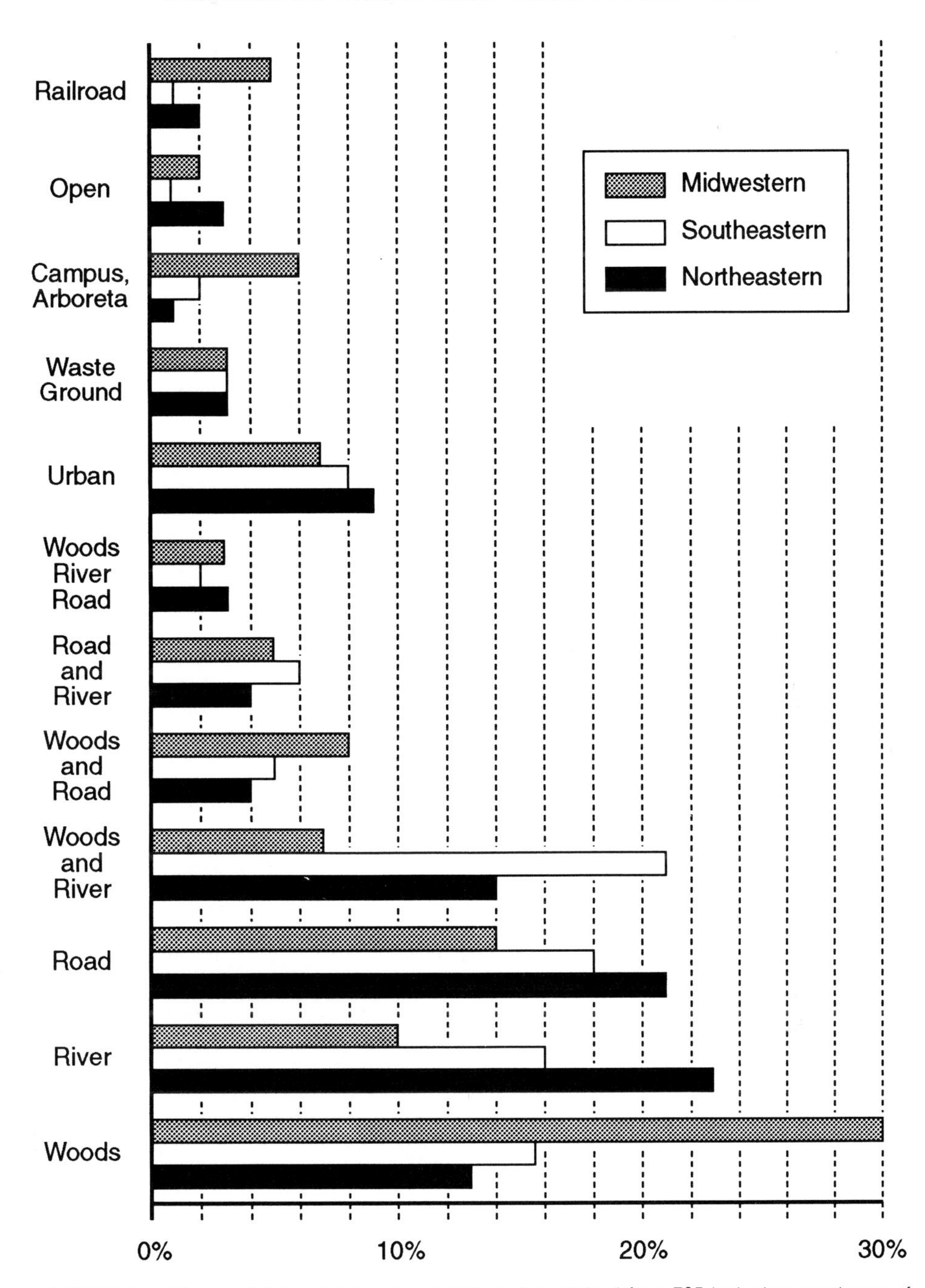


FIGURE 4.—*Alliaria petiolata* habitat by geographic region, derived from 705 herbarium specimens of *Alliaria petiolata* collected in the United States between 1868 and 1989. Northeastern states are Conn., Del., Mass., Md., Maine, N.H., N.J., N.Y., Pa., Vt., and the District of Columbia; southeastern states are Ky., N.C., Va., and W.Va.; Midwestern states are Ill., Ind., Iowa, Kans., Mich., Minn., Mo., Ohio, and Wis. Percentages based on total number of collections within each region.

Annual monitoring for and immediate removal of *Alliaria* will prevent establishment in individual natural areas. Once the species is well-established in a site, successful removal is unlikely without considerable expenditure of labor and money (*sensu* MacDonald et al. 1989; Coblentz 1990) over an extended period of time (*sensu* Usher 1988). In regions with multiple infestations, the frequency and abundance of *Alliaria* in unmanaged habitats limit the effectiveness of single-site management, as seeds are continually imported into the managed site.

Long-term effective control will require significant reduction or elimination of *Alliaria* populations from both public and private lands throughout the infested region. Auld et al. (1978) theorize that 20 percent of a weed's populations must be eliminated annually over a 20-year period in order to eradicate the species; elimination of 5 percent of the populations will not slow the rate of spread, and at least 15 percent of all infested sites must be managed to effect an immediate decrease in the total numbers of the weed species. Successful control of *Alliaria* at this level is unlikely with current management techniques and budgets. Development and implementation of biological control agents provide the most likely means to effectively reduce presence of *Alliaria* in the northcentral U.S. and adjacent Canada.

## ACKNOWLEDGMENTS

I would like to thank the curatorial staffs of the 77 contributing herbaria in the United States and Canada for making collection information available and the staffs of the more than 40 additional herbaria who searched for *Alliaria* collections. Many curators provided additional reference materials for which I am most grateful. I especially appreciate the considerable effort expended by field biologists who reported presence of *Alliaria.* I would also like to thank Bill McKnight and an anonymous reviewer for their helpful comments on an earlier version of this paper, and Matt Paulson for his meticulous preparation of the distribution map.

## REFERENCES

Auld, B. A. and B. G. Coote, 1980. A model of a spreading plant population. Oikos 34:287–92.

Auld, B. A., K. M. Menz and N. M. Monaghan. 1978. Dynamics of weed spread: implications for policies of public control. Protection Ecol. 1:141–48.

Babonjo, F., S. S. Dhillion and R. C. Anderson. 1990. Floral biology and breeding system of garlic mustard (*Alliaria petiolata*). Trans. Illinois State Acad. Sci., suppl. to vol. 83:32. (Abstract).

Barkley, T. M. 1977. Atlas of the flora of the Great Plains. Iowa State Univ., Ames.

Baskin, J. M. and C. C. Baskin. 1992. Seed germination biology of the weedy biennial *Alliaria petiolata.* (Draft manuscript).

Brooks, K. L. 1983. A Catskill flora and economic botany IV. (Part I.) Polypetalae, Chenopodiaceae through Capparidaceae. Univ. of the State of New York, Albany.

Byers, D. L. and J. A. Quinn. 1987. The effect of habitat variation in *Alliaria petiolata* on life history characteristics. Amer. J. Bot. 74:647. (Abstract).

———. 1988. Plant size as a factor in determining flowering time and reproductive output in *Alliaria petiolata.* Amer. J. Bot. 75:71. (Abstract).

Cavers, P. B., M. I. Heagy and R. F. Kokron. 1979. The biology of Canadian weeds. 35. *Alliaria petiolata* (M. Bieb.) Cavara and Grande. Can. J. Pl. Sci. 59:217–29.

Coblentz, B. E. 1990. Exotic organisms: a dilemma for conservation biology. Cons. Biol. 4:261–65.

Fernald, M. L. 1970. Gray's manual of botany. Van Nostrand Co., New York, N.Y.

Fernaldo, L. V. S. 1971. Selection and utilization of different food plants by *Pieris brassicae* (L.). Spolia Zeylon Bull. Natl. Mus. Ceylon 32:115–27.

Forcella, F. and S. J. Harvey. 1983. Relative abundance in an alien weed flora. Oecologia 59:292–95.

Furlow, J. 1991. unpublished distribution map.

Georgia, A. E. 1920. A manual of weeds. Macmillan Co., New York, N.Y.

Gleason, H. A. and A. Cronquist. 1963. Manual of vascular plants of northeastern United States and adjacent Canada. Van Nostrand Reinhold Co., New York, N.Y.

Gray, A., S. Watson and J. M. Coulter. 1889. Manual of the botany of the northern United States, including the district east of the Mississippi and north of North Carolina and Tennessee. American Book Co., New York, N.Y.

Kelley, T., S. Dhillion and R. Anderson. 1990. Aspects of the seed biology of garlic mustard (*Alliaria petiolata*). Trans. Illinois State Acad. Sci., suppl. to vol. 84:33. (Abstract).

Lacey, W. S. 1957. A comparison of the spread of *Galinsoga parviflora* and *G. ciliata* in Britain. Pages 109–15 *in* Progress in the study of the British Flora. J. E. Lousley (ed.). The Botanical Society of the British Isles Conference Report 5, London.

Lhotska, M. 1975. Notes on the ecology of germination of *Alliaria petiolata.* Folia Geobot. Phytotax. (Praha) 10:179–83.

MacDonald, I. A. W., L. L. Loope, M. B. Usher, et al. 1989. Wildlife conservation and the invasion of nature reserves by introduced species; a global perspective. Pages 215–55 *in* Biological invasions; a global perspective. J. White (ed.). SCOPE. John Wiley and Sons LTD, New York, N.Y.

Mack, R. N. 1981. Invasion of *Bromus tectorum* L. into western North America: an ecological chronicle. Agro-Ecosystems 7:145–65.

———. 1985. Invading plants: their potential contribution to population biology. Pages 127–42 *in* Studies on plant demography. J. White (ed.). Academic Press, London.

Murley, M. R. 1951. Seeds of the Cruciferae of Northeastern North America. Amer. Midl. Nat. 46:1–65.

Nuzzo, V. A. 1991a. Experimental control of garlic mustard *Alliaria petiolata* [(Bieb.) Cavara & Grande] in northern Illinois using fire, herbicide and cutting. Nat. Areas J. 11:158–67.

———. 1991b. Experimental control of garlic mustard (*Alliaria petiolata* [Bieb.] Cavara and Grande) in four natural communities in Illinois. Interim report to the Illinois Department of Energy and Natural Resources. Native Landscapes.

———. 1991c. Current and historic distribution of garlic mustard (*Alliaria petiolata*) in Illinois. Report to the Illinois Department of Conservation. Native Landscapes.

Patman, J. P. and H. H. Iltis. 1961. Preliminary reports on the flora of Wisconsin. No. 44 Cruciferae—mustard family. Wisconsin Acad. Sci., Arts and Letters 50:17–72.

Radford, A. E., H. E. Ahles and C. R. Bell. 1965. Atlas of the vascular flora of the Carolinas. Technical Bulletin No. 165. Univ. of North Carolina, Chapel Hill.

Rai, J. N., V. C. Saxena and J. P. Tewari. 1972. New leaf and stem spot diseases of Indian wild crucifers. Ind. Phytopathol. 25:253–56.

Roberts, H. A. and J. E. Boddrell. 1983. Seed survival and periodicity of seedling emergence in eight species of Cruciferae. Ann. Appl. Biol. 103:301–04.

Schwegman, J. 1989. Illinois garlic mustard alert. Illinois Department of Conservation, Division of Natural Heritage, Springfield.

Thompson, D. Q., R. L. Stuckey and E. B. Thompson. 1987. Spread, impact, and control of purple loosestrife (*Lythrum salicaria*) in North American wetlands. U.S. Fish and Wildlife Service, Fish and Wildlife Research 2. Washington, D.C.

Thompson, D. Q. 1991. History of purple loosestrife (*Lythrum salicaria* L.) biological control efforts. Nat. Areas J. 11:148–50.

Trimbur, T. J. 1973. An ecological life history of *Alliaria officinalis,* a deciduous forest "weed." M.S. Thesis. Ohio State Univ., Columbus.

Tutin, T. G., V. H. Heywood, N. A. Burges, et al. 1964. Flora Europaea, vol. 1. Cambridge University Press.

Usher, M. B. 1988. Biological invasions of nature reserves: a search for generalisations. Biol. Conserv. 44:119–35.

Voss, E. G. 1985. Michigan flora part II. Cranbrook Institute of Science and Univ. Michigan Herbarium. Bulletin 55. Ann Arbor.

Yatskievych, G. and J. Turner. 1990. Catalogue of the flora of Missouri. Missouri Bot. Gard., St. Louis.

# Control of the Ornamental Purple Loosestrife (*Lythrum salicaria*) by Exotic Organisms

Stephen D. Hight[1]

ABSTRACT: *Lythrum salicaria* (purple loosestrife) was introduced from its native Europe onto the eastern shores of North America nearly 200 years ago (Stuckey 1980). Since then this wetland plant has spread to every state in the northern half of the U.S. and much of adjacent Canada. An aggressive invader, it has displaced native vegetation and destroyed waterfowl habitat by forming dense, nearly monotypic stands (Thompson et al. 1987). Host specificity tests were completed with three species of natural enemies from Europe (Blossey and Schroeder 1991; Kok et al. 1992; Kok et al., in preparation): the root-feeding weevil, *Hylobius transversovittatus* Gaeze (Fig. 1), and the two leaf-feeding beetles, *Galerucella calmariensis* L. and *G. pusilla* Duft. Larvae of *H. transversovittatus* develop in the roots of *L. salicaria* and, with high numbers and/or several years of infestation, the plants are severely damaged (Fig. 2). Both species of *Galerucella* feed on the leaves and stems of *L. salicaria* and will completely defoliate plants. Like adult *H. transversovittatus,* adult *Galerucella* feed on aboveground parts of the plants, but the larvae cause the significant damage below ground. Approval was granted for the release of *H. transversovittatus* at a state wildlife area near Buffalo, New York, and at a federal wildlife area near Philadelphia, Pennsylvania. These locations will serve as nursery propagation sites to test overwintering capabilities of the agents, to evaluate various release methods, and to increase the numbers of insects for redistribution into new areas. In August 1991 a total of 516 adults and 15,208 eggs of *H. transversovittatus* from northern Germany were released at the two sites. Additional releases of *H. transversovittatus* and initial releases of *G. calamariensis* and *G. pusilla* are being requested for August 1992. To date, biological control offers the most promising control tactic (Hight and Drea 1991) as well as the most inexpensive method of control (Drea 1991). The insects selected are highly specific to the target plant, with no negative impact on non-target plants. The goal of the project is to reduce the level of *L. salicaria* infestation by 70–80 percent while causing minimal or no impact to the native biota.

## REFERENCES

Blossey, B. and D. Schroder. 1991. Final Report. Study and screening of potential biological control agents of purple loosestrife (*Lythrum salicaria* L.). European Stn. CIBC, Delémont.

Drea, J. J. 1991. The philosophy, procedures, and cost of developing a biological control of weeds project in the United States. Nat. Areas J. 11:143–47.

Hight, S. D. and J. J. Drea. 1991. Prospects for a classical biological control project against purple loosestrife (*Lythrum salicaria* L.). Nat. Areas J. 11:151–57.

Kok, L. T., T. J. McAvoy, R. A. Malecki, et al. 1992. Host specificity tests of *Hylobius transversovit-*

1. USDA, Agricultural Research Service, Insect Biocontrol Laboratory, Bldg. 406, BARC-East, 10300 Baltimore Ave., Beltsville, MD 20705-2350

*tatus* Goeze (Coleoptera: Curculionidae), a potential biological control agent of purple loosestrife, *Lythrum salicaria* L. (Lythraceae). Biological Control (in press).

———. Host specificity tests of *Galerucella calmariensis* (L.) and *G. pusilla* (Duft.) (Coleoptera: Chrysomelidae), potential biological control agents of purple loosestrife, *Lythrum salicaria* L. (Lythraceae). In preparation.

STUCKEY, R. L. 1980. Distributional history of *Lythrum salicaria* (purple loosestrife) in North America. Bartonia 47:3–20.

THOMPSON, D. Q., R. L. STUCKEY and E. B. THOMPSON. 1987. Spread, impact, and control of purple loosestrife (*Lythrum salicaria*) in North American wetlands. Fish and Wildlife Res. 2. U.S. Dept. Int., Washington, D.C.

FIGURE 1. (above left)—Adult *Hylobius transverovitlatus* on a leaf of purple loosestrife in laboratory in Germany. (photo by Bland Blossey) FIGURE 2. (above right)—Mature larvae of *Galerucella calmariensis* on shoot tip of purple loosestrife and their feeding drainage in Germany, 1989. (photo by Bland Blossey)

# The Himalayan Snowcock: North America's Newest Exotic Bird[1]

James D. Bland and Stanley A. Temple[2]

Until recently, snowcocks were found only in the mountains of central Asia. The few people who had encountered them—servants of khans and emperors, turn-of-the-century sportsmen and a handful of zoologists—provided only sketchy details of their biology (Jerdon 1864; Hume and Marshall 1878). Today as a result of Herculean efforts begun by game biologists in the 1950s, a new population of Himalayan snowcocks has been established in the Ruby-East Humboldt Range of Humboldt National Forest, northeast Nevada (Fig. 1).

In the 1950s the practice of exotic game introduction was in its heyday. Game biologists went to great lengths to acquire wild snowcocks in Pakistan, propagate them, and release their offspring in areas said by some to have too few game birds. Almost 20 years of effort resulted in the successful introduction of a new "trophy game bird" in Nevada, but a close look at the program reveals some lingering question about the wisdom of the program.

The purpose of this paper is to chronicle the history of snowcocks in America. We review the introduction effort, discuss some costs and benefits of the program, describe the present distribution and status of snowcocks in Nevada, and present some results of our recent studies of the species.

## HISTORY OF SNOWCOCK INTRODUCTIONS

The history of the Snowcock Introduction Program is an intriguing, seldom told story. In 1948 the United States Fish and Wildlife Service (USFWS) began a new program, Foreign Game Investigations (Bump 1951), whose stated objective was to seek out "new adaptable species possessing a high hunting resistance . . . so that . . . habitats thoroughly changed by man . . . or never fully occupied by native game . . . [could be stocked, and thus] provide greater hunting opportunities" (Bump 1968a). Program biologists conducted field studies on dozens of potential game birds in all corners of the globe. Informative leaflets were published for potential state collaborators, enabling them to match candidate species with local conditions (e.g., Bump 1973). Between 1960 and 1970 the program was responsible for releasing no fewer than 19 species of pheasants, partridges, quail, tinamou, and sandgrouse (Banks 1981) (Table 1). The program was a joint venture between the USFWS and participating state agencies, and the Wildlife Management Institute provided loans to purchase foreign birds. Federal funds were provided through the Pittman-Robertson Wildlife Restoration Act, which provides funds to states on a matching basis for *wildlife restoration*: generally land acquisition, research, development, and management. The federal Foreign Game Investigations Program was particularly popular in the Southeast, the arid Southwest and in Hawaii. In Nevada the objective was to enrich "60,000 square miles of [arid] habitats [54

1. An invited paper not presented at the symposium.
2. Department of Wildlife Ecology, University of Wisconsin-Madison, Madison, WI 53706

FIGURE 1.—Himalayan snowcock on Thomas Peak, Rudy Mountains, Humboldt National Forest, northeastern Nevada. (photo by Bland)

percent of the state] which were not permanently inhabited by upland game birds, [or were inhabited by one of] five [native] species [for which] hunting potential . . . is erratic" (Christensen 1963).

Snowcock introductions were attempted in five Nevada ranges and on Mauna Kea in Hawaii. Ironically snowcocks fell far short of the federal government's criteria of being adaptable, possessing high hunting resistance, and providing superior hunting opportunities. Furthermore, these strictly alpine birds had little chance of enriching Nevada's extensive deserts with game. But key players in the Foreign Game Investigations Program became enamored with this "giant cousin" of the chukar partridge (*Alectoris greca*). Bolstered by the overwhelming popularity of exotic chukars—by then the favorite quarry of Great

TABLE 1. A summary of birds successfully introduced outside of their natural geographic ranges in North America, primarily game birds for hunters.

| SPECIES | COMMON NAME | INITIAL DATE OF INTRODUCTION |
|---|---|---|
| *Alectoris greca* | *CHUKAR | 1893 |
| *Bonasa umbellus* | RUFFED GROUSE | 1956 |
| *Callipepla californicus* | CALIFORNIA QUAIL | 1857 |
| *C. gambelii* | GAMBEL'S QUAIL | 1885 |
| *C. squamata* | SCALED QUAIL | 1913 |
| *Colinus virginianus* | BOBWHITE QUAIL | 1910 |
| *Francolinus erkelii* | *ERKEL'S FRANCOLIN | 1957 |
| *F. francolinus* | *BLACK FRANCOLIN | 1950s |
| *F. pondicerianus* | *GREY FRANCOLIN | 1959 |
| *Gallus gallus* | *RED JUNGLEFOWL | 1500s |
| *Geopelia striata* | *ZELOZA DOVE | 1922 |
| *Lophura leucomelana* | *KALIJ PHEASANT | 1962 |
| *Meleagris gallopavo* | WILD TURKEY | 1880s |
| *Orerotyx pictus* | MOUNTAIN QUAIL | 1880s |
| *Ortalis vetula* | PLAIN CHACHALACA | 1923 |
| *Perdix perdix* | *GRAY PARTRIDGE | 1910 |
| *Phasianus colchicus* | *RING-NECKED PHEASANT | 1882 |
| *Pterocles exustus* | *CHESTNUT-BELLIED SANDGROUSE | 1959 |
| *Tetraogallus himalayensis* | *HIMALAYAN SNOWCOCK | 1933 |
| *Zenaida asiatica* | *WHITE-WINGED DOVE | 1959 |

*Denotes taxa not native to North America

Basin bird hunters—program participants set out to add the "king of chukars" to hunters' bags.

In 1961 a Reno trophy-hunter arranged to have six Himalayan snowcocks trapped near the town of Gilgit, Pakistan. The birds were transported by porter, pony, jeep, and airplane to a quarantine station in Honolulu, a journey of some 8,500 miles. Only a single bird survived the trip, but the bird so impressed the Nevada Game Commission that they requested 35 more. Nineteen of these survived the journey to Nevada and were released directly into the Ruby Mountains in April 1963. They vanished soon after. Nevada Department of Wildlife (NDW) then decided to establish a captive flock from which offspring would be released over successive years. A total of 107 wild snowcocks was imported from Pakistan. Between 1963 and 1979, 2,025 of their progeny were released in Nevada, 1,717 in the Ruby Mountains alone.

During the 1970s a series of policy shifts signaled a close to the era of exotic game bird introductions. In 1970 the USFWS terminated its Foreign Game Introductions Program. In 1977 President Carter issued Executive Order 11987, greatly restricting the use of federal funds, personnel, or lands for the introduction of exotic species. In 1979 NDW discontinued propagation and release of snowcocks. This is not to say the interest in exotic game has died. New subspecies of previously established species continue to be imported with impunity (Squibb 1987), and new populations of previously established species continue to be established by translocation within and between states.

In the mid-1980s Alberta Fish and Wildlife Division was on the verge of releasing descendants of Nevada's captive snowcocks into the Rocky Mountains, but professional and environmental groups convinced the agency to reconsider introducing the birds into a habitat that would have allowed them to spread widely (Bland 1989).

## STATUS AND DISTRIBUTION OF SNOWCOCKS

The status of Nevada's wild snowcock population has never been adequately documented. The introduction was first declared a success in 1971 (Abbott and Christensen 1971), although reports of reproduction were not substantiated until 1977 (Nevada Department of Wildlife 1980). Through the 1970s, research efforts consisted of a few searches on foot or horseback. More recently NDW biologists have begun to count snowcocks during sporadic helicopter and foot surveys for introduced mountain goats (*Oreamnos americanus*). In 1980 NDW opened a snowcock hunting season. Initially hunters were required to report the location and results of their hunt. This requirement could have provided rare data on Nevada's snowcocks, but it was soon eliminated. In 1985 NDW reported the snowcock population to number between 250 to 500 individuals (Stiver 1984). Since 1985 NDW has declined population estimates for lack of data.

The breeding range of snowcocks in the Ruby-East Humboldt Range is probably limited to elevations above 3,000 m (Bland and Temple 1990). Less than 50 $km^2$ of the Ruby-East Humboldt Range meet this basic criterion (when slope is not considered for areal calculations). Considering that snowcock breeding densities in China range from 1.3 to 2.0 individuals/$km^2$ (Huang et al. 1990; Liu et al. 1990), we can conclude the number of snowcocks in the Ruby-East Humboldt Range will never be great. In the Ruby Mountains, snowcocks appear to prefer deep glacial cirques rimmed with extensive moist meadows and sheer cliffs (Bland and Temple 1990). The spotty distribution of such cirques and alpine meadows limits the number of areas where large flocks can establish home ranges. The core of Nevada's snowcock population appears to inhabit in the Thomas Peak-Ruby Dome area of the Ruby Mountains, though coveys are regularly reported further north and south. Because Nevada's snowcocks are marooned on an "alpine island" at the center of the Great Basin, they are unlikely to disperse to other alpine habitats on their own.

## COSTS AND BENEFITS

The merits of exotic game bird introductions have long been debated. The biologist-in-charge of the Foreign Game Introductions Program once likened the critics of exotic game introductions to the "crowd of critics, of complainers, of commentators [that] darkened the face of learning [and brought about the fall of Rome]" (Bump 1968b). The Snowcock Introduction Program may well have "brightened the face of learning" with regard to snowcock biology, but the results of the program hardly reaffirm any wisdom in the introduction of exotics.

Since the snowcock introduction is nonetheless considered a success, an overall evaluation is warranted. On the positive side Nevada now has a trophy game bird that is a true challenge to bag, and one that can put a real meal on the table (snowcocks can weigh over 2.5 kg or 5.5 lb). In the words of a local snowcock hunter, snowcocks provide "a truly unique hunting opportunity." For the more hardy bird-watchers and hikers, snowcocks can also add unusual variety to an otherwise solemn mountain environment. The sight of a snowcock covey in flight, the alien cacophony of their calls, and the chance drama of a high-speed chase by an eagle are truly spectacular. Snowcocks generate notoriety and revenue for the state and for local businesses. Local hunting guides even offer specialized snowcock hunts. These benefits are valued highly by NDW, which depends on license sales and strives to fulfill the desires of the hunting public.

On the other hand, the Snowcock Introduction Program is said to have cost $750,000 (Stiver 1984). The figure would probably approach $1 million if private funds and funds not allocated directly to the project were included. Incredibly this sum was spent on snowcocks at a time when two native game birds—sage grouse (*Centrocercus urophasianus*) and sharp-tailed grouse (*Pedioecetes phasianellus*)—were in serious need of "restoration" in Nevada. The utilization of snowcocks by hunters has been limited, at best. Snowcocks are difficult to bag, and the few hunters who make the strenuous climb to snowcock habitat have collectively bagged an average of four birds each year—total. Many hunters try for snowcocks with hopes of acquiring a mountable specimen, but since snowcocks are usually shot in flight high over a deep, rocky canyon, they are often retrieved too badly mangled to warrant taxidermy (Fulton 1904, anonymous hunter pers. comm.).

Sadly, the fruits of Nevada's snowcock hunt had been foretold more than a century ago, when Hume and Marshall (1878) berated the species in their still-authoritative book *The Game Birds of India, Burma and Ceylon*. They noted:

> "With a gun they do not, as a rule, afford any sport. You may get them driven over you nicely at times [by coolies], and you might sometime stalk them—if it were worth the tremendous labour such stalks usually involve—but as a rule, whenever I have seen them, the rifle is the only weapon with which a bag can be made. I went in regularly for it. . . . though to me they seemed, after many trials, almost uneatable."

It is unlikely anyone will ever know the ecological impact introduced snowcocks have had on the unique Ruby Mountain environment or on the already diverse community of native herbivores. The small, isolated, alpine meadows on which Ruby Mountain wildlife congregate are fragile, as most alpine meadows are, and what's more, no other alpine plant community in the Great Basin is so rich with plant species (Loope 1969). Since the introduction of snowcocks went virtually unmonitored, with no prior assessment of the snowcocks' potential for harm, the condition of the environment before, during and after their introduction is largely unknown.

## RECENT STUDIES

In the opinion of one retired Nevada biologist who has observed the introduction program from its inception, the greatest merit of the project may be the opportunity it provided for studying snowcock ecology. Little was known of Nevada's wild snowcocks prior to 1981 when we began our field studies.

Our first study confirmed that snowcocks in the Ruby Mountains carry out the same daily elevational traverse that has been observed among Himalayan snowcocks in central Asia. This route takes the birds through a series of habitats, each well-suited for the behavior they engage in at particular times of day, be it foraging, loafing or roosting (Bland 1982).

Our initial observations led us to believe snowcocks face a serious ecological challenge in balancing conflicting demands for forage and cover. The alpine plants on which snowcocks feed (grasses, forbs, and sedges) are not highly nutritious, so snowcocks must consume large quantities. To do so they spend as much as 80 percent of daylight hours foraging (Bland 1982). Under these circumstances one might expect snowcocks to spend most of their time where good food is most abundant, but snowcocks are reluctant to forage where there is good food if the topographic setting leaves them vulnerable to golden eagles (*Aquila chrysaetos*) (Bland and Temple 1990). Snowcocks can best elude eagles—their principal predators in the Himalayas (Baker 1924; Whistler 1926) as well as in Nevada—by plummeting at high speed down steep slopes and outdistancing the eagles.

In 1985 we focused our attention on the relationships between snowcocks and their habitat, food and predators. To determine the foraging potential of various habitats, we closely observed the foraging behavior of tame, hand-reared snowcocks on plots with different vegetative characteristics (Bland and Temple 1987). To determine where snowcocks were most nervous about attacks by eagles, we observed the vigilance behavior—alert visual scanning—of wild birds foraging on level versus steep ground (Bland and Temple 1988). Wild birds have learned when and where extra wariness is necessary to prevent being trapped in a compromising situation by eagles.

Our tame birds indicated that snowcocks can feed most efficiently on level or slightly sloping meadows and in particular where grasses are abundant. But by observing wild snowcocks we found they avoid level terrain regardless of the quality of food there. Wild snowcocks were more alert and spent more time scanning for threats when foraging on level ground than when they foraged on steep ground.

In light of our findings regarding the effects of predation risk on habitat use, we were able to describe what might be considered ideal snowcock habitat and, more interestingly, explain why and how it best suits them. In retrospect the unique alpine meadows and glacial cirques of the Ruby-East Humboldt Range were the key to successful establishment of snowcocks. Those few meadows which are on or near steep cliffs or slopes or are nestled high in glacial cirques provide just the right combination of food and escape terrain for snowcocks to prosper.

Our work with Himalayan snowcocks in the Ruby Mountains has given us a unique opportunity to study this poorly known bird of central Asia. In some respects we now know more about the species in Nevada than in its native range. Moreover, the story of its introduction to America provides some examples of why new laws are needed to prohibit the introduction of exotic species. We now have a better understanding of why snowcocks are inexorably tied to the topography and vegetation of alpine environments, and why their distribution is clustered even in the Ruby-East Humboldt Range. They are generalist feeders that can make a meal out of just about any meadow, but their access to food resources is restricted by birds of prey. They must always have a quick escape route at their disposal.

We do not at this time believe it is likely that there would be much support for eradicating snowcocks from the Ruby-East Humboldt Range, even though much of the area has recently been designated as wilderness area. Snowcocks were established in Nevada for economic and recreational purposes, and many feel they should be enjoyed and utilized to those ends to the greatest degree possible. We do however ask that other agencies or individuals contemplating game bird introductions or translocations reflect on the snowcock story, review their reasoning a second and third time, and especially consider the competing demands for funds necessary to sustain native game species in viable, natural ecosystems.

## REFERENCES

ABBOTT, U. K. and G. C. CHRISTENSEN. 1971. Hatching and rearing the Himalayan snow partridge in captivity. J. Wildlife Manage. 35(2):301–06.

BAKER, E. C. S. 1924. The game birds of India, Burma and Ceylon, part XXXVIII. J. Bombay Nat. Hist. Soc. 30(1):1–11.

BANKS, R. C. 1981. Summary of foreign game liberations, 1969–78. U.S. Fish and Wildlife Service Special Sci. Rpt., Wildlife No. 239.

BLAND, J. D. 1982. Patterns of summer habitat use in Himalayan snowcocks introduced to Nevada, USA. Proc. 2nd Intnl. Pheasant Symp. Srinagar, Kashmir.

BLAND, J. D. and S. A. TEMPLE. 1990. The effects of predation-risk on habitat use in Himalayan snowcocks. Oecologia 82:187–91.

———. 1987. Using tame hand-reared birds in field studies. Proc. 2nd Intnl. Symp. on Breeding Birds in Captivity. Hollywood, Cal.

———. 1988. Effects of predator-risk on habitat use by Himalayan snowcocks. M.S. Thesis. Univ. Wisconsin, Madison.

BLAND, J. D. 1989. World Pheasant Association policies deter snowcock introduction in Canada. WPA International Newsletter, No. 26.

BUMP, G. 1951. Game introductions—when, where, and how. Trans. N. Amer. Wildl. Conf. 16:316–25.

———. 1968a. Foreign game investigation—a federal-state cooperative program. USDI, Bureau of Sport Fisheries and Wildlife. Washington, D.C.

———. 1968b. Exotics and the role of the state-federal Foreign Game Investigations Program. Pages 5–8 *in* Proceedings: Introduction of exotic animals: ecological and socioeconomic considerations. Caesar Kleberg Research Program in Wildlife Ecology. College Station, Tex.

———. 1973. The snowcocks. Foreign Game Leaflet No. 29. U.S. Fish and Wildlife Service, Washington, D.C.

CHRISTENSEN, G. C. 1963. Exotic game bird introductions into Nevada. Nevada Game and Fish Commission Bulletin, No. 3. Reno.

FULTON, H. T. 1904. Some notes on the birds of Chitral. J. Bombay Nat. Hist. Soc. 16(1):44–64.

HUANG, R-X, L. MA, H-G SHAO, et al. 1990. Preliminary studies on the ecology and biology of the Himalayan snowcock in Mt. Tian, Xinjiang, China. Pages 31–32 *in* Pheasants in Asia 1989. D. A. Hill, P. J. Garson and D. Jenkins (eds.). World Pheasant Association. Reading, U.K.

HUME, A. O. and C. H. T. MARSHALL. 1878. The game birds of India, Burma, and Ceylon, vol. 1. (Publ. by authors) Calcutta.

JERDON, T. C. 1864. Birds of India, vol. 2. G. Wyman and Co., Calcutta.

LIU, N., C. CHANG and X. WANG. 1990. Ecological studies of the Himalayan snowcock. Pages 33–36 *in* Pheasants in Asia 1989. D. A. Hill, P. J. Garson and D. Jenkins (eds.). World Pheasant Association. Reading, U.K.

LOOPE, L. L. 1969. Subalpine and alpine vegetation of northeast Nevada. Unpubl. Doctoral Dissertation. Duke University, Durham, N.C.

NEVADA DEPARTMENT OF WILDLIFE. 1980. Upland game, migratory game birds, furbearers, and mountain lion investigations and season recommendations. Nevada Department of Wildlife. Reno.

SQUIBB, P. 1987. How's our new pheasant? Michigan Natural Resources Magazine, Sept/Oct:26–29.

STIVER, S. J. 1984. Himalayan snowcocks—Nevada's newest upland game. Cal-Neva Trans., p. 55–58.

WHISTLER, H. 1926. Birds of the Kangra District of Punjab. Ibis 68:724–83.

# Philosophical and Ecological Perspectives of Highly Valued Exotic Animals: Case Studies of Domestic Cats and Game Birds

Richard E. Warner[1]

ABSTRACT: The domestic cat (*Felis domesticus*) and game birds are good examples of exotic animals in North America that are highly valued by humans. As exotic animals, game birds and domestic pets tend to have the following in common:

1. they are of great economic importance;
2. their popularity in North America has increased during the twentieth century, and
3. the negative ecological aspects of these animals have received relatively little attention from biologists and are poorly understood by the public at large. Although there are obvious reasons why such animals are popular, the emphasis here is on some of the more important negative aspects.

The introduction of game birds dates back to the late 1700s in North America. The propagation and liberation of exotic game birds became widespread in North America around 1900. During this era the demand for sport hunting was on the increase and indigenous game birds began to register declines related to the destruction of habitat and, in some localized settings, severe hunting pressure as well. To accommodate the demand for hunting, there have been numerous federal and state programs to identify and import potential game birds from other continents. Ecological aspects of game birds considered candidates for importation have generally been gleaned from a precursory study of their natural history in native environments, and such information has been used to infer their suitability for habitats in North America. Thus any negative aspects have typically not been detected until populations of exotic game birds have become well-established. Significant problems that have sometimes surfaced include:

1. the fact that exotics can serve as vectors for diseases that decimate endemic species;
2. inter-specific competition with native species can occur (such as between the native quail and prairie chicken and the exotic ring-necked pheasant) often leading to dominance of the exotic species, and
3. exotic game bird programs are expensive, sometimes diverting limited management resources away from benefiting native species.

Because any negative implications of such exotics have usually been identified after the fact—when range establishment is widespread and sport hunting occurs—remedial courses of action are difficult if not impossible. With the poten-

1. Center for Wildlife Ecology, Illinois Natural History Survey, 607 E. Peabody Dr., Champaign, IL 61820

tial for negative consequences to native species, introductions of exotic game birds in North American habitats should be curtailed.

Domestic cats pervade North America and so do cat-related problems. Numbers of cats have increased in recent decades as cat ownership has gained popularity; numbers of unwanted cats have likewise increased. An estimated 10–15 percent of the cats in the U.S. are euthanized annually, and these control efforts incur high costs that no one wants to pay. Further, free-ranging cats—including those that are well-cared-for as pets—are skilled predators. For example, free-ranging cats in both rural and urban settings kill a variety of small birds and mammals. The impact of cats on food webs and animal populations can be quite significant. Pet owners typically have little awareness of the well-documented problems associated with free-ranging pets (dogs as well as cats) and wildlife. Public awareness of such problems should be heightened along with specific emphasis on appropriate measures for curtailing such problems in the future.

# Exotic Weeds in North American and Hawaiian Natural Areas: The Nature Conservancy's Plan of Attack

John M. Randall[1]

## INTRODUCTION

The Nature Conservancy (TNC) is a non-profit conservation organization that operates the largest system of privately-owned preserves in the world. These areas are managed as part of TNC's mission to preserve plants, animals, and natural communities that represent the diversity of life on earth by protecting the lands and waters they need to survive. The presence and spread of non-native plants and animals is a problem on many of these sites and, in some cases it is the single greatest threat to the species or communities the preserves were designed to protect.

Techniques used to manage these problems on TNC lands vary from state to state and even from one preserve to the next, in keeping with TNC's decentralized approach to resource management. Recently, however, the realization that weed problems are so widespread and serious led TNC to create an invasive weed specialist position to help coordinate weed control efforts around the nation. Below, I will briefly describe The Nature Conservancy, outline its organization, focusing on the weed specialist's program and then give examples of control efforts on several preserves. The examples emphasize the importance of work done by volunteers and of considering weed control as part of a larger restoration program.

## HISTORY AND DESCRIPTION OF THE NATURE CONSERVANCY

The Nature Conservancy traces its roots to 1917 and the establishment of The Ecological Society of America's Committee for the Preservation of Natural Conditions (The Nature Conservancy 1985) but it was not until 1951 that TNC was incorporated as a non-profit organization for scientific and educational purposes. In 1954 TNC purchased land for its first preserve in the Mianus Gorge in Westchester Co., New York. From the beginning, TNC used a business-like, non-confrontational approach to acquire land and protect habitat. Over the years its mission and scope grew, gradually at first and far more rapidly in recent years. Today it is an international organization which cooperates with other conservation groups and non-profit organizations as well as with landowners, government, and industry. TNC now owns and manages some 1,500 preserves ranging in size from less than 0.5 ha (1 acre) to 130,191 ha (321,703 acres or 502.7 $mi^2$) and totalling more than 341,000 ha (843,000 acres) in the U.S. and Canada (Central Biological Conservation Data System, 30 September 1991). Many of the preserves receive few if any visitors, but several are heavily visited. Blowing Rocks

1. Exotic Species Program, The Nature Conservancy, 6500 Desmond Rd., Galt CA 95632 and Botany Department, University of California, Davis CA 95616

Preserve on the east coast of Florida, for example, receives 130,000 visitors a year, and Ramsey Canyon in southeastern Arizona, hosts 30,000 people annually, most of them birdwatchers.

TNC also acquires lands that are later transferred to other private conservation organizations or public agencies and has developed the use of conservation easements and other legal agreements to ensure that private landowners protect land from development. Through all of these activities TNC has been responsible for the protection of more than 2.2 million ha (5.5 million acres) in 50 U.S. states and 2 Canadian provinces. TNC has also helped like-minded organizations in Latin America and the Caribbean protect millions of acres and has begun a program to identify areas for protection in the Pacific and Indonesia.

## OUTLINE OF THE NATURE CONSERVANCY'S MANAGEMENT ACTIVITIES

Management of TNC preserves and cooperative projects is the responsibility of the Stewardship Program. As TNC's system of preserves has grown, this aspect of its work has taken on increasing importance. TNC stewards have become more aware that preservation is not accomplished by simply cordoning off areas and leaving them alone. Active management is needed because human activities, including those outside preserves, alter natural processes inside. For example, fire suppression on wildlands, and the firebreaking qualities of roads and other human developments have altered natural fire regimes over entire regions. Reestablishment of the natural regimes is thus a top management priority on many preserves. Stewardship created a fire management and research program to train and certify stewards to conduct prescribed burns under specific conditions in response to this need.

Invasions by non-native species of plants and animals have likewise resulted in tremendous changes even on preserves that were never developed and are distant from initial introduction sites. Responses of TNC stewards to these invasions differ depending on impacts of the invaders, the species involved, methods of control available, and environmental conditions at the site. In addition, personal preferences and experiences of the steward may determine the response, especially where the severity of the threat is not clear and no particular control technique is recognized as superior. TNC encourages this decentralized or multilocal approach to management. The realization that weed problems are so serious and widespread on TNC preserves, however, inspired the creation of an invasive weed specialist position at the national level in 1991. This was done to provide in-house expertise in weed control and biology, a center for the exchange of information on weeds, and a focal point around which policies on weed control, herbicide use, and the use of biocontrol agents can be developed. In brief, it is an attempt to coordinate efforts so that stewards faced with weed problems are not forced to reinvent the wheel at each turn.

I joined TNC as its first invasive weed specialist in July 1991. Four activities I have chosen to focus on initially are:

1. acting as a "switchboard" to facilitate the exchange of information on weeds among TNC stewards and with resource managers and researchers from other organizations and agencies;
2. coordinating efforts to develop policies on weed control;
3. conducting research on weed ecology and control and promoting research on these topics on TNC preserves; and

4. coordinating and updating TNC's data on weeds and control methods.

Wherever possible, I hope to encourage the attitude that control or eradication of alien or invasive plants is one aspect of an overall restoration effort. In this context the long-range goal of control is to eliminate the alien species as an ecological factor. This places an emphasis on consideration of the species that will replace the weed and acting to encourage desirable natives rather than simply hoping that another exotic species will not move in and become established. Thomas (1987, 1988), Hiebert (1990), Hester (1991), and others, have promoted this perspective in the National Park Service. In some instances, removal of weeds is the first and crucial step in a restoration program. For example, removal of tamarisks (*Tamarix ramosissima*) may be required before a desert spring will flow again, allowing restoration of an oasis community to proceed (Neill 1983). At the other end of the spectrum, some weed problems can best be addressed indirectly by restoring the natural hydrologic regime, fire, or native species to an area. At the Cosumnes River Preserve, plantings of native riparian forest species like valley oak (*Quercus lobata*), Fremont's cottonwood (*Populus fremontii*), and willows (*Salix spp.*) will shade out and largely eliminate alien annuals that now infest these areas.

I also encourage TNC stewards to consider the environmental costs of their weed control activities to ensure that they do not outweigh the benefits. As natural area managers it is good to keep in mind an injunction from the Hippocratic Oath: "First, do thy patient no harm."

The first policy issue I have addressed is that of whether or not to allow the release of biocontrol agents against invasive alien weeds. Current TNC policy simply states that intentional release of alien organisms is prohibited on lands owned or managed by TNC. Although it does not specifically mention them, this prohibition includes biocontrol agents not native to the area in question. A few preserve managers and stewards, however, opted to release alien biocontrol agents against alien weeds they judged to be immediate and serious threats to organisms or communities under their protection. In all cases the agents had been screened and approved for release by the USDA-Agricultural Research Service biological control labs. With the help of several TNC stewards and Dr. Charles E. Turner of the USDA biocontrol lab in Albany, California, I drafted a proposal to amend TNC policy to permit such releases under certain, carefully defined conditions. The proposal attempts to balance recognition of classical biocontrol's many virtues (c.f. DeLoach 1991) with an acknowledgment that it may also result in unanticipated problems as has been pointed out forcefully by Howarth (1991). The proposal is currently under review.

Much of TNC's data on the ecology and management of weeds that invade natural areas is currently contained in summaries of written and verbal information on selected species called Element Stewardship Abstracts (ESAs). Many ESAs cover rare or threatened species, but they are also available for over 60 species of weeds that invade natural areas, and more are currently in production (Table 1). They are similar to the Vegetation Management Guidelines produced by the Illinois Nature Preserves Commission and include data from numerous sources including journal articles, government agency reports, and researchers and managers actively working with the species. Each ESA includes a description of the species and information on its biology, habitat, and methods for monitoring and controlling it. In some cases, names and addresses of people who have experience with the species are included. The ESAs may be obtained through TNC's regional offices ( Table 1).

A database will be developed to supplement the ESAs. This database will contain names and addresses of managers and researchers who are willing to offer their expertise on weeds to others. The first edition should be available late in 1992. I also plan to populate a database containing lists of publications with information on the ecology and control of weeds that invade natural areas.

TABLE 1. Element Stewardship Abstracts (ESAs) on pest plants available from The Nature Conservancy.

| SPECIES | SPECIES CODE | LAST UPDATE |
|---|---|---|
| 1. *Acacia melanoxylon* | PDFAB020M0 | 06-17-91 |
| 2. *Ailanthus altissima* | PDSIM01010 | 11-30-88 |
| 3. *Ammophila arenaria* | PMPOA08010 | 11-22-88 |
| 4. *Anthoxanthum odoratum* | PMPOA0F020 | 11-04-88 |
| 5. *Artemisia absinthium* | PDAST0S020 | 06-12-87 |
| 6. *Arundo donax* | PMPOA0R010 | 09-03-86 |
| 7. *Brassica hyssopifolia* | PDCHEO6O2O | 09-03-86 |
| 8. *Bromus inermis* | PMPOA150L0 | 10-12-87 |
| 9. *Bromus rubens* | PMPOA15190 | 03-01-92 |
| 10. *Bromus tectorum* | PDPOA151H0 | 05-18-89 |
| 11. *Carduus nutans* | PDAST1S040 | 03-13-87 |
| 12. *Carduus pycnocephalus* | PDAST1S050 | 10-18-88 |
| 13. *Casuarina equisetifolia* | PDCAS01030 | 12-08-88 |
| 14. *Centauria maculosa* | PDAST1Y0C0 | 01-23-87 |
| 15. *Centauria solstitialis* | PDAST1Y0S0 | 02-04-86* |
| 16. *Cirsium arvense* | PDAST2E090 | 05-08-87 |
| 17. *Clidemia hirta* | PDMLS04020 | 06-17-91 |
| 18. *Conium maculatum* | PDAPI0Q010 | 05-31-89 |
| 19. *Convolvulus arvensis* | PDCON05020 | 10-17-88 |
| 20. *Cortaderia jubata* | PMPOA1P010 | 11-21-88 |
| 21. *Cynodon dactylon* | PMPOA1W020 | 03-01-92 |
| 22. *Cytisus scoparius*<br>*C. monspessulanus* | PDFAB18060<br>PDFAB18030 | 09-24-86 |
| 23. *Daucus carota* | PDAPI0X010 | 08-28-87 |
| 24. *Eleagnus umbellata* | PDELG01060 | 08-28-87 |
| 25. *Eucalyptus globulus* | PDMRT02020 | 03-22-89 |
| 26. *Euphorbia esula* | PDEUPOQOL0 | 07-28-87 |
| 27. *Foeniculum vulgare* | PDAPI12010 | 10-17-88 |
| 28. *Fraxinus uhdei* | PDOLE040J0 | 08-12-91 |
| 29. *Hedychium coronarium* | PMZIN04010 | 08-12-91 |
| 30. *Holcus lanatus* | PMPOA37010 | 11-21-88 |
| 31. *Juniperus virginiana* | PGCUP050E0 | 07-18-83 |
| 32. *Lonicera japonica* | PDCPR030G0 | 04-09-87 |
| 33. *Lonicera tatarica*<br>*L. morrowii*<br>+ their hybrid *L.* x *bella* | PDCPR030S0<br>PDCPR030K0<br>PDCPR03X10 | 07-18-84 |
| 34. *Lythrum salicaria* | PDLYT090B0 | 11-01-87 |
| 35. *Melilotus alba*<br>*M. officinalis* | PDFAB2H010<br>PDFAB2H060 | 07-27-87 |
| 36. *Neyraudia reynaudiana* | PDPOA4D010 | 04-30-90 |
| 37. *Pastinaca sativa* | PDAPI1M010 | 07-23-87 |
| 38. *Phalaris aquatica* | PMPOA4RO20 | 08-29-88 |

TABLE 1. (cont.)

| SPECIES | SPECIES CODE | LAST UPDATE |
|---|---|---|
| 39. *Phragmites australis* | PMPOA4V010 | 03-11-86* |
| 40. *Poa pratensis* | PMPOA4Z210 | 11-20-87 |
| *P. compressa* | PMPOA4Z0K0 | |
| 41. *Psidium cattleianum* | PDMRT0E020 | 08-12-91 |
| 42. *Rhamnus cathartica* | PDRHA0C050 | 08-07-84 |
| *R. frangula* | PDRHA0C080 | |
| 43. *Rhus glabra* | PDANA08030 | 06-29-84 |
| 44. *Robinia pseudoacacia* | PDFAB3G080 | 08-07-84 |
| 45. *Rosa multiflora* | PDROS1J0P0 | 07-24-87 |
| 46. *Rubus argutus* | PDROS1K0P0 | 08-12-91 |
| 47. *Rubus discolor* | PDROS1K1Y0 | 05-31-89 |
| 48. *Salsola iberica* (= *S. kali* var. *tenuifolia*) | PDCHEOKO2O | 11-30-88 |
| 49. *Schinus terebinthifolius* | PDANA09030 | 10-13-88 |
| 50. *Senecio jacobaea* | PDAST8H1U0 | 02-01-89 |
| 51. *Silybum marianum* | PDAST8M010 | 09-12-89 |
| 52. *Spartium junceum* | PDFAB3P010 | 03-22-89 |
| 53. *Taeniatherum caput-medusae* | PMPOA5Z010 | 09-21-88 |
| 54. *Tamarix ramosissima* | PDTAM01080 | 12-05-88 |
| *Tamarix chinensis* | PDTAM01050 | |
| 55. *Typha latifolia* | PMTYP01040 | 10-27-87 |
| *T. angustifolia* | PMTYP01020 | |
| *T. domingensis* | PMTYP01030 | |
| *T. x glauca* | PMTYP01X10 | |
| 56. *Ulex europaeus* | PDFAB42010 | 03-22-89 |
| 57. *Verbascum thapsus* | PDSCR1Z080 | 09-03-86 |
| 58. *Vinca major* | PDAPO0T020 | 10-17-88 |
| 59. *Xanthium spinosum* | PDAST9Z010 | 04-12-89 |
| 60. *Xanthium strumarium* | PDAST9Z020 | 02-01-89 |

California Regional Office, San Francisco CA
Eastern Regional Office, Boston MA
Florida Regional Office, Winter Park FL
Midwest Regional Office, Minneapolis MN
Pacific Regional Office, Honolulu HI
Southeastern Regional Office, Chapel Hill NC
Western Regional Office, Boulder CO
* = Revision in progress

## EXAMPLES OF WEED CONTROL EFFORTS ON NATURE CONSERVANCY PRESERVES

Efforts to control or eradicate weeds are under way on TNC preserves in every part of the nation (Fig. 1). The programs described below illustrate the diversity of approaches taken as well as the volunteer involvement and focus on restoration that are unifying themes in TNC stewardship.

### Blowing Rocks Preserve, Florida

Blowing Rocks Preserve encompasses 30 ha (73 acres) on Jupiter Island north of West Palm Beach on the Atlantic coast of Florida. It was established in 1969 largely to protect Blowing Rocks, the largest outcropping of Anastasia lime-

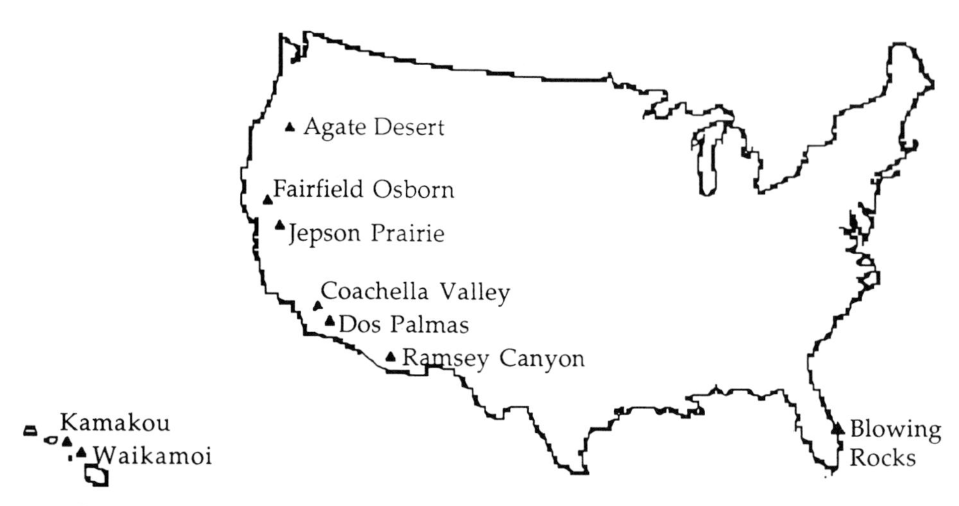


FIGURE 1.—Nature Conservancy preserves discussed in the text.

stone on the Atlantic coast. The formation's name derives from the plumes of saltwater that are forced up through fissures and solution holes in the limestone during severe winter storms. Marine turtles use the beach north of the rocks as a nesting area. The western edge, along Jupiter Sound, is an important feeding ground for West Indian manatees which winter in the area. The preserve encompasses shifting oceanfront dune, coastal strand, interior mangrove wetlands, and a small tropical hammock.

Invasive exotic plants dominated large portions of Blowing Rocks by the mid-1980s. Australian pine (*Casuarina equisetifolia*) was introduced to the island in 1916 for windbreaks and landscaping, and Brazilian pepper (*Schinus terebinthifolius*) was likely introduced to the area around the same time. These species and other exotics spread to areas that were covered with dredge spoil or otherwise disturbed during construction of the Intracoastal Waterway through Jupiter Sound in the 1910s and 1920s. Disturbances caused by mosquito control ditching on the island also encouraged the spread of exotics.

In 1985 efforts to control exotics on the preserve began when Australian pines in a 2.5 ha area along the dunes and coastal strand were cut down (Steve Morrison, pers. comm.). Two years later TNC contracted Post, Buckley, Schuh & Jernigan, Inc. (PBS&J) to assist with restoration of a 6 ha site on the east side of the preserve. PBS&J used chainsaws, a backhoe, and other heavy equipment to remove Australian pines and other exotics. Sea oats (*Uniola paniculata*) were planted on the seaward side of the first dune and mature saw palmettos (*Sabal palmetto*) taken from the site of a planned development nearby were planted in the dunes along with other native species. Mangrove seedlings were planted in wetlands and grasses, and other herbaceous species were established in adjacent areas.

Volunteers help maintain the restored area by removing exotics that attempt to reestablish there and elsewhere on the preserve (Norma Jeanne Byrd, pers. comm.). Species they encounter frequently include *Colubrina asiatica,* a woody vine in the Rhamnaceae, and *Wedelia trilobata,* a composite introduced as a ground cover. Recently a project to restore a 15 ha site on the western side of the preserve was begun. As with the earlier restoration effort, it involves removal of large numbers of aggressive exotics followed by plantings of native species. Volunteers now meet the first Saturday of each month, October through May, to accomplish these tasks. Others collect native plants from nearby areas slated for development. Still more help propagate native plants from local stock in the preserve's native plant nursery which was constructed in 1991 with funds from the Curtis & Edith Munson Foundation. In all, volunteers contributed over 2,400 hours of their time on this project in 1990 and 1991.

## Ramsey Canyon Preserve, Arizona

Infestations of *Vinca major* (periwinkle) are one of the most pressing problems in Ramsey Canyon, a 120 ha (300 acre) preserve in the Huachuca Mountains of southeastern Arizona (Tom Wood, pers. comm.). Many other exotic species including apples and other fruit trees were planted in the cool, shaded canyon during the late 1800s and early 1900s when it was used as a resort by miners and others who dwelt in hotter, lower areas nearby. These plants, however, are not spreading or preventing native species from reproducing as does *Vinca major.* This species, introduced as a ground cover around homesteads, spreads rhizomatously and grows so densely that other species cannot get established in areas it infests. It grows thickly along the banks of the creek that drains the canyon and may thereby encourage the creek to erode downward instead of into the banks themselves. This may cause the creek to become more deeply entrenched, altering the canyon's hydrology and eventually narrowing the zone of riparian vegetation along the canyon bottom (Anderson and Johnson 1991).

*Vinca major* covered 12 to 14 ha along the stream in 1974 when Dr. Nelson C. Bledsoe deeded the land to TNC. Attempts to clear some of patches by spraying it with herbicides were made and met with moderate success. Later, as more volunteer labor became available, a program of hand-grubbing the infestations was begun. Within the last few years roughly 4 ha have been cleared and the preserve manager, Tom Wood, predicts that the infestation will be under control within three years.

Ramsey Canyon is one of TNC's most heavily visited preserves and as such attracts large numbers of volunteers, many of whom have participated in this work. The bulk of the *Vinca* clearing, however, may be credited to one volunteer, Bill Gunterberg, who often comes out several times a week to work for a few hours. Like the other volunteers, he works from the edges of the patches using a hand-held cultivator to pull *Vinca* rhizomes from the ground. Chipping away at the infestations, this work proceeds at a pace that allows native species to move in and reclaim the area. This method has been advocated by the Bradleys, who attacked infestations of alien weeds in natural areas around Sydney, Australia (Bradley 1971, 1988; Fuller and Barbe 1985).

## Tamarisk on Coachella Valley and Dos Palmas Preserves, California

Tamarisk infestations were a severe problem in the palm oases and riparian areas of the 5,250 ha (13,000 acre) Coachella Valley Preserve near Palm Springs, California, until recently. Tamarisks, especially *Tamarix ramosissima* and *T. chinensis,* are notorious for their ability to reach supplies of groundwater up to 6 m below the surface (Kerpez and Smith 1987). They use this water inefficiently relative to native species like willows and cottonwoods in that they displace and may cause precipitous declines in groundwater levels and the cessation of surface seeps and flows (Gary and Campbell 1965; Kerpez and Smith 1987). These effects may in turn lead to the elimination of other native plants that can no longer reach the water table and of fish and other organisms dependent on surface flows

(Kerpez and Smith 1987; Howe and Knopf 1991). A single tamarisk can produce thousands of tiny wind-dispersed seeds, and the seedlings mature rapidly, often producing seed of their own by the end of their first year. Thus, an area that contains only a few tamarisks can be converted to an impenetrable thicket in less than a decade (Neill 1983). Native species usually fail to regenerate in these areas because their seedlings cannot tolerate the shade cast by tamarisks nor the deep layers of litter that they produce (Neill 1983). These thickets are also poor habitat for native animals, most notably birds and insects (Anderson and Ohmart 1976, 1984; Anderson et al. 1977; Cohan et al. 1978).

A tamarisk removal program that was sustained for nearly a decade finally resulted in the elimination of tamarisk as an ecological factor at the Coachella Valley in 1991 (Barrows, pers. comm.). This effort was a cooperative venture involving TNC, the Bureau of Land Management, the U.S. Fish and Wildlife Service, and the California Department of Fish and Game; all partners in the creation and continuing management of the preserve. Late in the fall of 1991 volunteers cleared the last heavily infested area on the preserve, a 1 km stretch of Pushwalla Canyon. Volunteers and staff continue to visit previously cleared areas to eliminate any resprouting or newly appearing tamarisk.

Strategies for eradicating *Tamarix ramosissima* from the preserve were developed in consultation with Bill Neill, a private citizen who has organized volunteer tamarisk removal projects on private, state, federal, and tribal lands throughout the Southwest for over ten years. Neill advocates identifying and attacking individual tamarisk infestations, concentrating on eliminating small outliers first, and then working from the edges inward on larger, core infestations (Neill, pers. comm.). He developed this stategy based on his own observations and experiences, but it is remarkably similar to strategies for invasive weed control advocated by the Bradleys (Bradley 1971, 1988; see above) and by Moody and Mack (1988). The technique currently regarded as the best for eliminating the trees involves cutting them with loppers or chainsaws and then quickly spraying the cut stumps with an ester-based solution of triclopyr (Garlon 4 or Pathfinder) (Neill 1991). Ideally the herbicide is mixed with a dye so that areas already sprayed can be readily identified.

The successes at Coachella Valley and at TNC's Hassayampa River Preserve in Arizona, where tamarisk was also eliminated as an ecological factor, have helped inspire programs at other sites. One such site is the Lower San Miguel River Preserve in southwestern Colorado where a contractor will be hired in the fall of 1992 to cut and treat the larger individuals and the densest infestations so that volunteers can concentrate their efforts in more accessible areas (Willits 1992). Upstream areas will be treated first in order to reduce the potential for reinfestation by waterborne seed.

A tamarisk control program has also been initiated at the Dos Palmas Preserve, approximately 40 km southeast of Coachella Valley (Fig. 1). Here, as at Coachella Valley in the past, tamarisk infestations threaten not only plant communities but the continued existence of pools that support populations of the threatened desert pupfish (*Cyprinodon macularius*). The core of Dos Palmas Preserve, consisting of two oases dominated by the fan palm (*Washingtonia filifera*), was purchased by TNC in 1989. Efforts to eliminate heavy infestations of tamarisk from the oases were begun shortly thereafter, again with the aid of Bill Neill and other volunteers.

Lands owned by the Bureau of Land Management (BLM) and the California Department of Fish and Game in the Salt Creek drainage surrounding the oases total over 1,620 ha (4,000 acres), nearly half of which is infested with tamarisk. The BLM prepared an environmental assessment for a project to eradicate tamarisk from this area in concert with efforts on the TNC property. Under this proposal the methods used will vary depending on the sizes and densities of tamarisk on a site, the percentage of native vegetation, and the proximity to aquatic resources (Bureau of Land Management 1992). The techniques used at Coachella Valley will be used in areas with large tamarisks, around existing native vegetation and along watercourses. In upland areas with open tamarisk stands and little or no native vegetation, a root plow will be pulled behind a bulldozer to sever tamarisk roots roughly 0.5 m below the soil surface. Severed brush will then be piled and burned on site and remaining stumps and resprouts hand-treated with triclopyr. A brush hog will be used in upland areas infested with smaller tamarisks. Due to the scale of the project, BLM plans to complete the work in phases and will proceed from west-to-east and north-to-south within the drainage in order to take advantage of the area's prevailing winds.

## Jepson Prairie Preserve, California

Jepson Prairie Preserve was established to protect a series of vernal pools and the relict California native bunchgrass prairie that surrounds them. Many species are unique to vernal pools, so-named because they fill with water during the winter rainy season and dry up in late spring or early summer. The pools at Jepson Prairie are relatively free of exotic plants. The prairie that forms the matrix around them is regarded as one of the best remaining stands of native grassland in the state. Nonetheless, large portions of it are dominated by annual grasses and forbs native to the Mediterranean region including *Avena barbata, A. fatua, Bromus diandrus, B. mollis, Erodium botrys, E. cicutarium, Hypochaeris glaba* and *Lolium perenne* (Barbour, unpubl. data). The same exotics have taken over nearly all of central California's grasslands which were once dominated by native perennial bunchgrasses such as *Melica californica, Poa scabrella* and *Stipa pulchra* (Barry 1972; Burcham 1981; Jackson 1985).

Many theories have been advanced to explain this remarkable replacement, which has taken place over the last two centuries. Several center on the role of grazing by cattle introduced first by Spanish settlers and expanded greatly in the years following the Gold Rush and statehood (Burcham 1981). Alteration of the natural fire regime is believed to have been important as well (Daubenmire 1968). TNC is currently supporting research to determine how seasonal grazing and prescribed fires influence competition between native bunchgrasses and introduced annuals at Jepson Prairie. The work is being conducted by Dr. John Menke and other researchers from the University of California, Davis, which shares management responsibility for the 634 ha (1,566 acre) preserve located roughly 65 km southeast of Sacramento.

The researchers have followed populations of native and introduced species inside paddocks that were treated with prescribed burns in late summer, grazed by sheep for short periods in early spring or summer, burned and grazed or left untreated as controls (Langstroth 1986; Fossum 1990). Their results thus far indicate that prescribed fires and grazing promote native bunchgrasses in different ways. Grazing reduces the heavy mulch produced by annual species which impedes germination and establishment by natives. Seed banks of the annuals are also reduced by grazing. Fire, on the other hand, breaks large bunchgrasses into smaller bunches which appear to be better able to spread vegetatively. In fact, it may be that this type of fire-promoted vegetative reproduction is more important than reproduction by seed in the establishment of dominance by bunchgrasses (Langstroth 1986; Fossum 1990).

## Fairfield Osborn Preserve, California

Yellow star thistle, (*Centauria solstitialis*) began invading the Fairfield Osborn Preserve 10 years ago (Larry Serpa, pers. comm.). Its first inroads were along trails through grasslands and oak savannas on the 85 ha (210 acre) site in the Sonoma Mountains about 100 km north of San Francisco. At the time, it appeared that this exotic annual, native to the Mediterranean, would not invade undisturbed areas. It soon turned up in undisturbed grasslands, however, and within a few years became dominant over large portions of the preserve. Fearing that yellow star thistle could outcompete some of the very species the preserve had been established to protect, the preserve manager began to look for control options (Serpa, pers. comm.).

Unfortunately, yellow star thistle is an extremely prolific seeder, and its long-lived seeds allow it to quickly re-establish after herbicidal control dissipates (Prather and Callihan 1991). The species has expanded its range in California at a roughly exponential rate since the late 1950s, increasing from 0.5–3.2 million ha (1.2 to 7.9 million acres) between 1958 and 1991 (Thompson et al. 1991). It is also a severe problem in Idaho, Oregon, and Washington. Because of its alarmingly rapid spread, significant resources have been dedicated to efforts to control it. For example, Dr. Charles Turner of the USDA-ARS biocontrol laboratory in Albany, California has focused on establishing a biocontrol program for yellow star thistle for the last several years. As a result of this work, six insect species have now been approved for release.

Efforts to control yellow star thistle at the Fairfield Osborn Preserve by hand-pulling were not successful. Volunteers were badly discouraged when areas they had helped clear one summer were thick with yellow star thistle the following year. After a few seasons they were, at best, reluctant to continue. This led the preserve manager to contact Dr. Turner to discuss the use of biocontrol. It was decided that yellow star thistle presented such a threat that release of one of the approved species, *Bangasternus orientalis,* was warranted. Roughly 225 of these seed-head-feeding beetles, collected from an established colony about 25 km north of the preserve, were released at two sites on the preserve in 1991. By autumn, healthy *B. orientalis* populations had become established, but it was too early to tell if star thistle populations were being affected. Unfortunately, yellow star thistle produces great quantities of seed which may persist for many years in the soil. Therefore, even if this insect and the others approved for release, all seed-head-feeders, become established, their activities alone may not noticeably reduce yellow star thistle populations for a decade or more.

Several TNC preserves in southern Oregon are also infested with yellow star thistle. The preserve manager responsible for the area, Darren Borgias, is considering use of biocontrol agents screened by Turner's lab because populations of several of the state's rarest plants are threatened by these infestations (Borgias, pers. comm.). One of the species of greatest concern is *Lomatium cookei,* a member of the Apiaceae, found only in the Agate Desert Preserve (Fig. 1) and a few other sites in southwestern Oregon. Borgias' decision will hinge, at least in part, on results of the releases at Fairfield Osborn.

## Hawaiian Preserves

The disruption of native communities and ecosystems by alien plants and animals has been nowhere more severe than in Hawaii (Vitousek et al. 1987a, b), where TNC manages preserves totalling 11,760 ha (29,070 acres) on five islands. On many preserves the threats posed by exotic ungulates, especially wild pigs, (*Sus scrofa*) and feral goats (*Capra hircus*) have been judged to be of overriding importance. The activities of these species facilitate the spread of many exotic plants and, until these mammals can be controlled, restoration attempts are doomed to failure (Stone 1985). On some preserves, ungulate control has reached a point that allows stewards to focus on control of the most troublesome plants.

Because there are so many species to contend with, TNC of Hawaii developed alien plant monitoring and management plans for several preserves with the help of a consultant (Rydell et al. 1990; Tunison 1991a, b). The plans prioritize the exotic species to be targeted. At Waikamoi Preserve on Maui, for example, 10 species are targeted including gorse (*Ulex europaeus*), Kahili ginger (*Hedychium gardnerianum*), and Florida prickly blackberry (*Rubus argutus*). The plan for Kamakou Preserve on Molokai proposes that weeds that have great potential to spread, including both exotics that are already widespread and those that are currently localized, be attacked first. Species in the former category include tropical ash (*Fraxinus uhdei*) and white ginger (*Hedychium coronarium*) and localized species include Clidemia (*Clidemia hirta*), strawberry guava (*Psidium cattleianum*), Florida prickly blackberry, and Brazilian pepper (*Schinus terebinthifolius*) also a problem in Florida as noted above. The plans also set specific annual goals for acreages to be cleared and provide estimates of the costs and worker hours that will be required. An excellent protocol for herbicide use was also developed, emphasizing practices to minimize damage to the environment and ensure applicator safety. The monitoring and management plans as well as the herbicide protocol may serve as models for exotic weed control programs in other TNC offices.

Exotic weed problems in Hawaii are already nearly overwhelming, but they could get worse. Simply stated, the invasion of the islands is not over. Over 4,600 species of plants have been introduced in the last two centuries and over 600 have become "naturalized," but more arrive every year (Saint John 1973; Smith 1985; Wagner et al. 1990). Hawaii Agriculture Alliance (1991) estimates that the state receives as many as 35 new alien animals and plants annually.

Recognizing that attempting to control exotics while ignoring new introductions is like shoveling sand against the tide, TNC of Hawaii and the Natural Resources Defense Council are preparing a background study on Hawaii's alien pest problem. The study is focused on identifying areas for improvement in current quarantine, inspection and control programs. Toward this end the two groups organized a workshop on the systems designed to prevent and control alien pest introductions in Hawaii. Representatives from Customs, the USDA Animal and Plant Health Inspection Service, the Hawaii Department of Agriculture, and other federal and state agencies involved with these systems attended. Organizations like the National Park Service that manage land affected by pest outbreaks also participated. One objective of the workshop was to improve contact among officials in all phases of the current system, allowing it to function more efficiently. Another objective was to generate ideas for improvements in the system so an outline of the approach taken in New Zealand was offered as a possible model. The outcome of the workshop will be represented in the background study report, due to be completed in 1992. The next goal is to use this report to produce a multi-agency cooperative action plan for improving Hawaii's protection against alien pest species.

## SUMMARY

Invasions by alien plants seriously threaten many of TNC's preserves throughout North America and Hawaii. In an attempt to coordinate weed control efforts within TNC the national stewardship office created an invasive-weed specialist position effective July 1991. Facilitating communication about the spread and control of alien weeds is an important part of this work, and I invite those with information on these topics to contact me. For those seeking information, Element Stewardship Abstracts (ESAs) covering over 60 pest plant species are available through TNC's Regional Offices.

The invasion process in Hawaii and North America has not stopped, and it is not enough to combat those species that are already pests on our shores. For this reason, TNC of Hawaii helped organize cooperative efforts to improve the systems designed to prevent and control exotic pest introductions in the islands. TNC hopes to promote and participate in similar cooperative work on the mainland.

In the field, TNC stewards attack weed problems with mechanical, chemical and/or biological methods, keeping in mind that the environmental benefits of their actions should clearly outweigh the costs. They are eager to improve on old methods and consider new ideas as well. I hope to help by promoting research on control methods and working to develop policies that will serve as guides to stewards while encouraging them to use their own knowledge and judgment.

The most successful programs view control or eradication of alien weeds as one part of an overall restoration program. They establish priorities and clear goals for weed control with carefully considered management plans, as was done for Kamakou and Waikamoi preserves in Hawaii. In some instances removal of weeds is the first and crucial step in a restoration program as it was at the Blowing Rocks Preserve. At the other end of the spectrum, some weed problems can best be addressed indirectly by restoring the natural hydrologic or fire regime, or reintroducing native species to an area.

Volunteers play extremely important roles throughout TNC's stewardship efforts. Weed control and broader restoration work account for the lion's share of volunteer time on many preserves. The skills and dedication that these people bring give us hope that we will control some of the most troublesome weeds on our preserves and restore much needed habitat for native species.

## ACKNOWLEDGMENTS

Norma Jeanne Byrd, Patrick Dunn, Greg Elliott, Doria Gordon, Alan Holt, Will Murray, and Rich Reiner contributed information and offered valuable editorial suggestions on an earlier draft. Cam Barrows, Darren Borgias, Steve Morrison, Larry Serpa, and Tom Wood contributed additional information.

## REFERENCES

Anderson, B. W., A. E. Higgins and R. D. Ohmart. 1977. Avian use of saltcedar communities along the lower Colorado River Valley. Pages 128–36 *in* Importance, preservation and management of riparian habitat: a symposium. R. R. Johnson and D. A. Jones (eds.). U.S. Forest Service General Technical Report RM-43.

Anderson, B. W. and R. D. Ohmart. 1976. A vegetation management study for the enhancement of wildlife along the lower Colorado River, annual report. In fulfillment of Bureau of Reclamation Contract #7-07-30-V0D09. Arizona State Univ., Tempe.

———. 1984. Vegetation management study for the enhancement of wildlife along the lower Colorado River. Comprehensive final report to the U.S. Bureau of Reclamation, Boulder City, Nev.

Anderson, C. and R. Johnson. 1991. 1991 riparian field tour: Sept. 3–7. Unpubl. manuscript, New Mexico Field Office of The Nature Conservancy, Santa Fe.

Barry, W. J. 1972. The central valley prairie. California prairie ecosystem, vol. 1. California Department of Parks and Recreation Report, Sacramento.

Bradley, J. 1971. Bush regeneration: the practical way to eliminate exotic plants from natural reserves. The Mosman Parklands and Ashton Park Association. Mosman, Sydney, Australia.

———. 1988. Bringing back the bush—The Bradley method of bush regeneration. Landowne Press, Sydney, Australia.

Burcham, L. T. 1981. California rangelands in historical perspective. Rangelands 3:95–104.

Bureau of Land Management. 1992. Salt Creek/Dos Palmas tamarisk eradication program environmental assessment. Palm Springs- South Coast Resource Area, California Desert District, Bureau of Land Management.

COHAN, D. R., B. W. ANDERSON and R. D. OHMART. 1978. Avian population responses to saltcedar along the lower Colorado River. Pages 371–81 *in* Strategies for protection and management of floodplain wetlands and other riparian ecosystems. R. R. Johnson and J. F. McCormick (eds.). U.S. Forest Service General Technical Report WO-120.

DAUBENMIRE, R. 1968. Ecology of fire in grasslands. Adv. Ecol. Res. 5:209–66.

DELOACH, C. J. 1991. Past successes and current prospects in biological control of weeds in the United States and Canada. Nat. Areas J. 11:129–42.

FOSSUM, H. C. 1990. Effects of prescribed burning and grazing on *Stipa pulchra* (Hitch.) seedling emergence and survival. M.S. Thesis, Univ. California, Davis.

FULLER, T. C. and G. D. BARBE. 1985. The Bradley method of eliminating exotic plants from natural reserves. Fremontia 13(2):24–25.

GARY, H. L. and C. J. CAMPBELL. 1965. Water table characteristics under tamarisk in Arizona. U.S. Forest Service Research Note RM 58.

HAWAII AGRICULTURE ALLIANCE. 1991. Introduced species: an overview of damages caused by introduction of some alien species to Hawaii. Draft report of the Hawaii Agriculture Alliance, Honolulu.

HESTER, F. E. 1991. The National Park Service experience with exotic species. Nat. Areas J. 11: 127–28.

HIEBERT, R. D. 1990. An ecological restoration model: application to razed residential sites. Nat. Areas J. 10(4):181–86.

HOWARTH, F. 1991. Environmental impacts of classical biological control. Ann. Rev. Entomol. 36:485–509.

HOWE, W. H. and F. L. KNOPF. 1991. On the imminent decline of Rio Grande cottonwoods in central New Mexico. Southwestern Natur. 36:218–24.

JACKSON, L. E. 1985. Ecological origins of California's Mediterranean grasses. J. Biogeo. 12:349–61.

KERPEZ, T. A. and N. S. SMITH. 1987. Saltcedar control for wildlife habitat improvement in the southwestern United States. U.S. Fish and Wildlife Service, Resource Publication 169.

LANGSTROTH, R. P. 1986. Fire and grazing ecology of *Stipa pulchra* grassland: a field study at Jepson Prairie, California. M.S. Thesis, Univ. California, Davis.

MOODY, M. E. and R. N. MACK. 1988. Controlling the spread of plant invasions: the importance of nascent foci. J. App. Ecol. 25:1009–21.

NEILL, W. 1983. The tamarisk invasion of desert riparian areas. Education Bulletin #83–84. Publication of the Education Foundation of the Desert Protective Council.

———. 1991. Tamarisk Newsletter. Published and distributed by W. M. Neill, Anaheim, Cal.

PRATHER, T. S. and R. H. CALLIHAN. 1991. Interference between yellow star thistle and pubescent wheatgrass during grass establishment. J. Range Manage. 44:443–47.

RYDELL, R., P. HIGASHINO, T. CASHMAN, et al. 1990. Waikamoi Preserve 5 year weed control plan. unpubl. manuscript, The Nature Conservancy of Hawaii, Honolulu.

SAINT JOHN, H. 1973. List and summary of the flowering plants in the Hawaiian Islands. Pacific Tropical Botanical Garden Mem. 1.

SMITH, C. W. 1985. Impact of alien plants on Hawaii's native biota. Pages 180–250 *in* Hawaii's terrestrial ecosystems preservation and management. C. P. Stone and J. M. Scott (eds.). Cooperative National Park Resources Studies Unit, Univ. Hawaii, Honolulu.

STONE, C. P. 1985. Alien animals in Hawaii's native ecosystems: toward controlling the adverse effects of introduced vertebrates. Pages 251–97 *in* Hawaii's terrestrial ecosystems preservation and management. C. P. Stone and J. M. Scott (eds.). Cooperative National Park Resources Studies Unit, Univ. of Hawaii, Honolulu. The Nature Conservancy. 1985. Employee Handbook. The Nature Conservancy, Arlington, Va.

THOMAS, L. K., JR. 1987. Experimental and technical management of exotic plants in the parks of the national capitol region: kudzu management *in* Abstracts of the thirteenth annual scientific research meeting, Great Smoky Mountains National Park, 21–22 May, 1987, J. D. Wood, Jr. (ed.). U.S. Department of the Interior, National Park Service, Science Publications Office, Atlanta, Ga.

———. 1988. Some principles of exotic species ecology and management and their interrelationships. Pages 96–110 *in* Proceedings of the Conference on Science in the National Parks, 1986, vol. 5: Management of exotic species in natural communities. 13–18 July, 1986, Fort Collins, Colo. L. K. Thomas (ed.). The U.S. National Park Service and the George Wright Society.

THOMPSON, C. D., M. E. ROBBINS and S. LARSON. 1991. Yellow star thistle control. Range Science Report No. 30. Department of Agronomy and Range Science, Univ. California, Davis.

TUNISON, T. 1991. Waikamoi Preserve: recommendations for managing and monitoring alien plants. unpubl. manuscript, The Nature Conservancy of Hawaii, Honolulu.

———. 1991. Kamakou Preserve: recommendations for managing and monitoring alien plants. unpubl. manuscript, The Nature Conservancy of Hawaii, Honolulu.

VITOUSEK, P. M., L. L. LOOPE and C. P. STONE. 1987. Introduced species in Hawaii: biological effects and opportunities for ecological research. Trends Ecol. Evol. 2:224–27.

VITOUSEK, P. M., L. R. WALKER, L. D. WHITEAKER, et al. 1987. Biological invasion by *Myrica faya* alters ecosystem development in Hawaii. Science 238:802–80.

WAGNER, W. L., D. R. HERBST and S. H. SOHMER. 1990. Manual of the flowering plants of Hawaii. University Press of Hawaii & Bishop Museum Press, Honolulu.

WILLITS, P. 1992. Tamarisk control at the Lower San Miguel River Preserve, Colorado. Unpubl. grant proposal for funding through The Nature Conservancy's Rodney Johnson & Katherine Ordway Endowments, The Nature Conservancy, Arlington, Va.

# The Ecological Impact and Management History of Three Invasive Alien Aquatic Plant Species in Florida

Don C. Schmitz,[1] Jeffrey D. Schardt,[1] Andrew J. Leslie,[1]
F. Allen Dray, Jr.,[2] John A. Osborne,[3] and Brian V. Nelson[1]

## INTRODUCTION

Florida's freshwater ecosystems were probably the first in the United States to experience invasions by alien vegetation. After the voyage of Christopher Columbus in 1492 to Hispaniola, the Spaniards and French extended their early explorations of the New World to include what is now called Florida (Tebeau 1980). By the late 1500s, the Spanish City of St. Augustine was established not far from the banks of the 435 km long St. John's River (Fig. 1). Within a few years, St. Augustine became a defense post that protected vital trade routes along the Florida coast and was an important port-of-call for international ship traffic from South America.

One of the earlier aquatic species introduced (possibly reintroduced into North America) was the South American floating plant waterlettuce (*Pistia stratiotes*). Its modern establishment may be linked to the arrival of the early Spanish settlers who established the city of St. Augustine (Stukey and Les 1984; Center et al. 1990). Even though waterlettuce presently has a cosmopolitan distribution, occurring on all continents except Antarctica (Holm et al. 1977), its geographic origin remains unclear. Fossil records place the genus *Pistia* in Siberia (Dorofeev 1955, 1958, 1963), northern Europe (Friis 1985), and North America (Stoddard 1989) prior to the Pleistocene Epoch. The dramatic climatic changes associated with this period, however, apparently eliminated waterlettuce from these regions.

By the 1880s, another South American native floating plant, the waterhyacinth (*Eichhornia crassipes*), was introduced into the St. John's River in northern Florida (Tabita and Woods 1962). Almost immediately, this floating pest spread throughout most of this river system with the assistance of cattlemen who held the mistaken belief that it made good cattle feed (United States Congress 1957). By the late 1950s, it was estimated that waterhyacinth populations occupied more than 51,000 ha of Florida's waterways (United States Congress 1965).

As waterhyacinth clogged Florida's waterways in the early 1950s, a submersed aggressive plant species (*Hydrilla verticillata*) from Ceylon (now Sri Lanka) was introduced into a drainage canal near the Tampa airport. Hydrilla originally was imported to St. Louis, Missouri, where it was sold as another species of *Anacharis* (now identified as the genus *Elodea*), and used as a popular aquarium plant (Schmitz et al. 1991). Within a few years, however, hydrilla es-

1. Florida Department of Natural Resources, 2051 East Dirac Drive, Tallahassee, FL 32310
2. University of Florida, IFAS, Ft. Lauderdale Research Education Center, Ft. Lauderdale, FL 33314
3. University of Central Florida, Department of Biological Sciences, Orlando, FL 32816

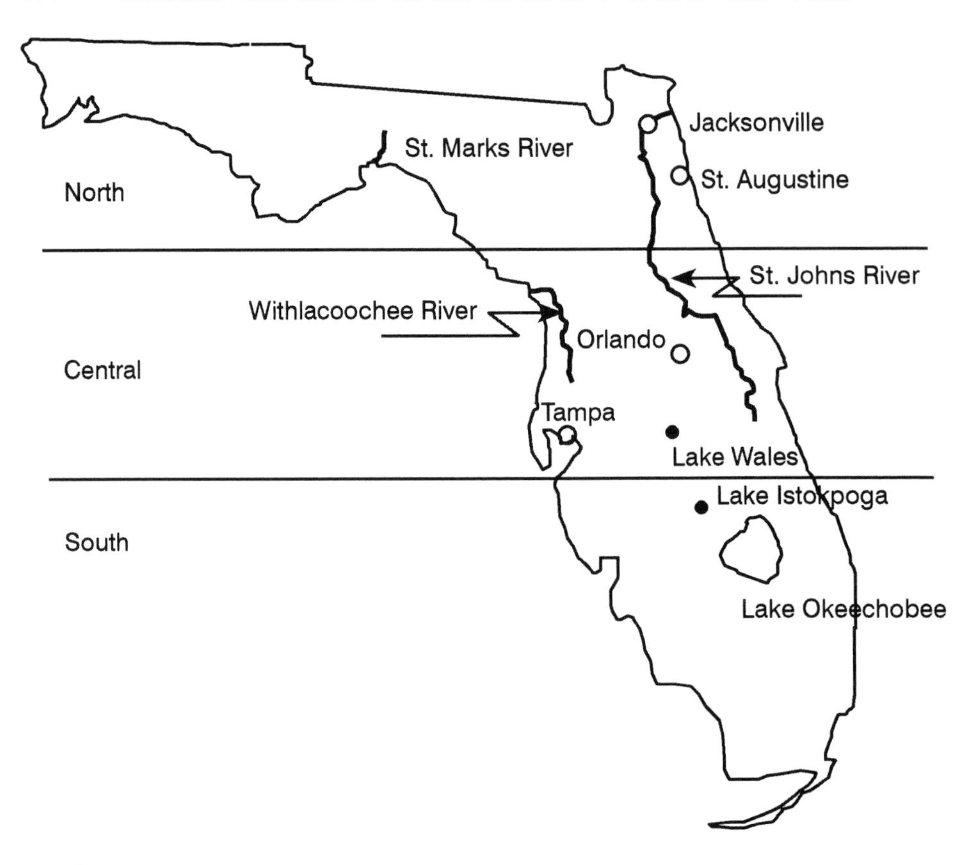


FIGURE 1.—Map of Florida.

caped the confines of the aquarium plant trade and began infesting central and south Florida lakes, rivers, and canals. Presently, the spread of hydrilla in Florida's waterways makes it one of the state's most aggressive alien plant species.

Vitousek (1986) argues that invading alien species can alter ecosystem properties. Their effects depend primarily on the ecological potential of the invader and the physical and biological properties of the ecosystem receiving the introduction (Taylor et al. 1984). This paper reviews the impacts of waterhyacinth, waterlettuce, and hydrilla on Florida's aquatic ecosystems. It also provides a historical perspective of efforts in Florida to manage these alien plant species.

## ALIEN AQUATIC PLANTS

### Waterhyacinth

Waterhyacinth is an impressive floating, erect plant with aerial leaves, bulbous petioles, and submerged roots and rhizomes. This alien plant is extremely productive creating dense floating mats of vegetation. Growth rates for waterhyacinth measured in terms of dry biomass, exceed that of any terrestrial, salt water, or fresh water vascular macrophyte (Wolverton and McDonald 1979; Hall 1980). Its rapid branching ability and vegetative reproduction enables this weed to quickly invade and overgrow an area, thereby precluding potential plant competition (Center and Spencer 1981) (Fig. 2). Biomass doubling times range be-

FIGURE 2.—The Ortega River bridge near Jacksonville, Florida, during the 1940s, showing waterhyacinth accumulation.

tween 6 and 18 days (Mitchell 1976; National Academy of Sciences 1976; Cornwell et al. 1977). Productivity decreases, however, as ambient air temperatures decrease, and approaches zero as temperatures near 10°C (Reddy and Bagnall 1981). Although this species ranges into temperate regions, it is not freeze-tolerant.

Large floating mats of waterhyacinth often degrade water quality and can lead to dramatic changes in plant and animal communities. Penfound and Earle (1948) reported that water temperatures were more uniform, acidity and dissolved carbon dioxide were higher, and dissolved oxygen was lower under a waterhyacinth mat when compared to an open water location in Louisiana. Increases in turbidity and color, have also been attributed to waterhyacinth infestations (Timmer and Weldon 1967). McVea and Boyd (1975) reported that total phosphorus concentrations decreased in order of increasing waterhyacinth cover in Alabama fish ponds. Filtering agricultural drainage waters through experimental reservoirs containing dense stands of waterhyacinth caused dissolved oxygen, temperature, pH, conductivity, turbidity, and bicarbonate alkalinity to decrease, and increased dissolved carbon dioxide (Reddy et al. 1983). Similar changes have been documented by Ultsch (1973), Rai and Datta Munshi (1979), and Schreiner (1980). Ultsch (1973) noted that the extent of these physicochemical changes was dependent on the time of day, location within the water column, and the season.

Waterhyacinth communities transpire large amounts of water. Evapotranspiration losses from waterhyacinth populations in Louisiana measured 3.2 times the rate from an adjacent open water area (Penfound and Earl, 1948). In Florida, evapotranspiration rates range from 3.7 to 6.0 times higher than adjacent open water areas (Timmer and Weldon 1967; Rogers and Davis 1972; Reddy and Tucker 1983). Whether such large losses can impact regional hydrologic cycles in Florida has not been studied.

Plant detritus production is a by-product of the natural aging of plants, biological or chemical control, and frost damage. Center and Spencer (1981) reported that leaves represent 60–70 percent of waterhyacinth biomass and that leaf turnover rate can range between 60–70 percent per month. One hectare of waterhyacinth can contain more than 1,600,000 plants and can deposit between 404 and 1,886 metric tons (wet weight) of detritus per year (Penfound and Earl 1948; Center and Spencer 1981). Joyce (1985) studied detrital accumulation over a one year period in waterhyacinth covered tanks. He found the plants produced the equivalent of 14.6 metric tons/ha (dry weight) and the organic portion of these sediments comprised 11.7 metric tons/ha (dry weight). The anoxia resulting from such detritus enhances the release of phosphorus into the water during decomposition (Reddy and Sacco 1981). With many of Florida's waterways experiencing increased sediment and nutrient loads due to agriculture and urban runoff, the additional detrital load generated from dense waterhyacinth mats is believed to overburden the nutrient loadings in these systems.

Heavy metal ions from water, such as cadmium, chromium, lead, mercury, and zinc have been shown to bioconcentrate in substantial quantities in waterhyacinth (Muramoto and Oki 1983; Kay et al. 1984). In addition, laboratory studies show this plant can remove significant quantities of synthetic phenolic organic compounds (common environmental pollutants) from water (O'Keeffe et al. 1987). Consequently, because of the rapid growth rate and high evapotranspiration rate of waterhyacinth, populations of this plant species have the potential to be a depository for heavy metals and probably for toxic organic compounds. Despite this removal of toxic substances from a waterway, any potential benefits are short-lived without the actual physical removal of the plants from a waterbody because detritus production will reintroduce these pollutants into the ecosystem.

Laboratory concentrations substantially higher than those reported for field collected plants of cadmium, copper, and lead in waterhyacinth leaf tissues fed to waterhyacinth weevils (*Neochetina eichhorniae*) had no biologically significant effects upon the mortality of adult weevils (Kay and Haller 1986). Lead and cadmium were shown to substantially bioaccumulate within the insect's tissues. The authors speculated that heavy-metal contaminated weevils could be a substantial source of lead and cadmium for predatory arthropods and insectivorous birds feeding in areas of heavy weevil infestation on waterhyacinths. Additionally, heavy metal and synthetic organic compound-contaminated waterhyacinth may pose some risk for the Florida populations of the West Indian manatee (*Trichechus manatus*), an endangered, herbivorous aquatic mammal species. Although, how much risk, if any, is not known.

McVea and Boyd (1975) noted that declining phytoplankton abundance in small fish ponds was related to shading and phosphorus uptake by waterhyacinth. One step up the food chain, studies on zooplankton densities and/or species diversity beneath waterhyacinth mats are sparse; none have been conducted within Florida. Studies conducted in Africa offer some conflicting clues to the possible impacts of waterhyacinth invasions on Florida plankton communities. Before (Rzoska et al. 1955) and after waterhyacinth invasion studies (Abu-Gideiri and Yousif 1974), conducted in the Sudan on phytoplankton and zooplankton communities in the White Nile, indicate the disappearance of three crustacean zooplankton species, a rotifer species, and one phytoplankton species. Abu-Gideri and Yousif (1974) also noted an apparent species shift and the addition of several new species as well as a large increase in phytoplankton and zooplankton densities. In contrast, Scott et al. (1979) and Ashton et al. (1979) reported declines in phytoplankton and zooplankton abundance after waterhyacinth invaded a waterway in South Africa. Clearly, further study is needed to determine overall ecological effects of waterhyacinth invasions on the dynamics of plankton communities.

The edges of waterhyacinth mats have been reported to support large numbers of invertebrates within its root system (O'Hara 1967; Schramm et al. 1987). Hansen et al. (1971) reported a partial food web of a waterhyacinth root community in a central Florida lake to have two producers (waterhyacinth and phytoplankton), five herbivores and eleven carnivores. The principle herbivore found within the root system of waterhyacinth was the amphipod *Hyalella azteca* (O'Hara 1967; Hansen et al. 1971). Amphipods, which often act as shredders of dead plant material, may substantially influence the rate of waterhyacinth breakdown in Florida waters (Bartodziej, in press).

At least seven species of fish are associated with waterhyacinth habitat in Florida (Hansen et al. 1971). In contrast, some authors have suggested that low dissolved oxygen levels associated with waterhyacinth infestations reduce acceptable fish spawning areas (Lynch et al. 1957; Achmad 1971). Gopal (1987) noted that waterhyacinth mats shade-out benthic communities and can nearly block the diffusion of oxygen through the water-atmosphere interface. Low oxygen concentrations beneath waterhyacinth mats have resulted in fish kills (Timmer and Weldon 1967). When waterhyacinth populations reached 10 and 25 percent vegetated cover in experimental ponds, McVea and Boyd (1975) reported that fish production was reduced. They speculated that this was due to lower phytoplankton densities from the shading produced by these floating macrophytes.

Drifting mats of waterhyacinth often smother beds of submersed vegetation and overwhelm marginal plants that are important to waterfowl (Tabita and Woods 1962; Chesnut and Barman 1974). Consequently, dense waterhyacinth populations can impact critical wildlife habitat areas. For example, declines in the population of the Florida Everglade kite (*Rostrhamus sociabilis*), an endangered bird species, have been attributed, in part, to waterhyacinth populations invading and destroying kite habitat in south Florida (Griffen 1989). Large waterhyacinth mats uproot stands of emergent native plant species making it difficult for the Everglade kite to locate their primary food source (apple snails) that normally utilize these emergent stands of vegetation (Griffen 1989).

## Waterlettuce

Waterlettuce is a free-floating plant whose weedy characteristics have enabled it to become a problem in Florida as well as many other parts of the world (Holm et al. 1977; Sculthorpe 1967) (Fig. 3). Its seasonal growth cycles are similar to those reported for waterhyacinth (Center and Spencer 1981; DeWald and Lounibos 1990). Waterlettuce propagates rapidly throughout the year through copious production of offsets (Datta and Biswas 1969; DeWald and Lounibos 1990). Seed production is also common in Florida (Dray and Center, 1989), and seedlings are particularly important as a source for re-infesting waterways in which waterlettuce has been reduced by control efforts or extreme cold weather. Prolific vegetative growth results in maximum standing crops of at least 700 g/m$^2$ (dry weight) (Hall and Okaii 1974) and plant densities exceeding 1,000 rosettes/m$^2$ (DeWald and Lounibos 1990).

Generally, waterhyacinth has been reported to out-compete waterlettuce (Chadwick and Obeid 1966; Bond and Roberts 1978; Tucker and DeBusk 1981; Sutton 1983; Agami and Reddi 1990). Consequently, large floating islands of waterlettuce that were encountered during the mid to late 18th century in Florida (Stuckey and Les 1984) became less prevalent with the spread of waterhyacinth throughout Florida (Schardt and Schmitz 1990).

The few studies that have been conducted concerning the impacts of waterlettuce on water quality indicate a level of ecological impact similar to those

FIGURE 3.—A waterlettuce infestation in one of the Everglades Conservation areas in 1991.

reported in the previous section on waterhyacinth (Sharma 1984). For example, Attionu (1976) reported that waterlettuce populations comprising only 350–500 rosettes/m$^2$ induced thermal stratification, depleted dissolved oxygen levels, and lowered the pH of affected waters in Lake Volta, Ghanna. In addition, Sharma (1984) reported that evapotranspiration by waterlettuce exceeds evaporation over open water sites by as much as tenfold. However, the question of whether or not these higher evapotranspiration rates have an impact on regional hydrologic cycles remains unanswered (see, however, Anonymous 1971). Like waterhyacinth, waterlettuce populations also can bioaccumulate considerable amounts of heavy metals (Sridhar 1986).

The long adventitious roots of both waterlettuce and waterhyacinth, with their copious lateral roots, permit large infestations to slow water velocities in rivers and streams, thereby causing increased siltation (Sculthorpe 1967; Anonymous 1971). Pollution studies have shown that increased siltation can degrade benthic substrates making them no longer suitable as nesting sites for fish (Beumer 1980) or as habitat for many macroinvertebrates (Roback 1974). However, increased siltation caused by waterlettuce mats has never been directly measured in Florida's flowing waters.

The impact of waterlettuce on native plant communities also has been poorly documented. The primary effect in waterways covered by waterlettuce mats for prolonged periods is the loss of native submersed species because of shading. The degree to which this occurs in Florida remains to be investigated, but Bua-ngam and Mercado (1976) reported that waterlettuce out-competes seedlings of cultivated rice, an aquatic grass (*Echinochloa crusgalli*), and the bulrush (*Scirpus maritimus*). These authors suggested that waterlettuce is better adapted than other species to utilize nutrients, particularly nitrogen.

Like waterhyacinth, waterlettuce harbors many macroinvertebrates of which the amphipod *Hyallela azteca* is the most abundant (Dray et al. 1988). The importance of these amphipods to the aquatic food chain in Florida waters has not been studied directly, but they probably play an ecological role similar to those found under waterhyacinth mats.

## Hydrilla

Hydrilla is a submersed plant species similar to the native Canadian waterweed (*Eloedea canadensis*) in appearance. But, this native of southeast Asia is more aggressive than Canadian waterweed, especially in warmer climates. Hydrilla, which branches profusely, can grow in water ranging from a few centimeters to 15 m deep (Yeo et al. 1981). Hydrilla is rooted in the hydrosoil but concentrates 70 percent of its biomass at the water surface, forming a dense canopy especially during the summer and early fall months. Hydrilla is only known to reproduce asexually in Florida. Detached stem fragments readily develop into new plants which attach themselves to the hydrosoil by unbranched adventitious roots (Cook and Luond 1982). Axillary buds (turions) and subterranean tubers are the primary vegetative propagules that are produced by hydrilla in Florida. Axillary buds, which are propagated primarily in the fall and winter, break loose, sink to the hydrosoil, and germinate during the spring and summer (Haller 1978). Tubers form at the ends of rhizomes usually during the fall and winter (Bowes et al. 1979). Germination is usually in the spring and summer, but also can occur year-round in south Florida waterways (Sutton and Portier 1985). Tuber densities in Florida lake bottoms infested with hydrilla have been estimated to be in the millions (Mitchell 1974; Bowes et al. 1979; Steward 1984).

The hydrilla populations in Florida are dioecious and only produce "female" flowers. But the recent introduction of the monoecious "male" hydrilla strain in the Potomac River (District of Columbia) increases the potential, through sexual reproduction, for genetic diversification and adaptation of this species to different habitats (Steward et al. 1984). If the northern strain of hydrilla, a more cold tolerant plant, spreads to Florida, this may lead to an even more aggressive hydrilla population, especially in colder north Florida waterways.

Hydrilla is well-adapted for distribution and growth in Florida's waterways. The most common mechanism for dispersal appears to be transportation of hydrilla fragments by boat trailers because new populations are often first noticed adjacent to boat ramps. Once established, boat traffic continues to fragment plants, enhancing dispersal (Schardt and Schmitz 1991).

Hydrilla out-competes native aquatic vegetation in Florida lakes and rivers because of its ability to adapt to low light levels (Bowes et al. 1977) and its ability to quickly produce a dense canopy near the surface of the water that shades-out other existing, native, submersed vegetation. Entire water bodies can be covered quickly by such canopies. For example, in August of 1987, approximately 1,618 ha of the 11,332 ha Lake Istokpoga (located in south-central Florida) were covered by hydrilla (Schardt and Schmitz 1990) (Fig. 4). By December of 1988, nearly 8,000 ha were covered by this alien plant species (Fig. 5). Hydrilla's ability to grow under low light levels also lengthens this species' growing season beyond that of native vegetation. Consequently, an entire statewide population reduction in the native aquatic plant eel grass (*Vallisneria*) in 1983 was attributed primarily to competition with hydrilla (van Dijk 1985).

The production of allelopathens by hydrilla has been shown to negatively effect the distribution of at least one native Florida species, *Ceratophyllum demersum* (Kulshreshtha and Gopal 1983). Yet culturing experiments have shown that at least one Florida native plant (*Sagittaria graminea*) was resistent to the establishment of hydrilla from sprouted tubers (Sutton 1986). However, hydrilla will likely be able to out-compete most submersed native plant species in

FIGURE 4.—Lake Istokpoga in 1987 showing Hydrilla coverage. (for details see Schardt and Schmitz 1990).

Florida because of hydrilla's ability to utilize extremely low light levels (Dr. Dave Sutton, Univ. Florida, pers. comm., 1992).

Florida's waterways have undergone increased loadings of nutrients since the turn of the century due to increased urbanization and agricultural runoff. Excessive growth of hydrilla in many lakes has been linked with this increased eutrophication (Canfield et al. 1983a). Changes in hydrilla biomass in Florida lakes have had profound impacts on water quality in highly productive lakes. Canfield et al. (1983a) reported that the effects of hydrilla on lake water chemistry are related to the percentage of the lake's volume infested with hydrilla and macrophyte standing crop. For example, during the growing season in a small (5.4 ha) eutrophic hydrilla-dominated lake located near Orlando in central Florida, total alkalinity, specific conductivity, dissolved oxygen, orthophosphate, water color, turbidity, and cholorphyll$_a$ decreased while pH and Secchi disc transparency increased (Schmitz and Osborne 1984). Specifically, chlorophyll$_a$ concentrations declined significantly from a high of 82 mg/m$^3$ when hydrilla biomass was low, to less than 6 mg/m$^3$ when hydrilla biomass was greatest. The dense submersed vegetation is known to inhibit phytoplankton growth (Hasler and Jones 1949). Also, dissolved oxygen concentrations in this small lake were similarly affected by hydrilla growth. When hydrilla biomass peaked during the fall, dissolved oxygen concentrations measured at the bottom of this Orlando lake were generally less than 2.0 mg/l and were probably related to reduced underwater light penetration and water circulation. Similar changes in water quality have been observed in other hydrilla-dominated Florida lakes (Canfield et al. 1983b; Osborne et al. 1983; Shireman et al. 1983b). Consequently, changes in lake trophic state classification based on water quality alone can be attributed to dense infestations of hydrilla, especially in small lakes. Because of an 80 percent coverage of hydrilla in a central Florida lake, Canfield et al. (1983a) recognized that large errors in state lake trophic assessment gave no consideration to the

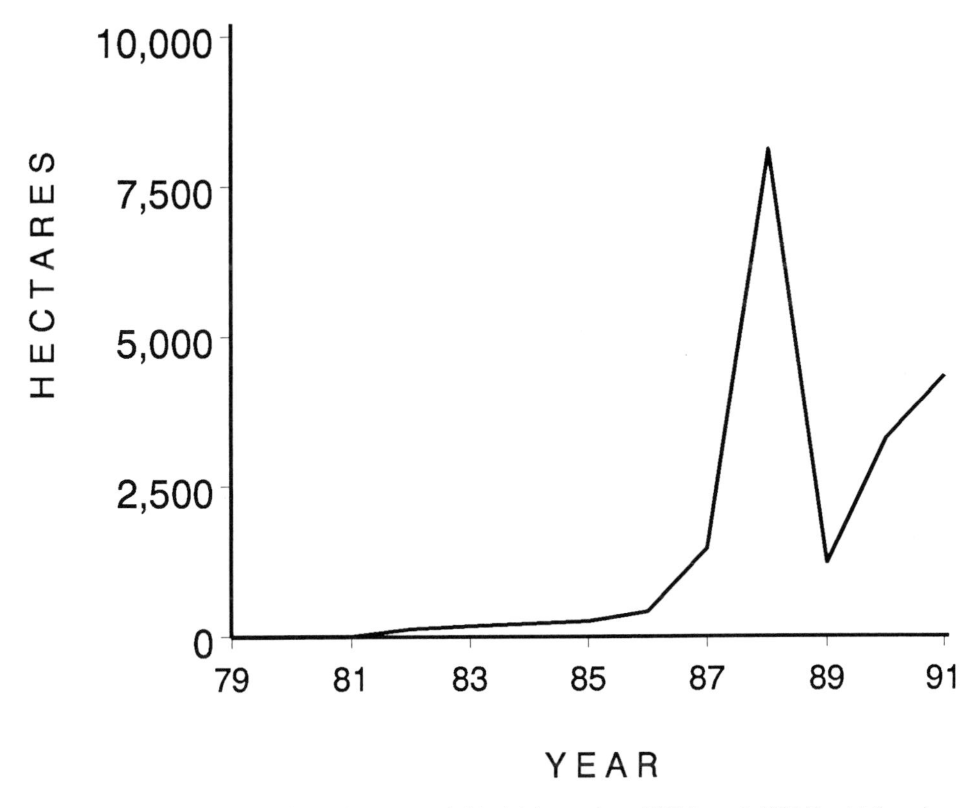


FIGURE 5.—The amount of hydrilla coverage in lake Istokpoga from 1979 through 1991 (for details on how this survey was conducted, see Schardt and Schmitz, 1990).

nutrients, plant biomass, or organic production associated with dense macrophyte populations.

Dense infestations of hydrilla in Florida lakes have been reported to decrease zooplankton abundance and increase the number of species (Schmitz and Osborne 1984; Richard et al. 1985). Shifts in zooplankton densities are often associated with changes in plant communities (Blancher 1979) and abundant aquatic vegetation may actually inhibit the density of zooplankton (Pennak 1966). Additionally, the dramatic decrease of the planktivore food-base (reduced phytoplankton abundance) by hydrilla expansion within Florida lakes has been suggested to influence changes in zooplankton composition and abundance (Schmitz and Osborne 1984; Richard et al. 1985).

Epiphytic (plant dwelling) and benthic macroinvertebrate populations and species diversity have been reported to increase in hydrilla-dominated areas of Florida lakes (Watkins et al. 1983; Schramm et al. 1987; Schramm and Jirka 1989). Higher benthic macroinvertebrate densities, particularly amphipods, have been linked to hydrilla creating a more desirable habitat condition by stabilizing the bottom sediments (Schramm and Jirka 1989). In addition, the greater leaf dissection of hydrilla has been suggested to afford more attachment sites for epiphytic and grazing macroinvertebrates (Watkins et al. 1983). In contrast, Bartodziej (in press) found epiphytic macroinvertebrate densities in the St. Marks River to be lower on hydrilla. Instead, these epiphytic macroinvertebrates tended to use simple structured native macrophytes, like the strap-leafed sagittaria (*Sagittaria kurziana*), as a habitat. In another study, Scott and Osborne

(1981) reported lower benthic macroinvertebrate densities and diversity in a hydrilla-dominated central Florida lake; likely related to seasonal insect emergence and low dissolved oxygen concentrations. Oligocheaetes and chironomids have been reported to dominate the macroinvertebrate fauna in lakes infested with hydrilla (Martin and Shireman 1976; Scott and Osborne 1981; Watkins et al. 1983). Although the studies on benthic macroinvertebrates and hydrilla in Florida are not extensive, it appears that dense hydrilla infestations can lead to higher benthic macroinvertebrate densities and diversity, with oligocheaetes and chironomids dominating the benthic fauna in lake systems when low dissolved oxygen concentrations are not a factor.

Biological effects on fish populations occur when plant coverage increases to the point where open-water feeding areas and spawning areas are no longer available (Shireman et al. 1983a). The near total lake coverage of hydrilla that occurs can reduce the bluegill (*Lepomis macrochirus*), redear (*Lepomis microlophus*), and black crappie (*Pomoxis nigromaculatus*) fishery (Colle et al. 1987). Populations of these species become skewed to smaller individuals due to insufficient predator cropping (Colle et al. 1987). Extensive hydrilla mats can delay age of recruitment to harvestable size from year two to year three or four for black crappie (Maceina and Shiremen 1985). Largemouth bass (*Micropterus salmoides*) avoid dense hydrilla mats, but such mats provide a haven for juvenile sport-fish and forage fishes resulting in lost condition (fatness) for harvestable largemouth bass and stunted panfish populations (Colle 1982; Shireman et al. 1983a).

Shireman and Maceina (1981) found that while aquatic vegetation is important for maintenance of sport-fish (Ware and Gasaway 1976; Miley et al. 1979; Leslie et al. 1987), the amount of vegetation needed for maximum production is undetermined. Hinkle (1986) reviewed literature dealing with fisheries and vegetation and concluded that aquatic plants should be managed at between 10 and 40 percent cover to maintain a high sustained yield of harvestable sport-fish. Colle and Shireman (1980) believe that complete elimination of hydrilla and other macrophytes would probably be detrimental to sport-fish communities. These authors found that hydrilla cover in excess of 30 percent resulted in low condition (fatness) of larger (> 250 mm TL) largemouth bass but smaller bass had high condition until cover exceeded 50 percent. Bluegill and redear condition was influenced more by the amount of plant material in the water column rather than percent of aerial cover. Condition factors of these species were not affected until the water column was nearly totally occupied by vegetation (78 percent cover of the entire lake bottom and 23 percent surface-matted vegetation).

Several southeastern workers have demonstrated the importance of aquatic vegetation to waterfowl. Florschutz (1972) found the alien submersed plant Eurasian watermilfoil (*Myriophyllum spicatum*), as well as several native species of macrophytes, were heavily grazed by waterfowl in North Carolina and Virginia. Duke and Chabreck (1976) classified common coontail, eelgrass (*Vallisneria americana*), and several species of pondweeds (*Potamogeton* spp.), among others, as important waterfowl food plants in Louisiana. Kerwin and Webb (1971) found that southern naiad (*Najas quadalupensis*) was the most important food for dabbling ducks while watershield (*Brasenia schreberi*) was most important for diving ducks. Watershield and floating heart (*Nymphoides aquaticum*) are considered good waterfowl foods in Florida (Chamberlain 1960). Krapu (1974) found a high occurrence of spikerush (*Eleocharis* spp.) in waterfowl diets. Montegut et al. (1976) found hydrilla to be heavily utilized by waterfowl in Lake Wales, Florida. Johnson and Montalbano (1987) believe there is a growing body of evidence indicating that hydrilla provides good waterfowl habitat. They reported that hydrilla was an important food for ducks, coots, and common moorhens in Florida. Further, they noted that waterfowl seemed to prefer hydrilla to the available native plant habitats, and that hydrilla supported the highest species diversity.

Because of the extensive wetland loss and degradation in Florida since the turn of the century (Fernald and Patton 1984), waterfowl managers have noted increased utilization of hydrilla-infested habitats. The growth habit of hydrilla appears to simulate the shallow marsh habitats that are being rapidly lost in Florida (Johnson 1987).

## MANAGEMENT HISTORY

### Waterhyacinth

Problems associated with waterhyacinth were reported along the St. Johns River as early as 1894 (Raynes 1964). One of the first records of waterhyacinth obstructions appeared in a February 9, 1895 letter from Mr. E.S. Crill of Palatka who noted that paddle wheel steamers were frequently hindered and entirely blocked by waterhyacinth. By 1897 or 1898, waterhyacinth had spread south to Lake Okeechobee (Will 1965).

In March, 1899, an aquatic plant management program in Florida began after the 55th Congress passed the Rivers and Harbors Act authorizing the U.S. Army Corps of Engineers (Corps) to crush, divert, or remove waterhyacinth from access areas of the St. Johns River. On May 11, 1899, Florida enacted Chapter 4753 (now S.861.04 Florida Statutes) to prohibit the further placing of waterhyacinth in streams and waters of the state and to prescribe penalties for violations. At the turn of the century, only the Corps attempted waterhyacinth control. Passive devices, such as log booms and barriers, were placed across infested creeks and streams to prevent discharge of floating waterhyacinth plants into the St. Johns River. Physical labor was employed to simply break up small mats which had collected among snags and pilings which freed these mats to float down the river and out into the Atlantic Ocean (Zeiger 1962). Mechanical apparatus, such as conveyor belts, derricks, and grapples, were used along shorelines and bridges but usually had little effect on retarding waterhyacinth production for several reasons. First, machines could not operate effectively in shallow water along shorelines and thus, these devices left a continuous source of plant material for re-infestation. Second, the sheer mass and volume of waterhyacinth was an impediment to success (Fig. 6). In 1973, the Corps estimated that one hectare of waterhyacinth weighs between 191 and 415 metric tons with an average volume of 168 cubic meters. Finally, the growth rate of waterhyacinth was so great, that these turn-of-the-century machines could not keep pace, despite, in some cases, around the clock operations.

In 1902, the Rivers and Harbors Act was amended to allow for the "extermination" of waterhyacinth by mechanical and chemical or any other means. Toxic sodium arsenite was used in Florida until 1905 when it was discontinued after problems arose concerning the poisoning of cattle. Other chemicals were tested including sulfuric and carbolic acids and kerosene. All of these chemicals were rejected because of their ineffectiveness or extreme toxicity to non-target organisms (Buker, 1982). In 1931, the State of Florida passed Chapter 1465 (now S.342.05 Florida Statute) to allow the use of . . . "any poisonous substance, chemical or spray in killing waterhyacinth . . . provided no such poisonous substance, chemical or spray shall be used which might injure fish life or human or other animal life. . . . "

This resulted in waterhyacinth control operations conducted almost exclu-

FIGURE 6.—Removing waterhyacinths from a south Florida bridge using mechanical means during the 1930s.

sively by mechanical means (crusher-boats, saw-boats, and harvesters) until the late 1940s. The crusher-boats, used most successfully in Louisiana, hauled plants aboard, crushed them under 40,000 pounds of pressure per square inch and discharged them back into the water or catapulted them to shore. Disposal in water was the preferred method since shoreline disposal often met with protests from property owners as levees of crushed waterhyacinth reached as high as 4.6 m along river banks (Wunderlich 1964). In other cases, shorelines were not accessible due to shallow water and dense tree growth. Hauling plants to ramps and then trucking material to dumps was deemed to be economically and logistically undesirable.

Saw-boats had a shallower draft than the more cumbersome crusher-boats and were able to operate closer to shore. With a 3.03 m wide bank of saw blades operating at the water surface, these boats shred waterhyacinth without first lifting them from water thus improving efficiency (Wunderlich 1967). In spite of their increased efficiency, saw-boats were not able to access shorelines and left a continual renewable source of waterhyacinths (Gallagher 1962).

As the Corps continued their search for more effective controls, a reported "blighted worm infested waterhyacinth" population was found in the Withlacoochee River in the early 1940s. The cause of the damage was investigated and affected plants were placed throughout the river system in an attempt to spread the infestation to other waterhyacinths (Buker 1982).This apparently was the first attempt at using an organism for control of waterhyacinth populations in Florida waterways. However, these efforts were considered to be unsuccessful; the organism appears not to have been identified and thus remains a mystery.

## Use of Modern Day Herbicide

In 1941, the herbicidal properties of 2,4-D, were evaluated, followed by terrestrial and aquatic field trials in the mid-1940s. Waterhyacinth was found to be highly susceptible to it and by 1948, the Corps began using 2,4-D operationally

(Buker 1982). Ironically, as waterhyacinth populations decreased, there was a rapid expansion of the 2,4-D resistant alien aquatic plant, alligatorweed (*Alternanthera philoxeroides*), into areas previously occupied by waterhyacinth (Coulson 1977).

On April 3, 1952, with federal funding (Dingell Johnson Federal Aid to Fisheries Fund), the Florida Game and Freshwater Fish Commission (GFC) began a limited state program controlling waterhyacinth with 2,4-D (Woods 1963). Dense mats of waterhyacinth were sprayed by an aircraft while a surface water craft operated in areas inaccessible to aerial spraying (narrow streams with tree canopies and near agricultural lands). Local interests were required to supply chemicals and other materials while the GFC provided labor and equipment (Tabita and Woods 1962). In 1955, the Florida Legislature appropriated $226,500 while the GFC added another $100,000 for a two-year waterhyacinth control program in Florida waters. The GFC and the Corps worked in tandem using 2,4-D and the newly available herbicide diquat along with an array of mechanical harvesters. At this time, local interests no longer had to provide control costs as long as the waterbody was open to the public. Consequently, this union slowed but did not halt waterhyacinth expansion in Florida.

In 1957, the Corps completed a study on the necessity of allowing the control of additional plant species not included in the Rivers and Harbors Act of 1899. The program, referred to as the Expanded Project for Aquatic Plant Control was approved by Congress in 1958. It approved a five-year pilot program to test the feasibility of control and progressive eradication of waterhyacinth, alligatorweed, and other objectionable aquatic plant species from the waters of eight states (U.S. Congress 1965). One of the provisions of the project required that states receiving federal revenues must provide 30 percent of the operating funds. In 1959, the GFC was designated as the official state agency to represent Florida in operations with the Corps on the Expanded Project (Tabita and Woods 1962). Due to fiscal restrictions and policies within the Corps, their funds were limited to the purpose of navigation. As the Expanded Project began in March, 1960, the Corps estimated 50,587 ha of waterhyacinth-infested Florida waters with a projected 70,822 ha possible without improved control techniques. It should be noted that an important finding of the project was that waterhyacinth eradication was not feasible, but management was possible if done on a consistent basis (U.S. Congress 1965).

At the beginning of the Expanded Project, the herbicide 2,4-D was sprayed on waterhyacinth populations from outboard or specially designed inboard motorboats. It was soon realized that, like the mechanical harvesters, these types of boats could not operate in extremely shallow waters where the alien plants multiplied and continued to re-infest lakes and rivers. A study confirmed that airboats could operate most efficiently and could traverse waterhyacinth mats, jams, and submersed obstructions even better than saw boats. Airboats emerged as the choice of conveyance for aquatic plant managers (Gallagher 1962).

## Hydrilla

By the early 1960s, a new threat was noticed in Florida's beleaguered waterways. Hydrilla was beginning to infest the flood control canals of southeastern Florida and there was little knowledge about how to control its fast spread. Initially, control measures were adapted from the management of the introduced elodea (*Egeria densa*) and native plant species. Mechanical devises included harvesters, draglines, and booms and chains that were dragged along the bottom of waterways to sheer plants loose. However, these control measures could not keep up with hydrilla expansion in large waterways. Additionally, harvesting proved to be too expensive for large-scale management efforts (Thayer and Ramey 1986).

Although there were some herbicides available, there was little knowledge about their effective use against submersed plants such as hydrilla. Some 800 herbicides were screened and evaluated to control hydrilla of which 37 warranted field evaluations (Blackburn and Weldon 1970). Unfortunately, few of these herbicides provided cost-effective control along with a safety margin for fish and wildlife. One of the herbicides that did provide reasonable control and environmental safety was copper sulfate mixed with diquat and endothall. In the late 1960s and early 1970s, chelated copper products, which did not precipitate in hard Florida waters, were used alone or in combination with diquat. During the 1980s, the low toxicity herbicide fluridone was tested and registered for aquatic use giving managers an effective method to control hydrilla on a large-scale while also being reasonably selective.

In 1970, the Florida Legislature passed the Aquatic Weed Control Act authorizing the Florida Department of Natural Resources (FDNR) to direct the control, eradication, and regulation of noxious aquatic weeds and to direct the research and planning related to said activities. In response, the Corps decreased their role in Florida aquatic plant management activities while FDNR provided funding to encourage local governments to establish their own aquatic plant control programs under the coordination of the FDNR. Also, the Legislature mandated that alien aquatic plant species be controlled under maintenance programs.

## Maintenance Control

The concept of maintenance control of alien vegetation in Florida's natural waterways was developed by the Corps upon completion of Operation Clean Sweep on the St. Johns River in 1973 (Buker 1982). Prior to this time, waterhyacinth was allowed to reach problem levels before control measures were implemented. These efforts resulted in the killing of large amounts of floating plant biomass which resulted in severe negative environmental impacts by detrital loading. It was realized that the only way to avoid severe environmental disturbances associated with such waterhyacinth control operations was to aggressively prevent these alien plants from attaining large population sizes. Joyce (1985) reported that maintaining waterhyacinth at 5 percent cover (or less) reduced herbicide usage by a factor up to 2.6, reduced organic sedimentation from live plants by a factor up to 4.0, reduced dissolved oxygen depressions, and even reduced the self-insulating nature of large waterhyacinth mats. Consequently, environmental impacts and economical costs could be kept to a minimum if the concept of maintenance control is employed.

Since its inception, the FDNR objective has been to achieve maintenance control of hydrilla, waterhyacinth, and waterlettuce using biological controls and better mechanical controls integrated with herbicides; control operations are evaluated through a permitting program for general environmental management oversight (Burkhalter 1972). The United States Department of Agriculture and the Corps began research to find biocontrol agents for waterhyacinth during the early 1960s and expanded to target waterlettuce. Host-specific insects and plant pathogens have been released in Florida to increase waterhyacinth leaf mortality, decrease plant size, and reduce overall population expansion (Goyer and Stark 1981, 1984; Center and Van 1989; Center et al. 1990; Grodowitz et al. 1991). However, these agents may take many months or years (Harley 1990) to reach densities that will reduce waterhyacinth populations. Therefore, because waterhyacinth grows so quickly, it must be managed with herbicides or harvesters in high-use waterbodies. Insects were recently released to control waterlettuce in

Florida show promise for long-term control (Center and Dray 1990; Dray et al. 1990; Dray and Center 1992), but they may not be able to provide rapid reductions of large floating mats.

The search for effective biological control agents to control hydrilla began in the late 1960s. Various hybrids and chromosome morphs of the Asian grass carp (*Ctenopharyngodon idella*) have been tested for hydrilla control since the early 1970s. Although grass carp have proven to be very effective in controlling hydrilla, the fish are not host-specific and consume native aquatic plant species, particularly when a hydrilla population declines (Van Dyke et al. 1984). This can be a positive management outcome for drainage canal managers but a problem for fishery and waterfowl managers who depend upon healthy native plant populations. Low stocking rates of grass carp along with applications of herbicides to significantly reduce hydrilla have been tested in Florida lakes to control hydrilla regrowth from tubers, but have led to inconsistent results. Consequently, the search continues for effective biological control methods which can reduce hydrilla plant populations while allowing native vegetation to flourish. Recently, several host-specific insects have been tested and released but their overall effect is inconclusive as of this date.

Despite maintenance control efforts, hydrilla continues to expand its range in Florida. Yet if detected early, before tubers are established, eradication is possible. Therefore, all of Florida's freshwater navigable rivers and lakes that are accessible to the public through boat ramps are inspected each year for new alien plant populations. Boat ramp inspections are emphasized as boat trailers seem to be the primary vector in the spread of hydrilla into public waterways. During 1990, a total of 19 pioneer hydrilla infestations were detected at boat ramps and eradication programs were begun immediately.

In addition to the Corps, FDNR contracts plant management activities with 18 state and local government entities to manage alien aquatic plant species in 325 public waterbodies which cover more than 404,000 ha of fresh water. More than 95 percent of the federal, state and local funds expended managing aquatic plants in Florida's waterways is for the control of hydrilla, waterhyacinth and waterlettuce. In 1992, approximately $7.2 million were available for this purpose.

While there are techniques available to manage invasive alien aquatic plant species, in most instances, government funding has not kept pace with management needs. Since 1986, there has been sufficient funding to address only the most urgent infestations. As a result, management priorities have been established based on plant type, degree of infestation, type and amount of public use of each waterbody, and the probability for successful management.

Control of floating alien plants remains the highest priority because of the rapid growth and mobility of these species. From 1980 through 1991, more than $43 million was spent managing waterhyacinth and waterlettuce in Florida public waters (Table 1). During this period, floating alien plant populations were reduced from a high of 9,600 ha (1983) to fewer than 2,400 ha in 1991 (Table 2). An average of 13,760 ha of floating alien plants have been controlled from 1988 through 1991 to cope with their rapid growth rates. Although floating alien plant populations occasionally grow out of control in some Florida locations, the overall statewide infestation was considered to be under maintenance control at the end of 1991.

From 1980 through 1991, more than $55 million was spent conducting hydrilla management (Table 1). Whereas an estimated $7.5 million is necessary for appropriate hydrilla control in Florida public waters per year, an average of only $3.9 million has been spent from 1988–1991. Hydrilla is quite difficult to manage in large lake systems. For example, after hydrilla rapidly spread in Lake

TABLE 1. Dollars spent to manage floating plants (waterhyacinth and waterlettuce) and hydrilla in Florida from 1980 through 1991.

| YEAR(S) | HYDRILLA | FLOATING PLANTS |
|---|---|---|
| 1980–88[a] | $43,572,000 | $35,668,000 |
| 1989[b] | 4,493,000 | 2,632,000 |
| 1990[b] | 4,142,000 | 2,016,000 |
| 1991[b] | 3,146,000 | 2,872,000 |
| Total | $55,352,000 | $43,188,000 |

[a] Estimated cost of operations in all waters except those exempt from permitting and reporting requirements.
[b] Dollars spent under state and federal funding programs in public waters.

Istokpoga in 1987 (Fig. 2), more than $1.4 million was spent on fluridone applications in 1988 and 1989 to return hydrilla to maintenance control levels. However, despite management efforts to control hydrilla in Lake Istokpoga, this invasive alien aquatic plant species increased again in 1989 and occupied more than 4,000 ha in 1991 (Fig. 4). As a result, hydrilla continues to expand in Florida and required control costs continue to spiral upward. In 1991, this species was established in 41 percent of Florida's public waters and infests some 26,960 ha (Table 2).

## SUMMARY

Large populations of invasive alien aquatic vegetation in Florida have been reported to substantially alter Florida's aquatic ecosystems. Specifically, changes in the physiochemical environment beneath hydrilla, waterhyacinth, and waterlettuce mats can lead to uniform water temperatures, cause lower pH and specific conductivity, increase dissolved carbon dioxide concentrations, lower dissolved oxygen and phosphorus concentrations, and decrease phytoplankton populations. Overall changes in state lake trophic classification based on water quality alone can be linked to large standing hydrilla populations. Both waterhyacinth and waterlettuce infestations increase the evapotranspiration losses and increase the detrital load in a waterbody. Heavy metal accumulation by waterhyacinth has been speculated to be a source of metal compounds to some species of wildlife.

The edges of waterhyacinth mats can support large numbers of invertebrates. The principle herbivores found within the root system of both waterhyacinth and waterlettuce populations are amphipods. Dense infestations of hydrilla in Florida lakes have been reported to decrease zooplankton abundance and increase the number of species. Epiphytic and benthic macroinvertebrate populations and species diversity have been reported to increase in hydrilla-dominated areas when low dissolved oxygen concentrations are not a factor. Oligocheaetes and chironomids have been reported to dominate the macroinvertebrate fauna in lakes infested with hydrilla. Fish kills and lower fish production has been linked to dense stands of waterhyacinths. Similarly, large populations of hydrilla can lead to reduced or stunted fish populations, especially when hydrilla infests more than 50 percent of a waterbody. However, hydrilla stands have been reported to provide a habitat for juvenile sport-fish. Waterhyacinth mats in Florida have

TABLE 2. Estimated hectares of Florida public waters infested with hydrilla, waterhyacinth and waterlettuce from 1982 through 1991. Estimated water surface area occupied is from annual surveys of public waterways conducted by the Florida Department of Natural Resources (for details on how these surveys were conducted, see Schardt and Schmitz 1990).

| | HECTARES INFESTED | | |
|---|---|---|---|
| YEAR | WATERHYACINTH | WATERLETTUCE | HYDRILLA |
| 1982 | 2,530 | 1,360 | 5,320 |
| 1983 | 7,610 | 1,990 | 16,150 |
| 1984 | 3,340 | 2,670 | 16,830 |
| 1985 | 1,810 | * | 19,740 |
| 1986 | 2,800 | 2,050 | 14,470 |
| 1987 | 2,360 | 920 | 14,650 |
| 1988 | 1,200 | 830 | 21,310 |
| 1989 | 880 | 840 | 16,840 |
| 1990 | 440 | 640 | 23,090 |
| 1991 | 890 | 1,460 | 26,960 |

* waterlettuce populations not surveyed

been reported to out-compete and destroy stands of native submersed and emergent vegetation. The growth habit of hydrilla has been reported to simulate native waterfowl shallow marsh habitats that are being rapidly lost in Florida.

Waterhyacinth control in Florida has been attempted since 1899. Throughout the first fifty years of this century, a number of control methods have been employed including passive barriers, crusher-boats, saw-boats, harvesters, and applications of unsafe chemicals. It was not until the early 1950s, when the herbicide 2,4-D was used operationally to control waterhyacinths, that a limited state program began. The Expanded Project for Aquatic Plant Control that began in 1960 provided a five-year pilot program to test the feasibility of control and progressive eradication of waterhyacinth, alligatorweed, and other objectionable aquatic plant species from the waters of eight states. An important finding of this project was that waterhyacinth eradication was not feasible, but management was possible if done on a consistent basis.

By the early 1960s, hydrilla began to infest the flood control canals of southeastern Florida. Initial hydrilla control efforts consisted of methods previously used to manage other submersed aquatic plant species, but were mostly unsuccessful in large waterbodies. Herbicides that provided reasonable hydrilla control and environmental safety were copper sulfate mixed with diquat and endothall. During the 1980s, the low-toxicity herbicide fluridone was found to be reasonably selective in removing hydrilla and provided large-scale control.

The concept of maintenance control was developed in the early 1970s after it was realized that the only way to avoid serious environmental disturbances associated with waterhyacinth control operations was to aggressively prevent them from attaining large population sizes. Presently, a state-wide program using biological controls, better mechanical controls, and integrated with herbicides has resulted in the maintenance control of waterhyacinth and watterlettuce. However, despite maintenance control efforts, hydrilla continues to expand its range in Florida and occupied some 26,960 ha of public waters in 1991. From 1980 through 1991, more than $43 million and $55 million were spent managing floating alien vegetation and hydrilla, respectively, in Florida.

## REFERENCES

Abu-Gideiri, Y. B. and A. M. Yousif. 1974. The influence of *Eichhornia crassipes* Solms. on planktonic development in the White Nile. Arch. Hydrobiologia 74:463–67.

Achmad, S. 1971. Problems and control of aquatic weeds in Indonesian open waters. Pages 107–13 *in* Tropical weeds: some problems, biology and control. M. Soerjani (ed.). Proc. First Indonesian Weed Sci. Conf., Biotrop. Bull. 2. SEAMEO-BIOTROP, Bogor.

Agami, M. and K. R. Reddy. 1990. Competition for space between *Eichhornia crassipes* (Mart). Solms and *Pistia stratiotes* L. cultured in nutrient-enriched water. Aquatic Bot. 38:195–208.

Anonymous. 1971. Economic damage caused by aquatic weeds. Tech. Rpt. TA/OST/71-5, Office of Science and Technology, U. S. Agency for International Development, Washington, D.C. 13 p.

Ashton, P. J., W. E. Scott and D. J. Steyn. 1979. The chemical control programme against the water hyacinth (*Eichhornia crassipes* (Mart.) Solms) on Hartbeespoort dam: historical and practical aspects. South African J. Sci. 75:303–06.

Attionu, R. H. 1976. Some effects of water lettuce (*Pistia stratiotes* L.) on its habitat. Hydrobiologia 50(3):245–54.

Bartodziej, W. In press. Epiphytic invertebrate populations on *Hydrilla verticillata* and *Egeria densa* versus native submersed macrophytes. 2nd Annual Proc. Florida Lake Manage. Soc.

Beumer, J. P. 1980. Hydrology and fish diversity of a North Queensland tropical stream. Austral. J. Ecol. 5:159–86.

Blackburn, R. D. and L. W. Weldon. 1970. Control of *Hydrilla verticillata*. Hyacinth Contr. J. 8:4–9.

Blake, N. M. 1980. Land into water—water into land: a history of water management in Florida. Univ. Presses of Florida, Tallahassee.

Blancher, E. C. 1979. Lake Conway, Florida: nutrient dynamics, trophic state, zooplankton relationships. Ph. D. Thesis. Univ. of Florida, Gainesville.

Bond, W. J. and M. G. Roberts. 1978. The colonization of Cabora Bassa Mozambique, a new man-made lake, by floating aquatic macrophytes. Hydrobiologia 60:243–59.

Bowes, G., T. K. Van, L. A. Garrard et al. 1977. Adaptation to low light levels by hydrilla. J. Aquatic Plant Manage. 15:32–35.

Bowes, G. A., S. Holaday and W. T. Haller. 1979. Seasonal variation in the biomass, tuber density, and photosynthetic metabolism of hydrilla in three Florida lakes. J. Aquatic Plant Manage. 15:61–65.

Bua-ngam, T. and B. L. Mercado. 1976. Competition of water lettuce (*Pistia stratiotes*) with rice and commonly associated weed species. Philippine Agric. 60:22–30.

Buker, G. E. 1982. Engineers vs. Florida's green menace. Florida Hist. Quart., April, p. 413–27.

Burkhalter, A. P. 1972. Florida Department of Natural Resources policies and plans as related to aquatic weed control and research. Hyacinth Contr. J. 10:2–4.

Canfield, D. E., Jr., K. A. Langeland, M. J. Maceina, et al. 1983a. Trophic state classifications of lakes with aquatic macrophytes. Can. J. Fish. Aquatic Sci. 40:1713–18.

Canfield, D. E., Jr., M. J. Maceina and J. V. Shireman. 1983b. Effects of hydrilla and grass carp on water quality in a Florida lake. Water Res. Bull. 19(5):773–78.

Center, T. D., A. F. Cofrancesco and J. K. Balciunas. 1990. Biological control of aquatic and wetland weeds in the southeastern United States. Pages 239-62 *in* Delfosse, E. S. (ed.). Proc., VII Int. Symp. Biol. Contr. Weeds, 6–11 March 1988, Rome, Italy, 1st. Sper. Patol. Veg. (MAF), Rome, Italy.

Center, T. D. and F. A. Dray, Jr. 1990. Release, establishment, and evaluation of insect biocontrol agents for aquatic plant control. Pages 39–49 *in* Proc., 24th Ann. Meeting Aquat. Plant Contr. Res. Prog., 13–16 November 1989, Huntsville, Alabama. U. S. Army Corps of Engineers, Vicksburg, Mississippi.

Center, T. D. and N. R. Spencer. 1981. The phenology and growth of water hyacinth (*Eichhornia crassipes* (Mart.) Solms) in a eutrophic north-central Florida lake. Aquat. Bot. 10:1–32.

Center, T. D. and T. K. Van. 1989. Alteration of waterhyacinth (*Eichhornia crassipes* (Mart.) Solms) leaf dynamics and phyto-chemistry by insect damage and plant density. Aquat. Bot. 35:181–95.

Chadwick, M. J. and M. Obeid. 1966. A comparison of the growth of *Eichhornia crassipes* Solms. and *Pistia stratiotes* L. in water culture. J. Ecol. 54:563–75.

Chamberlain, E. B. 1960. Florida waterfowl populations habitat and management. Florida Game and Freshwater Fish Commission Tech. Bull., no. 7, Tallahassee.

Chesnut, T. L. and E. H. Barman, Jr. 1974. Aquatic vascular plants of Lake Apopka, Florida. Florida Sci. 37:60–64.

Colle, D. E. 1982. Hydrilla—miracle or migraine for Florida's sportfish. Aquatics 4:6.

Colle, D. E. and J. V. Shireman. 1980. Coefficients of condition for largemouth bass, bluegill, and redear sunfish in hydrilla-infested lakes. Trans. Amer. Fish. Soc. 109:521–31.

Colle, D. E., J. V. Shireman, W. T. Haller, et al. 1987. Influence of hydrilla on harvestable sport-

fish populations, angler use, and angler expenditures at Orange Lake, Florida. North Amer. J. Fish. Manage. 7:410–17.
Cook, D. K. and R. Luond. 1982. A revision of the genus *Hydrilla* (Hydrocharitaceae). Aquatic Bot. 13:485–504.
Cornwell, D. A., J. Zoltec, Jr., C. D. Patrinely, T. et al. 1977. Nutrient removal by waterhyacinth. J. Water Poll. Contr. Fed. 49:57–65.
Coulson, J. R. 1977. Biological control of alligatorweed, 1959–1972. A review and evaluation. USDA Agricultural Res. Serv., Tech. Bull. No. 1547, 98 p.
Datta, S. C. and K. D. Biswas. 1969. Physiology of germination in *Pistia stratiotes*. Biologia, Bratislava. 24:1.
DeWald, L. B. and L. P. Lounibos. 1990. Seasonal growth of *Pistia stratiotes* seeds. Biologia Bratislava 24(1):70–79.
Dorofeev, P. I. 1955. Ob ostatkach rastenij iz treticnych otloz enij v. rajone s. Novonikolskogo na Irtyse v Zapadnoj Sibiri, Doki. Akad. nauk SSSR. 101:941–44.
Dorofeev, P. I. 1958. Novye dannye ob oligocenovoj flore d. Belojarki na r, Tavde v Zapadnoj Sibiri. Doki. Akad. nauk SSSR. 123:543–45.
Dorofeev, P. I. 1963. Treticnye fiory zapadnoj Sibiri Izd. Akad. nauk SSR. Moskva and Leningrad.
Dray, F. A., Jr. and T. D. Center. 1989. Seed production of *Pistia stratiotes* L. (water lettuce) in the United States. Aquat. Bot. 33:155–60.
Dray, F. A., Jr. and T. D. Center. 1992. Waterlettuce biocontrol agents: releases of *Namangana pectinicornis* and dispersal by *Neohydronomus affinis*. Pages 237–43 *in* Proc., 26th Ann. Meeting Aquat. Plant Contr. Res. Prog., 18–22 November 1991, Dallas, Texas. U. S. Army Corps of Engineers, Vicksburg, Mississippi.
Dray, F. A., Jr., C. R. Thompson, D. H. Habeck, et al. 1988. A survey of the fauna associated with *Pistia stratiotes* L. (waterlettuce) in the United States. USACE Technical Report A-88-6, Vicksburg, Ms.
Dray, F. A., Jr., T. D. Center, D. H. Habeck, et al. 1990. Release and establishment in the southeastern U. S. of *Neohydronomus affinis* (Coleoptera: Curculionidae), an herbivore of waterlettuce (*Pistia stratiotes*). Environ. Entomol. 19(3):799–803.
Duke, R. W. and R. H. Chabreck. 1976. Waterfowl habitat in lakes of the Atchafalaya Basin, Louisiana. Proc. Southeast. Assoc. Game Fish Comm. 29:501–12.
Fernald, E. A. and Patton, D. J., (eds.). 1984. Water Resources Atlas of Florida. Inst. Sci. Public Aff., Florida State Univ., Tallahassee.
Florschutz, O. 1972. The importance of Eurasian milfoil (*Myriophyllum spicatum*) as a waterfowl food. Proc. Southeast Assoc. Game Fish Comm. 26:189–94.
Friis, E. M. 1985. Angiosperm fruits and seeds from the middle Miocene of Jutland (Denmark), Det. Kongellge Danske Videnskaberne Selskab Biologiske Skriffer 24:3.
Gallagher, J. E. 1962. Waterhyacinth control with amitrol-T. Hyacinth Contr. J. 1:17–18.
Gopal, B. 1987. Waterhyacinth. Elsevier Science Publishers. Amsterdam, The Netherlands. 471 p.
Goyer, R. A. and J. D. Stark. 1981. Suppressing waterhyacinth with an imported weevil. Ornamental South 3(6):21–22.
Goyer, R. A. and J. D. Stark. 1984. The impact of *Neochetina eichhorniae* on waterhyacinth in southern Louisiana. J. Aquat. Plant Manage. 22:57–61.
Griffen, J. 1989. The Everglades kite. Aquatics 11(3):17–19.
Grodowitz, M. J., R. M. Stewart and A. F. Cofrancesco. 1991. Population dynamics of waterhyacinth and the biological control agent *Neochetina eichhorniae* (Coleoptera: Curculionidae) at a southeast Texas location. Environ. Entomol. 20(2):652–60.
Hall, J. B. and D. U. U. Okali. 1974. Phenology and productivity of *Pistia stratiotes* L. on the Volta Lake, Ghanna. J. Appl. Ecol. 11:709–26.
Hall, D. O. 1980. Biological and agricultural systems: an over-view. Pages 1–30 *in* Biochemical and photo-synthetic aspects of energy production. A. San Pietro (ed.). Academic Press, N. Y.
Haller, W. T. 1978. Hydrilla: a new and rapidly spreading aquatic weed problem. Univ. Florida, Gainesville, Circular S-245.
Hansen, K. L., E. G. Ruby and R. L. Thompson. 1971. Trophic relationships in the water hyacinth community. Quart. J. Florida Acad. Sci. 34:107–13.
Harley, K. L. S. 1990. The role of biological control in the management of water hyacinth, *Eichhornia crassipes*. Biocontrol News and Information 11(1):11–22.
Hasler, A. D. and E. Jones. 1949. Demonstration of the antagonistic action of large aquatic plants on algae and rotifers. Ecology 30:359–64.
Hinkle, J. 1986. A preliminary literature review on vegetation and fisheries with emphasis on the largemouth bass, bluegill, and hydrilla. Aquatics 8:9–14.
Holm, L. G., D. L. Plucknett, J. V. Pancho et al. 1977. The world's worst weeds: distribution and biology. University Press of Hawaii, Honolulu.
Johnson, F. A. 1987. Lake Okeechobee's waterfowl habitat: problems and possibilities. Aquatics 9 (2):20–21.

JOHNSON, F. A. and F. MONTALBANO. 1987. Considering waterfowl habitat in hydrilla control policies. Wildl. Soc. Bull. 15:466–69.

JOYCE, J. C. 1985. Benefits of maintenance control of waterhyacinth. Aquatics 7:11–13.

KAY, S. H. and W. T. HALLER. 1986. Heavy metal bioaccumulation and effects on waterhyacinth weevils, *Neochetina eichhorniae*, feeding on waterhyacinth, *Eichhornia crassipes*. Bull. Environ. Contam. Toxicol. 37:239-45.

KAY, S. H., W. T. HALLER and L. A. GARRARD. 1984. Effects of heavy metals on water hyacinths [*Eichhornia crassipes* (mart.) Solms]. Aquat. Toxicol. 5:117–28.

KERWIN, J. A. and L. G. WEBB. 1971. Foods of ducks wintering in coastal South Carolina, 1965–1967. Proc. Southeast Assoc. Game Fish Comm. 25:223–45.

KRAPU, G. L. 1974. Foods of breeding pintails in North Dakota. J. Wildl. Manage. 38:408–17.

KULSHRESHTHA, M. and B. GOPAL. 1983. Allelopathic influence of *Hydrilla verticillata* (L. F.) Royle on the distribution of *Ceratophyllum* species. Aquatic Bot. 16:207–09.

LESLIE, A. J., J. M. VAN DYKE, R. S. HESTAND et al. 1987. Management of aquatic plants in multi-use lakes with grass carp (*Ctenopharyngodon idella*). Lake and Reservoir Manage. 3:266–76.

LYNCH, J. J., J. E. KING, T. K. CHAMBERLAIN et al. 1957. Effects of aquatic weed infestations on the fish and wildlife of the Gulf states. Spec. Sci. Rep., U. S. Fish Wildl. Serv. 39:1–77.

MACEINA, M. J. and J. V. SHIREMAN. 1985. Influence of dense hydrilla infestation on black crappie growth. Proc. Ann. Conf. Southeast. Assoc. Fish and Wildlife Agencies. 36:394–402.

MARTIN, R. G. and J. V. SHIREMAN. 1976. A quantitative sampling method for hydrilla-inhabiting macroinvertebrates. J. Aquatic Plant Manage. 14:16–19.

MCVEA, C. and C. E. BOYD. 1975. Effects of waterhyacinth cover on water chemistry, phytoplankton, and fish in ponds. J. Environ. Qual. 4(3):375–78.

MILEY, W. W., A. J. LESLIE and J. M. VAN DYKE. 1979. The effects of the grass carp (*Ctenopharyngodon idella* Val.) on vegetation and water quality in three central Florida lakes. Florida Dept. of Natural Resources Technical Report, Tallahassee, 119 p.

MITCHELL, D. S. 1974. The development of excessive populations of aquatic plants. Pages 38–49 *in* Mitchell, D. S. (ed.), Aquatic vegetation and its use and control. UNESCO, New York, N.Y.

MITCHELL, D. S. 1976. The growth and management of *Eichhornia crassipes* and *Salvinia* spp. in their native environment and in alien situations. Pages 167–75 *in* Aquatic Weeds in South East Asia. C. K. Varshney and J. Rzoska (ed).

MONTEGUT, R. S., R. D. GASAWAY, D. F. DURANT et al. 1976. An ecological evaluation of the effects of grass carp (*Ctenopharyngodon idella*) introduction in Lake Wales, Florida. Florida Game and Freshwater Fish Commission Report, Tallahassee, 106 p.

MURAMOTO, S. and Y. OKI. 1983. Removal of some heavy metals from polluted water by water hyacinth (*Eichhornia crassipes*). Bull. Env. Cont. Tox. 39:170–77.

NATIONAL ACADEMY OF SCIENCES. 1976. Making aquatic weeds useful: some perspectives for developing countries. National Academy of Sciences, Report of an Ad Hoc Panel of the Advisory Committee on Technology Innovation, Board on Science and Technology for International Development, and the Commission on International Relations. Washington, D.C. 175 p.

O'HARA, J. 1967. Invertebrates found in water hyacinth mats. Quart. J. Florida Acad. Sci. 30:73–80.

O'KEEFFE, D. H., T. E. WIESE and M. R. BENJAMIN. 1987. Effects of positional isomerism on the uptake of monosubstituted phenols by the water hyacinth. Pages 505–12 *in* Aquatic plants for water treatment and resource recovery, Reddy, K. R. and W. H. Smith (eds.). Magnolia Pub., Orlando, Florida.

OSBORNE, J. A., D. I. RICHARD and J. W. SMALL, JR. 1983. Environmental effect and vegetation control by grass carp (*Ctenopharyngodon idella* Val.) and herbicides in four Florida lakes. Final Report, Florida Department of Natural Resources, Tallahassee, 431 p.

PENFOUND, W. T. and T. T. EARLE. 1948. The biology of the water hyacinth. Ecol. Monogr. 18:448–72.

PENNAK, R. W. 1966. Structure of zooplankton populations in the littoral macrophyte zone of some Colorado lakes. Trans. Amer. Microsc. Soc. 85:329–49.

RAI, D. N. and J. DATTA MUNSHI. 1979. The influence of thick floating vegetation (water hyacinth: *Eichhornia crassipes*) on the physico-chemical environment of a fresh water wetland. Hydrobiologia 62:65–69.

RAYNES, J. J. 1964. Aquatic plant control. Hyacinth Contr. J. 3:2–4.

REDDY, K. R. and L. O. BAGNALL. 1981. Biomass production of aquatic plants used in agricultural drainage water treatment. Pages 376–90 *in* 1981 Proc. of the International Gas Res. Conf.

REDDY, K. R. and P. D. SACCO. 1981. Decomposition of waterhyacinth in agricultural drainage water. J. Environ. Qual. 10:228–34.

REDDY, K. R. and J. C. TUCKER. 1983. Effect of nitrogen source on productivity and nutrient uptake of waterhyacinth (*Eichhornia crassipes*). Econ. Bot. 37:236–46.

REDDY, K. R., P. D. SACCO, D. A. GRAETZ, et al. 1983. Effect of aquatic macrophytes on physico-chemical parameters of agricultural drainage water. J. Aquat. Plant Manage. 21:1–7.

RICHARD, D. I., J. W. SMALL and J. A. OSBORNE. 1985. Response of zooplankton to the reduction and elimination of submerged vegetation by grass carp and herbicide in four Florida lakes. Hydrobiologia 123:97–108.

ROBACK, S. S. 1974. Insects (Arthropoda: Insecta). Pages 313–76 *in* Pollution ecology of freshwater invertebrates. C. W. Hart, Jr. and S. L. H. Fuller (eds.), Academic Press, New York, N.Y.

ROGERS, H. H. and D. E. DAVIS. 1972. Nutrient removal by water-hyacinth. Weed Sci. 20:423–28.

RZOSKA, J., A. J. BROOK and G. A. PROWSE. 1955. Seasonal plankton development in the White Nile and Blue Nile near Khartoum. Verh. Int. Verein. Limnol. 12:327–34.

SCHARDT, J. D. and D. C. SCHMITZ. 1990. 1989 Florida Aquatic Plant Survey. Florida Department of Natural Resources, Tallahassee, Technical Report 90-CGA, 67 p.

SCHARDT, J. D. and D. C. SCHMITZ. 1991. 1990 Florida Aquatic Plant Survey, the exotic aquatic plants. Florida Department of Natural Resources, Tallahassee, Technical Report 91-CGA, 89 p.

SCHMITZ, D. C., B. V. NELSON, L. E. NALL, et al. 1991 Chapter 21: Exotic aquatic plants in Florida: a historical perspective and review of the present aquatic plant regulation program. Pages 303-27 *in* Proc. Symp. Exotic Pest Plants, held on Nov. 2–4, 1988, Miami, Florida, NPS/NREVER/NRTR-91/06, National Park Service, Denver, Colo.

SCHMITZ, D.C. and J. A. OSBORNE. 1984. Zooplankton densities in a hydrilla infested lake. Hydrobiologia 111:127–32.

SCHRAMM, H. L., JR., K. J. JIRKA and M. V. HOYER. 1987. Epiphytic macroinvertebrates on dominated macrophytes in two central Florida lakes. J. Freshwater Ecol. 4(2):151–61.

SCHRAMM, H. L., JR., and K. J. JIRKA. 1989. Effects of aquatic on benthic macroinvertebrates in two Florida lakes. J. Freshwater Ecol. 5(1):1–12.

SCHREINER, S. P. 1980. Effects of waterhyacinth on the physico-chemistry of a south Georgia pond. J. Aquat. Plant Manage. 18:9–12.

SCOTT, S. L. and J. A. OSBORNE. 1981. Benthic macroinvertebrates of a hydrilla infested central Florida lake. J. Freshwater Ecol. 1(1):41–49.

SCOTT, W. E., P. J. ASHTON and D. J. STEYN. 1979. The chemical control of water hyacinth on Hartbeespoort Dam. Water Research Commission, Pretoria. 84 p.

SCULTHORPE, C. D. 1967. The biology of aquatic vascular plants. London: Edward Arnold Publ. Ltd. 1985 reprint, Konigstein, West Germany: Koeltz Scientific Books, 610 p.

SHARMA, B. M. 1984. Ecophysiological studies on water lettuce in a polluted lake. J. Aquatic Plant Manage. 22:19–21.

SHIREMAN, J. V., W. T. HALLER, D. E. COLLE et al. 1983a. Effects of aquatic macrophytes on native sportfish populations in Florida. Pages 208–14 *in* Proc. Int. Symp. Aquatic Macrophytes, September 18–23, 1983, Nijmegen, The Netherlands.

SHIREMAM, J. V., W. T. HALLER, D. E. COLLE, et al. 1983b. The ecological impact of integrated chemical and biological aquatic weed control. Final Report. Submitted to United States Environmental Protection Agency, Environmental Research Laboratory Sabine Island, Gulf Breeze, Florida, 333 p.

SHIREMAN, J. V. and M. J. MACEINA. 1981. The utilization of grass carp, *Ctenopharyngodon idella* Val., for hydrilla control in Lake Baldwin, Florida. J. Fish Biol. 19:629–36.

SRIDHAR, M. K. C. 1986. Trace element composition of *Pistia stratiotes* L. in a polluted lake in Nigeria. Hydrobiologia 131:273–76.

STEWARD, K. K. 1984. Growth of hydrilla (*Hydrilla verticillata*) in hydrosoils of different composition. Weed Sci. 32:371–75.

STEWARD, K. K., T. K. VAN, V. CARTER et al. 1984. Hydrilla invades Washington, D.C. and the Potomac. Amer. J. Bot. 7(1):162–63.

STODDARD, A. A., III. 1989. The phytogeography and paleofloristics of *Pistia stratiotes* L. Aquatics 11:21–24.

STUCKEY, R. L. and D. H. LES. 1984. *Pistia stratiotes* (waterlettuce) recorded from Florida in Bartrams's Travels, 1765–74. Aquaphyte 4(2):6.

SUTTON, D. L. 1983. Aquatic plant competition. Aquatics 5(10):12–14.

SUTTON, D. L. 1986. Growth of hydrilla in established stands of spikerush and slender arrowhead. J. Aquatic Plant Manage. 24:16–20.

SUTTON, D. L. and K. M. PORTIER. 1985. Density of tubers and turions of hydrilla in south Florida. J. Aquatic Plant Manage. 23:64–67.

TABITA, A. and J. W. WOODS. 1962. History of waterhyacinth control in Florida. Hyacinth Contr. J. 1:19–23.

TAYLOR, J. N., W. R. COURTENAY, JR. and J. A. MCCANN. 1984. Known impacts of exotic fishes in the continental United States. Pages 332–73 *in* Distribution, biology, and management of exotic fishes, W. R. Courtenay, Jr. and J. R. Stauffer, Jr. (ed.). The John Hopkins Univ. Press, Baltimore, Md.

TEBEAU, C. W. 1980. A history of Florida. Univ. Miami Press, Coral Gables.

THAYER, D. and V. RAMEY. 1986. Mechanical harvesting of aquatic weeds: literature review. Florida Dept. of Natural Resources Tech. Report, Tallahassee, 21 p.

TIMMER, C. E. and L. W. WELDON. 1967. Evapotranspiration and pollution of water by water hyacinth. Hyacinth Contr. J. 6:34–37.

TUCKER, C. S. and T. A. DEBUSK. 1981. Productivity and nutritive value of *Pistia stratiotes* and *Eichhornia.* J. Aquatic Plant Manage. 19:61–63.

ULTSCH, G. R. 1973. The effects of waterhyacinths (*Eichhornia crassipes*) on the microenvironment of aquatic communities. Arch. Hydrobiol. 12(4):460–73.

UNITED STATES CONGRESS. 1957. Water hyacinth obstructions in the waters of the gulf and south Atlantic states: letter from the Secretary of the Army. 35th Congress, Document no. 37.

UNITED STATES CONGRESS. 1965. Expanded project for aquatic plant control: letter from the Secretary of the Army. 89th Congress, Document no. 251.

VAN DIJIK, G. 1985. *Vallisneria* and its interactions with other species. Aquatics 7(3):6–10.

VAN DYKE, J. M., A. J. LESLIE, JR. and L. E. NALL. 1984. The effects of grass carp on the aquatic macrophytes of four Florida lakes. J. Aquatic Plant Manage. 22:87–89.

VITOUSEK, P. M. 1986. Biological invasions and ecosystem properties: can species make a difference? Pages 163–76 *in* Ecology of biological invasions of North America and Hawaii, H. A. Mooney and J. A. Drake (ed.). Springer-Verlag, New York, N.Y.

WARE, F. J. and R. D. GASAWAY. 1976. Effects of grass carp on native fish populations in two Florida lakes. Proc. Southeast. Assoc. Game Fish Comm. 30:324–34.

WATKINS, C. E. II, J. V. SHIREMAN and W. T. HALLER. 1983. The influence of aquatic vegetation upon zooplankton and benthic macroinveretebrates in Orange Lake, Florida. J. Aquatic Plant Manage. 21:78–83.

WILL, L. E. 1965. Okeechobee boats and skippers. Great Outdoors Publishing Co.

WOLVERTON, B. C. and R. C. MCDONALD. 1979. Waterhyacinth (*Eichhornia crassipes*) productivity and harvesting studies. Econ. Botany 33:1–10.

WOODS, J. W. 1963. Aerial application of herbicides. Hyacinth Contr. J. 2:20.

WUNDERLICH, W. E. 1964. Waterhyacinth control in Louisiana. Hyacinth Contr. J. 3:4–7.

WUNDERLICH, W. E. 1967. The use of machinery in the control of aquatic vegetation. Hyacinth Contr. J. 6:22–24.

YEO, R. R., R. H. FALK and J. R. THURSTON. 1981. The morphology of hydrilla (*Hydrilla verticillata* (L. f.) Royle). J. Aquat. Plant Manage. 22:1–17.

ZEIGER, C. F. 1962. Hyacinth-obstruction to navigation. Hyacinth Contr. J. 1:16–17.

# How Illinois Kicked the Exotic Habit

Francis M. Harty[1]

## INTRODUCTION

For the purpose of this paper, an exotic species is defined as "a plant or animal not native to North America." The history of folly surrounding the premeditated and accidental introduction of exotic animals has been well-documented (De Vos et al. 1956; Elton 1958; Hall 1963; Laycock 1966; Ehrenfeld 1970; Bratton 1974, 1975; Howe and Bratton 1976; Moyle 1976; Courtenay 1978; Coblentz 1978; Iverson 1978; Weller 1981; Bratton 1982; Vale 1982; Savidge 1987).

In 1963 Dr. E. Raymond Hall wrote, "Introducing exotic species of vertebrates is unscientific, economically wasteful, politically shortsighted, and biologically wrong." Naturalizing exotic species are living time bombs, but no one knows for sure how much time we have. For example, the ring-necked pheasant (*Phasianus colchicus*), touted as the Midwestern example of a good exotic introduction, has recently developed a nefarious relationship with the greater prairie chicken (*Tympanuchus cupido*) in Illinois (Fig. 1). Parasitism of prairie chicken nests by hen pheasants and harassment of displaying male chickens by cock pheasants are contributing to the decline of prairie chickens in Illinois (Vance and Westemeier 1979). The interspecific competition between the exotic pheasant (which is expanding its range in Illinois) and the native prairie chicken (which is an endangered species in Illinois) may be the final factor causing the extirpation of the prairie chicken from Illinois; it has already been extirpated from neighboring Indiana.

In 1953 Klimstra and Hankla wrote, "In connection with the development of a pheasant adapted to southern conditions, the compatability of pheasants and quail (*Colinus virginianus*) needs to be evaluated. It would be unwise to establish a game bird that would compete with another desirable species." It has been recently discovered that ring-necked pheasants also parasitize quail nests (Westemeier et al. 1989). Management at Illinois' prairie chicken preserves now includes pheasant control 12 months of the year.

Michigan is apparently not moved by the potential threats that exotic animals pose (Huggler 1991). They recently released the Sichuan or "brush" pheasant (*Phasianus colchicus strauchi*) into the wild, and Indiana officials are considering the same move. The preferred habitat of the Sichuan pheasant and the ruffed grouse (*Bonasa umbellus*) overlap. Consequently if the "brush" pheasant becomes fully naturalized, it may become a serious competitor to the native ruffed grouse.

The raccoon dog (*Nytereutes procynoides*), a member of the dog family, apparently is native to Asia (Fig. 2). Between 1927 and 1957 they were introduced to Europe both intentionally and accidentally as escapees from fur farms. As a result of their high reproduction rate, omnivorous feeding habits, and general lack of enemies, they have spread rapidly throughout northern and western Europe. Later, like the nutria (*Myocuster coypus*), they were brought to the United

1. Division of Natural Heritage, Illinois Department of Conservation, 8 Henson Pl., Champaign IL 61820

FIGURE 1.—Ring-necked pheasant harassing greater prairie chicken on booming ground in Marion County, Illinois. (photo by Richard Day).

States as fur-bearing stock. Raccoon dogs prefer forested, riparian habitats, marshes, and dense cover surrounding lakes. They are opportunistic, eating a wide variety of plants and animals, including eggs, fish and carrion. They are the only canid that hibernates, but they are not "deep sleepers." On warm days they forage and in the southern parts of their range they do not hibernate at all (Ward and Wurster-Hill 1990). If raccoon dogs were to escape captivity and become fully naturalized, they have all the characteristics to become a serious predator of many native species of wildlife.

On July 23, 1983, it became illegal to possess, propagate, or release a raccoon dog in Illinois. But three separate fur farms had purchased raccoon dogs before the law was passed. On July 11, 1984, the Illinois Department of Conservation paid $41,000 to buy the raccoon dogs from those fur farmers.

The rusty crayfish (*Orconectus rusticus*), originally sold as fish bait, is an aggressive exotic species that replaces native crayfish (Page 1985). Four species of crayfish, are currently listed on Illinois' endangered species list. The establishment of rusty crayfish in Illinois would be a serious threat to these native species. Consequently, as of June 29, 1990, the importation, possession, and sale of live rusty crayfish was banned in Illinois.

## THE ISSUE OF EXOTIC PLANT MATERIALS

Dr. Hall was right about the dangers of introducing exotic vertebrates, and his analysis applies to the introduction of exotic plant material as well (Reed 1977). Unfortunately, less is appreciated about the tremendous damage that is occurring to our continent's ecosystems due to the escape and naturalization of exotic plants (Bratton 1982; Harty 1986; Mooney and Drake 1986). In Illinois, as elsewhere, the perceived merits versus the perceived impacts associated with introducing exotic plant species is argued as a matter of philosophy among wildlife biologists, soil conservationists, foresters, landscapers, and ecologists. However, evidence is mounting to indicate that the introduction of exotic plant species has resulted in major ecological damage and caused serious management problems.

FIGURE 2.—Raccoon dog (*Nyctereutes procyonoides*); photo by M. Cukierski, with permission of the American Society Mammalogists Mammal Slide Library.

Multiflora rose (*Rosa multiflora*) is the classic Midwestern example of an exotic species run amuck, aggressively overgrowing pastures and abandoned farm ground. It was originally promoted in the 1940s for use as a living fence, erosion control, and wildlife food and cover, with the added assurance during its initial promotion that it would not spread or become a nuisance. These claims seem naive in retrospect. Nevertheless, variations of the same scenario have been used to promote autumn olive (*Elaeagnus umbellata*), bush honeysuckle (*Lonicera tatarica*), amur honeysuckle (*L. maackii*), and many other exotic species. Klimstra (1956) was one of the first to point out the problems associated with the widespread planting of multiflora rose; moreover, he questioned the *real* versus the *perceived* value of multiflora rose for wildlife habitat planting.

In Illinois 811 species, or 29 percent, of the state's flora are naturalized from foreign countries (Henry and Scott 1980). Not all these species can be classified as problem species today, but Tartarian honeysuckle, amur honeysuckle, tree-of-heaven (*Ailanthus altissima*), thistle (*Carduus nutans*), Canada thistle (*Cirsium arvense*), crown vetch (*Coronilla varia*), giant teasel (*Dipsacus laciniatus*), European beach grass (*Elymus arenarius*), tall fescue (*Festuca pratensis*), sericea lespedeza (*Lespedeza cuneata*), multiflora rose, Japanese honeysuckle (*Lonicera japonica*), purple loosestrife (*Lythrum salicaria*), white poplar (*Populus alba*), smooth and shining buckthorn (*Rhamnus cathartica* and *R. frangula*), and Johnson grass (*Sorghum halepense*), are just a few examples of exotic plant introductions causing farmers, foresters, land-managers and grounds-keepers considerable problems in various regions of the state (West 1984; Schwegman 1988).

Moreover, autumn olive, osage orange (*Maclura pomifera*), and winged-euonymus (*Euonymus alatus*), three of the long-term "neutrals" in the game of exotic roulette, have now adapted sufficiently to Illinois' conditions that they, too, are becoming naturalized weeds, spreading from plantings into the landscape (Nyboer and Ebinger 1978; Ebinger and Lehnen 1981; Ebinger et al. 1984; Nestleroad et al. 1987). Ebinger (1983) summarizes the problems that naturalized exotic shrubs (multiflora rose, Japanese honeysuckle, autumn olive, winged-euonymus, and blunt-leaved privet [*Ligustrum obtusifolium*]) are causing managers of natural areas in Illinois. Additionally, climbing euonymous (*Euonymous fortunei*) and oriental bittersweet (*Celastrus orbiculatus*), two popular ornamental vines, are becoming invaders in southern Illinois forests (Schwegman 1991, pers. comm.).

## WHY DID WE PLANT EXOTICS IN THE FIRST PLACE?

The common refrain associated with the promotion of exotic plant materials is that "they are hardy, disease free, have few if any insect pests, reproduce or propagate easily, and provide food or cover for wildlife." These characteristics are precisely what makes exotic species such serious competitors when released into a new ecosystem or habitat.

Two recent examples of this scenario are the release of "Elsmo" lace-bark elm (*Ulmus parvifolia* Jacq.) and "Redstone" Cornelian cherry dogwood (*Cornus mas* L.) by the Soil Conservation Service (Plants for Conservation 3(1) January 1991 and Plants for Conservation 3(2) July 1991). Siberian elm (*Ulmus pumila*) was promoted to replace American elm (*Ulmus americana*) as a street tree after the American elm was devastated by an introduced pathogen that caused Dutch Elm disease. Siberian elm failed in its original purpose but was successful at naturalizing into many parts of the country. Now lace bark elm is being promoted to replace Siberian elm. From an ecological standpoint we seem to be trapped on a devil's merry-go-round. *Cornus mas* has been in the landscaping trade for years, known as Cornelian cherry (Rehder 1960); it seems unfortunate that we would now be promoting it for conservation purposes. There are at least 11 species of shrubby dogwoods native to the eastern United States. These native species offer equivalent soil conservation benefits and an abundance of fruits for wildlife with no ecological risks.

Sawtooth oak (*Quercus acutissima*), a species from Asia, provides another example of promoting an exotic species as an alternative wildlife food plant (Hopkins and Huntley 1979). Thirty-six years ago, Klimstra (1956) pointed out the potential problems associated with planting multiflora rose. Similarly, Coblentz (1981) has pointed out the lack of foresight and, more importantly, the lack of hindsight in promoting the exotic sawtooth oak over the 37 species of oaks native to the southeastern United States for mast production for wildlife.

In spite of the mounting evidence of the ecological dangers associated with exotics and the skyrocketing costs of controlling them, exotic species continue to be tested and promoted for the same worn-out reasons:

1. wildlife habitat plantings (autumn olive and bush honeysuckle);
2. landscaping purposes (blackthorn and purple loosestrife);
3. wood and fiber production (Princess tree [*Paulownia tomentosa*] and tree-of-heaven);
4. soil conservation practices (crown vetch and multiflora rose); and
5. forage improvement (Johnson grass and tall fescue).

## WHAT IS AT RISK BECAUSE OF EXOTIC PLANTS?

Entire plant communities such as fens, bogs and marshes can be significantly altered by invasive plant species such as purple loosestrife (Thompson et. al. 1987). Endangered species such as Kankakee mallow (*Iliamna remota*) may be crowded out of its last habitat by multiflora rose and the bush honeysuckles (Glass 1986, pers. comm.). Common plants, such as bluebells (*Mertensia virgin-*

*ica*) are being crowded out of forests by garlic mustard (Iverson et al. 1991). The native high-bush cranberry (*Viburnum trilobum*) is known to hybridize with the ornamental, *Viburnum opulus*. This may result in the loss of the native genotype, or it could result in creating an aggressive hybrid species similar to the case of *Spartina anglica* (Thompson 1991).

Oak reproduction is a major concern of foresters, ecologists, wildlife biologists, and natural area managers, and naturalized exotic shrubs and vines are now being identified as serious competitors to oak regeneration. A recent example is oriental bittersweet which is becoming a serious pest on many hardwood regeneration sites in the Appalachians (McNab and Meeker 1987).

## WHAT ARE THE COSTS ASSOCIATED WITH EXOTIC INTRODUCTIONS?

Although the economic cost of controlling exotic introductions can be calculated, the ecological damage cannot be measured in dollars. For example, Brandenburg Bog in northeastern Illinois was purchased to preserve, protect, and perpetuate a rare calcareous fen community. Purple loosestrife is invading the fen, and it may be beyond eradication (Heidorn 1986, pers. comm.). The direct cost of this exotic species to the State of Illinois in this example is at least $379,000, the cost of purchase in 1973. However, "Brandenburg Bog is the premier calcareous fen in the state and as such is irreplaceable" (Schwegman 1988).

## HOW ILLINOIS KICKED THE EXOTIC HABIT—A CASE HISTORY

The Illinois Department of Conservation nurseries began producing autumn olive in 1964. By 1982 our nurseries were distributing more than 1,000,000 autumn olive seedlings a year, which represented about 20 percent of the state nursery's production of all species combined (Sternberg 1982). We also produced large numbers of bush honeysuckles.

In 1983 our Seedling Needs Committee met to review the needs of the department relative to seedling production. This is a standing committee comprised of representatives from the divisions of Wildlife Resources, Forestry, Public Lands, Planning, and Natural Heritage. The issue of exotics and the role of the state nurseries in their production was addressed by the committee. The committee agreed that further production of exotic plant materials in our nurseries was not necessary if suitable native species could be grown as substitutes for the exotics. The concept of substituting native species for exotic species is compelling when one considers that:

1. Native species comprise 99 percent of the wildlife species we manage habitat for, and they evolved with native plant species. Furthermore, there is no hard evidence to support the contention that exotic plant materials are superior to native species for wildlife (Martin et al. 1951);
2. There is no reason to believe that native species of trees and shrubs cannot be grown in nurseries using techniques similar to those we use to grow exotics (Schopmeyer 1974);
3. When developing landscaping plans for state parks, conservation areas, and other Department of Conservation facilities, it seems more appropriate to use native plant materials in keeping with the natural setting (Hightshoe 1988); and
4. Future management problems caused by introducing new exotic plant

materials could be reduced if we promoted and planted native species for conservation purposes.

Today our nurseries produce 67 species of native trees and shrubs for use in developing wildlife habitat, reclamation projects, and community restorations (Fig. 3, Table 1). The seeds necessary to propagate these native species are collected from state parks and conservation areas by our wildlife biologists, foresters, natural heritage biologists, site superintendents, maintenance workers and volunteers.

TABLE 1. Native trees and shrubs grown at Illinois Department of Conservation nurseries.*

| | |
|---|---|
| *Acer rubrum* | RED MAPLE |
| *Acer saccharinum* | SILVER MAPLE |
| *Acer saccharum* | SUGAR MAPLE |
| *Aronia melanocarpa* | BLACK CHOKEBERRY |
| *Betula nigra* | RIVER BIRCH |
| *Carya illinoensis* | PECAN |
| *Carya ovata* | SHAGBARK HICKORY |
| *Carya ovalis, C. cordiformis, C. glabra, C. tomentosa* | (various HICKORY SPECIES) |
| *Carya texana* | BLACK HICKORY |
| *Celtis occidentalis* | HACKBERRY |
| *Cephalanthus occidentalis* | BUTTONBUSH |
| *Cercis canadensis* | REDBUD |
| *Cornus obliqua* | PALE DOGWOOD |
| *Cornus racemosa* | GRAY DOGWOOD |
| *Cornus stolonifera* | RED OSIER DOGWOOD |
| *Corylus americana* | HAZELNUT |
| *Crataegus crus-galli* | COCK-SPUR THORN |
| *Crataegus phaenopyrum* | WASHINGTON HAWTHORN |
| *Diospyros virginiana* | COMMON PERSIMMON |
| *Fraxinus americana* | WHITE ASH |
| *Fraxinus pennsylvanica* | GREEN ASH |
| *Gymnocladus dioica* | KENTUCKY COFFEE-TREE |
| *Ilex decidua* | SWAMP HOLLY |
| *Juglans nigra* | BLACK WALNUT |
| *Juniperus virginiana* | RED CEDAR |
| *Liquidambar styraciflua* | SWEET GUM |
| *Liriodendron tulipifera* | TULIP TREE |
| *Malus ioensis* | IOWA CRAB APPLE |
| *Morus rubra* | RED MULBERRY |
| *Nyssa sylvatica* | SOUR GUM |
| *Pinus resinosa* | RED PINE |
| *Pinus strobus* | WHITE PINE |
| *Pinus taeda* | LOBLOLLY PINE |
| *Platanus occidentalis* | SYCAMORE |
| *Prunus americana* | WILD PLUM |
| *Prunus serotina* | WILD BLACK CHERRY |
| *Quercus alba* | WHITE OAK |
| *Quercus bicolor* | SWAMP WHITE OAK |
| *Quercus imbricaria* | SHINGLE OAK |
| *Quercus lyrata* | OVERCUP OAK |

TABLE 1. (cont.)

| | |
|---|---|
| *Quercus macrocarpa* | BUR OAK |
| *Quercus marilandica* | BLACKJACK OAK |
| *Quercus michauxii* | BASKET OAK |
| *Quercus nuttalii* | NUTTALL'S OAK |
| *Quercus pagoda* | CHERRY-BARK OAK |
| *Quercus palustris* | PIN OAK |
| *Quercus prinoides* var. *acuminata* | YELLOW CHESTNUT OAK |
| *Quercus rubra* | RED OAK |
| *Quercus shumardii* | SHUMARD'S OAK |
| *Quercus stellata* | POST OAK |
| *Quercus velutina* | BLACK OAK |
| *Rhus aromatica* | FRAGRANT SUMAC |
| *Rhus copallina* | DWARF SUMAC |
| *Rhus glabra* | SMOOTH SUMAC |
| *Rhus typhina* | STAGHORN SUMAC |
| *Robinia pseudoacacia* | BLACK LOCUST |
| *Rubus allegheniensis* | COMMON BLACKBERRY |
| *Sambucus canadensis* | ELDERBERRY |
| *Symphoricarpos orbiculatus* | CORALBERRY |
| *Taxodium distichum* | BALD CYPRESS |
| *Vaccinium arboreum* | FARKLEBERRY |
| *Viburnum lentago* | NANNYBERRY |
| *Viburnum recognitum* | SMOOTH ARROWWOOD |
| *Viburnum trilobum* | HIGH-BUSH CRANBERRY |

*Nomenclature follows Mohlenbrock (1986).

FIGURE 3.—Native plants being grown at the Illinois Department of Conservation Mason Tree Nursery. (photo provided by S. Pequignot)

In 1977 the Illinois nursery system moved forward once more by producing big bluestem (*Andropogon gerardii*) and Indian grass (*Sorghastrum nutans*) seed for prairie reconstructions. By 1980 our Mason Tree Nursery had expanded its operation to include 37 different species of prairie forbs. In 1983, 35,000 prairie forbs were obtained from 596 $m^2$ of bed space (Wallace et al. 1986).

The grass and forb program has been very successful. Production for 1991 included 54 forb species and 7 grass species (Fig. 4 and Table 2); 293,457 prairie forb rootstocks were grown in 1,900 $m^2$ of bed space. In addition, attempts to grow woodland herbaceous species have been initiated with 13 species currently involved (Table 3). Approximately $4 million in capital improvements at the Mason Nursery has increased seed bed space from 16 ha to 40 ha and built a 279 $m^2$ greenhouse for herbaceous production. The facility is also equipped with a grass seed cleaning and processing facility and a center pivot irrigation system which will allow expansion of the grass seed collection area to 16 hectares (Pequignot 1992, pers. comm.).

Besides attempts to eliminate exotic species from our nursery operations, educational articles discussing the problems with exotic plants and animals have been published in our department's official publication, *Outdoor Highlights* (Harty 1985; Schwegman 1985, 1988). Moreover, a colorful flier was prepared that explained the problems associated with planting purple loosestrife and recommended measures for its control. Species-specific alert fliers were produced for garlic mustard (*Alliaria petiolata*), rudd (*Scardineus erythrophthalmus*), rusty crayfish, and zebra mussels (*Dreissena polymorpha*). A slide tape program describing the problems associated with exotic plant species in Illinois was also produced for use in educating the general public.

TABLE 2. Native prairie forb and grass species grown at the Mason Tree Nursery, Topeka, Ill.*

| FORBS | |
|---|---|
| *Allium cernuum* | NODDING ONION |
| *Amorpha canescens* | LEADPLANT |
| *Anemone cylindrica* | THIMBLEWEED |
| *Asclepias sullivantii* | PRAIRIE MILKWEED |
| *Asclepias tuberosa* | BUTTERFLY-WEED |
| *Aster laevis* | SMOOTH ASTER |
| *Aster novae-angliae* | NEW ENGLAND ASTER |
| *Astragalus tennesseensis* | GROUND PLUM |
| *Baptisia lactea* | WHITE WILD INDIGO |
| *Baptisia leucophaea* | CREAM WILD INDIGO |
| *Boltonia decurrens* | DECURRENT FALSE ASTER |
| *Calcalia plantaginea* | PRAIRIE INDIAN PLANTAIN |
| *Camassia scilloides* | WILD HYACINTH |
| *Ceanothus americanus* | NEW JERSEY TEA |
| *Coreopsis lanceolata* | TICKSEED COREOPSIS |
| *Coreopsis palmata* | PRAIRIE COREOPSIS |
| *Coreopsis tripteris* | TALL TICKSEED |
| *Dalea candida* | WHITE PRAIRIE CLOVER |
| *Dalea foliosa* | LEAFY PRAIRIE CLOVER |
| *Dalea purpurea* | PURPLE PRAIRIE CLOVER |
| *Desmanthus illinoensis* | ILLINOIS MIMOSA |
| *Desmodium canadense* | SHOWY TICK TREFOIL |
| *Desmodium illinoense* | ILLINOIS TICK TREFOIL |
| *Dodecatheon meadia* | SHOOTING-STAR |

TABLE 2. (cont.)

| | |
|---|---|
| *Echinacea pallida* | PALE CONEFLOWER |
| *Eryngium yuccifolium* | RATTLESNAKE MASTER |
| *Helianthus occidentalis* | WESTERN SUNFLOWER |
| *Heliopsis helianthoides* | FALSE SUNFLOWER |
| *Heuchera richardsonii* | PRAIRIE ALUMROOT |
| *Hieracium longipilum* | HAIRY HAWKWEED |
| *Iliamna remota* | KANKAKEE MALLOW |
| *Iris shrevei* | WILD BLUE IRIS |
| *Lespedeza capitata* | ROUND-HEADED BUSH CLOVER |
| *Lespedeza leptostachya* | PRAIRIE BUSH CLOVER |
| *Liatris aspera* | ROUGH BLAZING STAR |
| *Liatris pycnostachya* | PRAIRIE BLAZING STAR |
| *Liatris spicata* | MARSH BLAZING STAR |
| *Monarda fistulosa* | WILD BERGAMONT |
| *Napaea dioica* | GLADE MALLOW |
| *Parthenium integrifolium* | AMERICAN FEVERFEW |
| *Physostegia virginiana* | FALSE DRAGONHEAD |
| *Polytaenia nuttalli* | PRAIRIE PARSLEY |
| *Potentilla arguta* | PRAIRIE CINQUEFOIL |
| *Prenanthes aspera* | ROUGH WHITE LETTUCE |
| *Ratibida pinnata* | DROOPING CONEFLOWER |
| *Rosa carolina* | PASTURE ROSE |
| *Rudbeckia subtomentosa* | FRAGRANT CONEFLOWER |
| *Silene regia* | ROYAL CATCHFLY |
| *Silphium integrifolium* | ROSINWEED |
| *Silphium laciniatum* | COMPASS-PLANT |
| *Silphium terebinthinaceum* | PRAIRIE-DOCK |
| *Solidago rigida* | RIGID GOLDENROD |
| *Tradescantia ohiensis* | SPIDERWORT |
| *Zizia aurea* | GOLDEN ALEXANDERS |
| **GRASSES** | |
| *Andropogon gerardii* | BIG BLUESTEM |
| *Panicum virgatum* | SWITCH GRASS |
| *Schizachryium scoparium* | LITTLE BLUESTEM |
| *Sorghastrum nutans* | INDIAN GRASS |
| *Spartina pectinata* | CORD GRASS |
| *Sporobolus heterolepis* | PRAIRIE DROPSEED |
| *Stipa spartea* | PORCUPINE GRASS |

*Nomenclature follows Mohlenbrock (1986).

Thirty-four vegetation management circulars were prepared by various authors for the Illinois Nature Preserves Commission. These circulars provide information about specific exotic plant species and management recommendations for their control or eradication. Many of these circulars have been published in the Natural Areas Journal.

In addition to these efforts, two other publications, *Landscaping for Wildlife* and *Illinois Prairie: Past and Future—A Restoration Guide,* promote the use of native species and point out the problems associated with exotic plant species.

Another significant step forward was the development of a windbreak manual for Illinois (Bolin et al. 1987). This is a cooperative effort by the University of

FIGURE 4.—Illinois Department of Conservation employees harvesting seed of native species being grown at Mason Tree Nursery. (photo provided by S. Pequignot)

Illinois Department of Forestry, Cooperative Extension Service, the USDA Soil Conservation Service, and the Illinois Department of Conservation. The issue of exotics was addressed early in the planning of this manual, and the committee, which is comprised of inter-agency foresters, wildlife biologists, and natural-heritage biologists, recommended 30 native trees and shrubs and 3 nonnative species as suitable for use for windbreaks and snow trips in Illinois (Table 4). The three nonnative species to Illinois, Norway spruce (*Picea abies*), blue spruce (*Picea pungens*), and Douglas-fir (*Pseudotsuga mensiesii*), have been planted throughout Illinois for many years and have not been found to reproduce spontaneously from seed. This has proved to be a prudent compromise.

TABLE 3. Native woodland herbaceous species grown at the Mason Tree Nursery, Topeka, Ill.*

| | |
|---|---|
| *Actaea pachypoda* | DOLL'S EYES |
| *Arisaema triphyllum* | JACK-IN-THE-PULPIT |
| *Asarum canadense* | WILD GINGER |
| *Claytonia virginica* | SPRING BEAUTY |
| *Dentaria laciniata* | TOOTHWORT |
| *Dicentra cucullaria* | DUTCHMAN'S-BREECHES |
| *Isopyrum biternatum* | FALSE RUE ANEMONE |
| *Jeffersonia diphylla* | TWINLEAF |
| *Panax quinquefolius* | GINSENG |
| *Polygonatum commutatum* | SOLOMON'S SEAL |
| *Sanguinaria canadensis* | BLOODROOT |
| *Smilacina racemosa* | FALSE SOLOMON'S SEAL |

*Nomenclature follows Mohlenbrock (1986).

The Illinois Department of Transportation has also been quite cooperative regarding management of exotics along right-of-ways and cloverleafs adjacent to Department of Conservation properties. For the past three years the Department of Conservation and the Soil Conservation Service (SCS) have been working cooperatively to collect seeds from native shrubs for the SCS to evaluate at their Elsberry Plant Improvement Center in Missouri. Seventeen species are currently being evaluated for their wildlife food and cover value (Table 5).

On January 1, 1988, Illinois passed the Illinois Exotic Weed Act. Exotic weeds are defined as " . . . plants not native to North America which, when planted, either spread vegetatively or naturalize and degrade natural communities, reduce the value of fish and wildlife habitat, or threaten an Illinois endangered or threatened species." There are currently three listed species—Japanese honeysuckle, multiflora rose, and purple loosestrife. It is unlawful to sell, buy,

TABLE 4. Native tree and shrub species recommended for windbreaks and snow trips in Illinois (Bolin et al. 1987).*

| | |
|---|---|
| *Amelanchier arborea* | SHADBUSH |
| *Aronia melanocarpa* | BLACK CHOKEBERRY |
| *Cornus alternifolia* | ALTERNATIVE-LEAVED DOGWOOD |
| *Cornus drummondii* | ROUGH-LEAVED DOGWOOD |
| *Cornus obliqua* | PALE DOGWOOD |
| *Cornus racemosa* | GRAY DOGWOOD |
| *Cornus stolonifera* | RED OSIER DOGWOOD |
| *Corylus americana* | HAZELNUT |
| *Crataegus crus-galli* | COCK-SPUR THORN |
| *Crataegus mollis* | RED HAW |
| *Crataegus phaenopyrum* | WASHINGTON HAWTHORN |
| *Hamamelis virginiana* | WITCH-HAZEL |
| *Ilex verticillata* | WINTERBERRY |
| *Juniperus communis* | COMMON JUNIPER |
| *Juniperus virginiana* | RED CEDAR |
| *Malus ioensis* | IOWA CRAB APPLE |
| *Picea abies*** | NORWAY SPRUCE |
| *Picea pungens*** | BLUE SPRUCE |
| *Pinus strobus* | WHITE PINE |
| *Prunus americana* | WILD PLUM |
| *Prunus virginiana* | COMMON CHOKECHERRY |
| *Pseudotsuga menziesii*** | DOUGLAS-FIR |
| *Symphoricarpos orbiculatus* | CORALBERRY |
| *Taxus canadensis* | CANADA YEW |
| *Thuja occidentalis* | ARBOR VITAE |
| *Viburnum acerifolium* | MAPLE-LEAVED ARROWWOOD |
| *Viburnum lentago* | NANNYBERRY |
| *Viburnum prunifolium* | BLACK HAW |
| *Viburnum rafinesquianum* | DOWNY ARROWWOOD |
| *Viburnum recognitum (V. dentatum)* | SMOOTH ARROWWOOD |
| *Viburnum rufidulum* | SOUTHERN BLACK HAW |
| *Viburnum trilobum* | HIGH-BUSH CRANBERRY |

* Nomenclature follows Mohlenbrock (1986).
** Not native to Illinois

TABLE 5. Native shrubs being evaluated as food and cover plants by the Soil Conservation Service's Plant Improvement Center, Elsberry, Mo.*

| | |
|---|---|
| *Amelanchier arborea* | SHADBUSH |
| *Aronia melanocarpa* | BLACK CHOKEBERRY |
| *Cornus alternifolia* | ALTERNATE-LEAVED DOGWOOD |
| *Cornus drummondii* | ROUGH-LEAVED DOGWOOD |
| *Cornus obliqua* | PALE DOGWOOD |
| *Cornus racemosa* | GRAY DOGWOOD |
| *Corylus americana* | HAZELNUT |
| *Juniperus virginiana* | RED CEDAR |
| *Prunus americana* | WILD PLUM |
| *Prunus virginiana* | COMMON CHOKECHERRY |
| *Ribes americana* | WILD BLACK CURRANT |
| *Thuja accidentalis* | ARBOR VITAE |
| *Viburnum lentago* | NANNYBERRY |
| *Viburnum prunifolium* | BLACK HAW |
| *Viburnum recognitum (V. dentatum)* | SMOOTH ARROWWOOD |
| *Viburnum trilobum* | HIGH-BUSH CRANBERRY |

* Nomenclature follows Mohlenbrock (1986).

offer for sale, or distribute seeds, plants, or plant parts without a permit issued by the Illinois Department of Conservation. A violation of the act is a Class B misdemeanor, and listed plants are confiscated and destroyed.

On May 25, 1989, the Director of the Illinois Department of Conservation signed a department policy on the planting and removal of exotic plant species. The policy lists 12 species which cannot be used on DOC property and lists 5 species which can be used only for short rotation, research or erosion control (Table 6).

## SUMMARY

Once exotics become naturalized, they often change community species composition, alter structure, and reduce natural diversity of native plant and animal communities. Moreover, if an exotic becomes naturalized and spreads throughout a system, getting it out of that system is like trying to unscramble an egg.

It is the responsibility of all natural resource professionals to provide proper and prudent management advice to private and public landowners and managers. To continue to ignore the documented consequences associated with introducing exotic species in the name of soil conservation, wildlife management, or reforestation would fall short of this obligation.

A giant step forward is necessary to head off the invasion of exotic plant materials into the natural landscape. We must redirect our reforestation and wildlife habitat restoration efforts away from exotics and toward the utilization of native plant species that are compatible with native ecosystems.

Illinois is extremely fortunate to have natural resource agencies and resource professionals who have taken decisive action in addressing the issue of exotic species.

Laycock (1966) described the pursuit of exotic species as a "perpetual relay race with one generation passing the stick to the next." I am happy to report that

TABLE 6. List of exotic plant species and their permissible uses on Illinois Department of Conservation properties as authorized by Policy Manual Code No. 2450 dated May 25, 1989.

| SCIENTIFIC NAME | COMMON NAME | PERMISSIBLE USES |
|---|---|---|
| *Celastrus orbiculatus* | ROUND-LEAVED BITTERSWEET | None |
| *Coronilla varia* | CROWN VETCH | None |
| *Elaeagnus umbellata* | AUTUMN OLIVE | None |
| *Euonymus alatus* | WINGED EUONYMUS | None |
| *Euonymus fortunei* | CLIMBING EUONYMUS | None |
| *Festuca elatior* | TALL FESCUE | Critical area erosion |
| *Lespedeza cuneata* | SERECIA LESPEDEZA | Cover crop and nitrogen fixation |
| *Lonicera japonica* | JAPANESE HONEYSUCKLE | None |
| *Lonicera maackii* | AMUR HONEYSUCKLE | None |
| *Lonicera tartarica* | TARTARIAN HONEYSUCKLE | None |
| *Lythrum salicaria* | PURPLE LOOSESTRIFE | None |
| *Melilotus alba* | WHITE SWEET CLOVER | Short rotation cropland |
| *Melilotus officinalis* | YELLOW SWEET CLOVER | Short rotation cropland |
| *Pueraria lobata* | KUDZU | None |
| *Rhamnus frangula* | GLOSSY BUCKTHORN | None |
| *Robinia pseudoacacia*** | BLACK LOCUST | Strip mine reclamation, nurse crop in black walnut plantation, mixed with 34 other species in forest application with unfavorable site conditions. |
| *Rosa multiflora* | MULTIFLORA ROSE | None |

* Nomenclature follows Mohlenbrock (1986)
**Black locust is native to extreme southeastern Illinois. It is not an exotic species, but it has become an invasive species in Illinois especially on sandy soils.

the Illinois Department of Conservation has dropped the baton. It is the author's hope that this paper will stimulate activity in other states to address the issue of exotic species within their jurisdiction.

## ACKNOWLEDGMENTS

I wish to acknowledge the following people and their agencies because they are the ones who worked together so successfully to address the issue of exotic plant species in Illinois. Richard Oliver, Steve Brady, Ray Herman and Gene Barickman, Soil Conservation Service; Mike Bolin, University of Illinois Extension Service; Gary Rolfe, Department of Forestry, University of Illinois; Al Mickelson, Stewart Pequignot, and Dick Little, Division of Forestry, Illinois Department of Conservation; John Schwegman and Carl Becker, Division of Natural Heritage, Illinois Department of Conservation, and Guy Sternberg, Division of Special Services, Illinois Department of Conservation.

As always a special thank you to Arlene Gallagher for typing the manuscript and thanks to John Schwegman, Robert Szafoni, Bill McKnight, Stewart Pequignot, and an anonymous reviewer for improving the manuscript.

## REFERENCES

Bolin, M., R. Oliver and S. Brady. 1987. Illinois windbreak manual. Illinois Department of Conservation, Springfield.

Bratton, S. P. 1974. The effect of the European wild boar (*Sus scrofa*) on high-elevation vernal flora in Great Smoky Mountains National Park. Bull. Torrey Bot. Club 101:198–206.

———. 1975. The effect of the European wild boar, *Sus scrofa*, on Grey Beech Forest in the Great Smoky Mountains. Ecology 56:1356–66.

——— 1982. The effects of exotic plant and animal species on nature preserves. Nat. Areas J. 2(3): 3–13.

Coblentz, B. E. 1978. The effects of feral goats (*Capra hircus*) on island ecosystems. Biol. Conserv. 13:279–86.

———. 1981. Possible dangers of introducing sawtooth oak. Wildlife Soc. Bull. 9 (2):136–38.

Courtenay, W. R., Jr. 1978. The introduction of exotic organisms. Pages 237–52 *in* Wildlife and America, H. P. Brokaw (ed.). Council on Environmental Quality, U.S. Fish and Wildlife Service, U.S. Forest Service, and National Oceanic and Atmospheric Administration.

De Vos, A., R. H. Manville and R. G. Van Gelder. 1956. Introduced mammals and their influence on native biota. Zoologica 41(4):163–94.

Ebinger, J. E. and L. Lehnen, Jr. 1981. Naturalized autumn olive in Illinois. Trans. Illinois State Acad. Sci. 74(3/4):83–85.

Ebinger, J. E. 1983. Exotic shrubs a potential problem in natural area management in Illinois. Nat. Areas J. 3(1):3–6.

Ebinger, J. E., J. Newman and R. Nyboer. 1984. Naturalized winged wahoo in Illinois. Nat. Areas J. 4(2):26–29.

Ehrenfeld, D. W. 1970. Biological conservation. Holt, Rinehart and Winston, New York, N.Y.

Elton, C. 1958. The ecology of invasions by animals and plants. Butler and Tanner, London.

Hall, E. R. 1963. Introduction of exotic species of mammals. Trans. Kansas Acad. Sci. 66(3):516–18.

Harty, F. M. 1985. Foreigners. Outdoor Highlights 13(4):8–9.

Harty, F. M. 1986. Exotics and their ecological ramifications. Nat. Areas J. 6(4):20–26.

Henry, R. D. and A. R. Scott. 1980. Some aspects of the alien component of the spontaneous Illinois vascular flora. Trans. Illinois State Acad. Sci. 73(4):35–40.

Hightshoe, G. L. 1988. Native trees, shrubs, and vines for urban and rural America. A planting design manual for environmental designers. Van Nostrand Reinhold, New York, N.Y.

Hopkins, C. R. and J. C. Huntley. 1979. Establishment of sawtooth oak as a mast source for wildlife. Wildlife Soc. Bull. 7(4):253–58.

Howe, T. D. and S. P. Bratton. 1976. Winter rooting activity of the European wild boar in the Great Smoky Mountains National Park. Castanea 41:256–64.

Huggler, T. 1991. New gamebird on the block. Gun Dog Mag. Dec.–Jan. 12–15.

Iverson, J. B. 1978. The impact of feral cats and dogs on populations of the West Indian rock iguana, *Cyclura carinata*. Biol. Conserv. 14:63–73.

Iverson, L. R., G. L. Rolfe, T. J. Jacob, et al. 1991. Forests of Illinois. Illinois Council on Forestry Development, Urbana, and Illinois Natural History Survey, Champaign.

Klimstra, W. D. 1956. Problems in the use of multiflora rose. Trans. Illinois State Acad. Sci. 48: 66–72.

Klimstra, W. D. and D. Hankla. 1953. Preliminary report on pheasant stocking in southern Illinois. Trans. Illinois State Acad. Sci. 46:235–39.

Laycock, G. 1966. The alien animals. Natural History Press, Garden City, N.Y.

Martin, A. C., H. S. Zim and A. L. Nelson. 1951. American wildlife and plants: a guide to wildlife food plants. Dover Publications, Inc., New York, N.Y.

McNab, W. H. and M. Meeker. 1987. Oriental bittersweet: a growing threat to hardwood silviculture in the Appalachians. Northern J. Appl. For. 4(4):174–77.

Mohlenbrock, R. H. 1986. Guide to the vascular flora of Illinois. Southern Illinois University Press, Carbondale.

Mooney, H. A. and J. A. Drake (eds.). 1986. Ecology of biological invasions of North America and Hawaii. Ecol. Stud. 58:322.

Moyle, P. B. 1976. Fish introductions in California: history and impact on native fishes. Biol. Conserv. 9:101–18.

Nestleroad, J., D. Zimmerman and J. E. Ebinger. 1987. Autumn olive reproduction in three Illinois state parks. Trans. Illinois Acad. Sci. 80(1/2):33–39.

Nyboer, R. W. and J. E. Ebinger. 1978. *Maclura pomifera* (Raf.) Schneid. in Coles County, Illinois. Trans. Illinois State Acad. Sci. 71(4):389–91.

Page, L. M. 1985. The crayfishes and shrimps (*Decapoda*) of Illinois. Illinois Nat. Hist. Surv. Bull. 33(4):448.

Reed, C. F. 1977. Economically important foreign weeds—Potential problems in the United States.

USDA. Agricultural Research Service Animal and Plant Health Inspection Service. Agriculture Handbook No. 498.

Rehder, A. 1960. Manual of cultivated trees and shrubs hardy in North America. The Macmillan Co., New York, N.Y.

Savidge, J. A. 1987. Extinction of an island forest avifauna by an introduced snake. Ecology 68 (3):660–68.

Schopmeyer, C. S. 1974. Seeds of woody plants in the United States. USDA Forest Service, Agriculture Handbook No. 450. Washington, D.C.

Schwegman, J. E. 1985. Purple plague. Outdoor Highlights 13(21):10–11.

———. 1988. Exotic invaders. Outdoor Highlights 16(6):6–11.

Soil Conservation Service. 1991. Plants for Conservation 3(1), January 1991, Newsletter. U.S. Department of Agriculture, Elsberry Plant Materials Center, Elsberry, Mo.

———. 1991. Plants for Conservation 3(2), July 1991 Newsletter. U.S. Department of Agriculture, Elsberry Plant Materials Center, Elsberry, Mo.

Sternberg, G. 1982. Autumn olive in Illinois conservation practice. Illinois Department of Conservation, Division of Planning, Springfield.

Thompson, D. Q., R. L. Stuckey and E. B. Thompson. 1987. Spread, impact, and control of purple loosestrife (*Lythrum salicaria*) in North American wetlands. U.S. Fish and Wildlife Service, Fish and Wildlife Research No. 2. U.S. Department of Interior, Washington, D.C.

Thompson, J. D. 1991. The biology of an invasive plant. What makes *Spartina anglica* so successful? BioScience 41(6):393–401.

Vale, T. R. 1982. Plants and people (vegetation change in North America). Association of American Geographers, Washington, D.C.

Vance, D. R. and R. L. Westemeier. 1979. Interactions of pheasants and prairie chickens in Illinois. Wildlife Soc. Bull. 7(4):221–25.

Wallace, V. K., S. Pequignot and W. Yoder. 1986. The role of state forest nurseries in prairie plant propagation. Pages 201–03 *in* The prairie—past, present, and future: Proceedings of the ninth North American prairie conference, G. K. Clambey and R. H. Pemble (eds.). Tri-College University, North Dakota State Univ., Fargo.

Ward, O. G. and D. H. Wurster-Hill. 1990. *Nyctereutes procynoides.* Mammalian Species No. 358. American Society of Mammalogists, New York, N.Y.

Weller, M. W. 1981. Freshwater marshes, ecology and wildlife management. Univ. Minnesota Press, Minneapolis.

West, K. A. 1984. Major pest species listed, control measures summarized at the (tenth Midwest) Natural Areas Workshop. Restor. Manage. Notes 2(1):34–35.

Westemeier, R. L., T. L. Esker and S. A. Simpson. 1989. An unsuccessful clutch of northern bobwhites with hatched pheasant eggs. Wilson Bull. 10(4):640–42.

# Prioritizing Patches for Control of Invasive Plant Species: A Case Study with Amur Honeysuckle

James O. Luken[1]

## INTRODUCTION

Managers of nature reserves are frequently faced with two dilemmas when the problem of invasive plant species is considered. First, invasive plants may not be limited solely to specific habitats or sites but instead may be widely distributed, occupying a range of habitats and showing various levels of performance. Such a pattern is common in invasive plants because of generalist habitat requirements (Bazzaz 1986) and also because invasion success may be associated with historic disturbance regimens (Orians 1986). Second, measures that resource managers can take to control invasive species are expensive while budgets for managing nature reserves are limited. In the Cape of Good Hope Nature Reserve in South Africa, 39 percent of the management budget is presently devoted to the control of invasive woody plants (Macdonald et al. 1988).

One potential solution to these dilemmas is a prioritization process that considers both the performance (productivity) and resilience (reestablishment of performance after a stress event) of invasive plant species. Such a prioritization process is perhaps best-suited to invasive woody plants occurring in patchy nature reserves but may also be applied to grasses and herbaceous species as well. Prioritization of management effort is based on the assumption that performance and resilience of invasive plant species vary in a predictable fashion relative to plant age and relative to gradients of environmental factors (Luken 1988; Luken and Mattimiro 1991). If management unit patches can be identified relative to these factors, then prioritization can occur that will guide the application of control measures.

## CASE STUDY

Concepts presented here for the control of invasive species are based largely on data collected in the exurban landscape of northern Kentucky. Here, Amur honeysuckle (*Lonicera maackii* (Rupr.) Maxim.), an upright, deciduous, multistemmed shrub (Fig. 1) native to northeastern Asia, dominates habitats ranging from recently disturbed soil to mature forest.

Research on the performance of this shrub in different habitats indicates that light is an important environmental factor affecting production, reproduction, population structure, demography, and resilience (Table 1). Specifically, open-grown adult shrubs have higher performance when undisturbed and greater resilience when cut than do forest-grown shrubs of similar age. These data indicate that forests are not optimal habitats for Amur honeysuckle.

Further observational data suggest that forest patches with complete canopies can resist invasion by Amur honeysuckle. However, if gaps are created or if

1. Department of Biological Sciences, Northern Kentucky University, Highland Heights, KY 41099-0400

FIGURE 1.—*Lonicera maackii* with fruit. (photo by B. N. McKnight)

forests pass through a developmental stage where canopy gaps exist, then Amur honeysuckle will invade and it will persist. Efforts to eradicate adult plants of Amur honeysuckle will be most successful when the shrubs are young or when they are growing in light-limited habitats (i.e., closed-canopy forests).

Forest and open sites presently exist that are essentially Amur honeysuckle thickets. These dense thickets are associated with a near complete absence of ground cover species. This greatly complicates the task of control because removal of the shrub component exposes the site to both erosion and colonization by other invasive species.

## PRIORITIZATION PROCESS

The landscape element of interest in the development of a prioritization process for invasive species control is the management unit patch. This is a single area of land that is managed to manipulate the community development path (Luken 1990a). In invasive species control, the management goal is commonly to eliminate invasive species while allowing the gradual release or establishment of plants from the native flora.

Management unit patches can be classified in terms of performance and re-

TABLE 1. Characteristics of open- and forest-grown Amur honeysuckle populations when undisturbed and after cutting. Data from Luken (1988) and Luken and Mattimiro (1991).

| PARAMETER | OPEN-GROWN | FOREST-GROWN |
|---|---|---|
| Above ground production | high | low |
| Leaf/stem ratio | high | low |
| Standing dead stems | low | high |
| Seed production | high | low |
| Stem release after cutting | high | high |
| Stem survival after release | high | low |
| Shrub survival after cutting | high | low |

silience of the invading plant. The best indicator of performance or importance is net primary production, because this is a direct measure of resource utilization (Whittaker 1975). The best indicator of resilience is measurement of plant performance after an artificially imposed stress regimen (e.g., repeated clipping, burning or herbicide spraying). However, more easily measured parameters may be used as performance and resilience indices. For example, in Amur honeysuckle, both performance and resilience increase with shrub population age and then decrease as trees invade the shrub populations and forests develop (Luken 1988; Luken and Mattimiro 1991). Thus population age and successional development of management unit patches produce a gradient of performance and resilience (Fig. 2). The successional phases presented in Fig. 2 represent convenient categories for grouping patches. Yet in any successional phase where invasion is recent, shrubs will be less resilient than in phases where shrubs are older.

The gradient presented in Fig. 2 is complicated by the fact that patch size (edge effects) and canopy gaps may create deviations from the general trend. Small forest patches (strong edge effects) and those with high canopy disturbance may need to be treated as special cases.

Mid-successional sites such as pastures, scrub, small forest patches with canopy gaps, and all sites where shrub populations are old (> 3 years) have low priority for management because of cost associated with shrub removal and low efficiency of eradication methods. Furthermore, removal of the shrub canopy in nonforested sites where invasive performance is high would likely open the area to erosion and colonization by other invasive species. Thus, in order to restore such sites, the manager would need to immediately protect the soil and supplement the propagule supply.

Recently disturbed sites, mature forests lacking canopy gaps, and all sites where shrubs are young (i.e., < 3 years old) should be given highest priority for management because complete eradication would be most easily achieved with the least money invested and the least amount of environmental disturbance. When shrubs are young or are stressed by light limitation they can be easily pulled or killed. Removal of the shrub layer in older forests can occur without colonization by other alien plants and with some native tree regeneration taking place after management (Luken 1990b).

## OTHER CONSIDERATIONS

In addition to performance and resilience of invasive species, the presence or potential presence of native species are important factors. Specifically, sites where invasives are threatening rare or endangered species or where diversity is

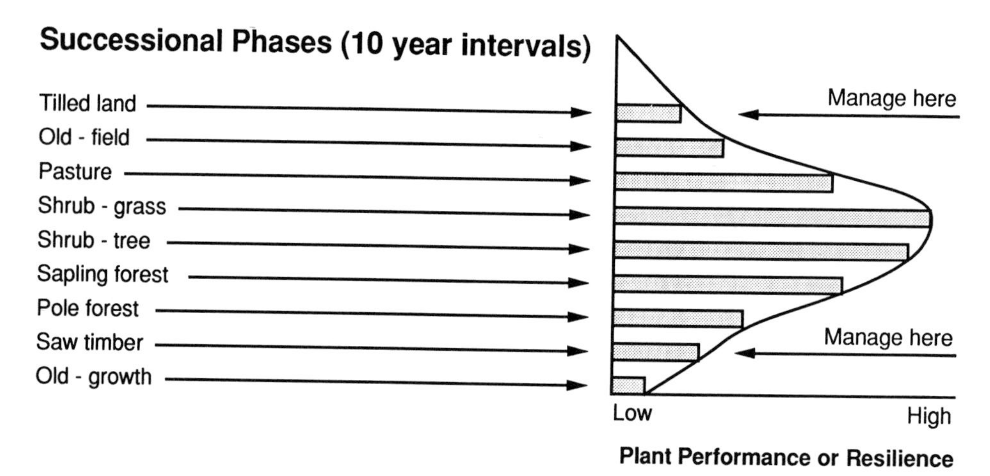


FIGURE 2.—A gradient of performance and resilience created by successional processes in patchy landscapes. Shrubs participating in succession from time zero will show increased performance and resilience as populations age and then will show decreased performance and resilience as forests develop.

being reduced can be given higher priority for plant removal. However, the greater the performance or importance of an invasive plant on a site, the greater will be the disturbance or environmental change as a result of shrub removal. Such conditions favor colonization by other invasive species. As such, resource managers should be prepared for an ongoing process of community restoration in these sites after invasives are eliminated.

It is unknown whether the prioritization process presented here has utility beyond the scope of light-limited invasive shrubs. However, all plant species have varying performance and resilience relative to environmental gradients. Thus it may be necessary to measure and assess the important environmental factor (or factors) controlling performance and resilience for each plant species and then prioritize patches based on this information.

## REFERENCES

BAZZAZ, F. A. 1986. Life history of colonizing plants: some demographic, genetic, and physiological features. Pages 96–110 *in* Ecology of biological invasions of North America and Hawaii. H. A. Mooney and J. A. Drake (eds.). Springer-Verlag, New York, N.Y.

LUKEN, J. O. 1988. Population structure and biomass allocation of the naturalized shrub *Lonicera maackii* (Rupr.) Maxim. in forest and open habitats. Amer. Midl. Nat. 119:258–67.

———. 1990a. Directing ecological succession. Chapman and Hall, London.

———. 1990b. Forest and pasture communities respond differently to cutting of exotic Amur honeysuckle (Kentucky). Restor. Manage. Notes 8:122–23.

———. 1991. Habitat-specific resilience of the invasive shrub Amur honeysuckle (*Lonicera maackii*) during repeated clipping. Ecol. Appl. 1:104–09.

MACDONALD, I. A. W., D. M. GRABER, D. DEBENEDETTI, et al. 1988. Introduced species in nature reserves in Mediterranean type climatic regions of the world. Biol. Conserv. 44:37–66.

ORIANS, G. H. 1986. Site characteristics favoring invasions. Pages 133–48 *in* Ecology of biological invasions of North America and Hawaii. H. A. Mooney and J. A. Drake (eds.). Springer-Verlag, New York, N.Y.

WHITTAKER, R. H. 1975. Communities and ecosystems. Macmillan Publishing, New York, N.Y.

# Classical Biological Control—An Endangered Discipline?

John J. Drea[1]

## WHAT IS CLASSICAL BIOLOGICAL CONTROL?

Unlike other papers presented at this symposium that vividly describe various aspects of biotic pollution, my contribution is more a cry of alarm. Actually the title of my presentation should be "Classical Biological Control—Endangered or Dangerous?" However, before considering these two distinct but, in this case, related concepts, we should look at classical biocontrol itself to see what it is and if it is worthy of worry.

Biological control has as many definitions as there are proponents of the field. In general, any definition refers to the use of living organisms or their by-products for the control of pests, mainly insects and weeds. The fundamental concept is that living organisms are the basis of the control. Biological control may include the use of biorationals, such as pheromones and kairomones for trapping or disruption of the pest, the creation of genetically engineered organisms to meet specific needs, the distribution of sterilized insects to sabotage mating, or the use of just about any other combination of an organism and its excretions, odors, toxins, etc. that can be imagined. Some consider the development of plant resistance as biocontrol. The range of definitions appears to be unlimited. However, it is generally accepted that the practice of intentionally introducing exotic insects for the control of pests of exotic, or rarely native, origin is usually referred to as "classical biological control." For this discussion the important concept is the *intentional introduction of exotic living organisms into new areas* as a method of pest control.

Should we be concerned if classical biological control is an endangered species or if it is becoming a dangerous species? YES! Is biological pollution a possible consequence of the pursuit of control by the introduction of natural enemies? YES!

Experience tells us that the successful invasion of any exotic organism will have an impact in varying degrees on the environment. The effect may be innocuous or extremely detrimental and highly polluting. Consider the exotic ornamental flowers that are now noxious weeds, exotic diseases that are decimating our trees, exotic insects that devastate our crops, and exotic vertebrates that compete with livestock or that eat rare plants and degrade habitat.

The intentional introduction and establishment of an organism to control a pest will have a definite effect on agriculture and various components of the environment. This effect must be beneficial and nonpolluting or we are defeating the purpose of the introduction. Unfortunately, in recent years unfavorable conditions have developed, including insufficient funds and reductions in personnel, that are causing a malaise in classical biological control. This decline, in my opinion, may lead to serious biological pollution.

---

1. 4607 Marie St., Beltsville, MD 20705. Retired; formerly with USDA, Agricultural Research Service, Insect Biocontrol Laboratory, Beltsville, MD 20705

## THE VALUE OF CLASSICAL BIOLOGICAL CONTROL

Is classical biological control worthy of consideration? Absolutely. Do the results of the past indicate a potential for the future? I think they do. Biological control has been around for many years. It started in antiquity, but most of us are unaware of these early beginnings. The ancient Chinese manipulated ants in citrus groves to control insect pests. It was just over 100 years ago that a lady beetle, the Vedalia, introduced from New Zealand and Australia, saved the citrus industry in California from the ravages of an introduced pest, the exotic cottony-cushion scale. The spectacular control achieved by the Vedalia laid the groundwork for additional biocontrol projects in the United States. Over the years some were successful, some were not. Some introduced beneficials gave complete control, others partial control, and others were total failures. Despite the failures, overall the science of introducing natural enemies has been effective.

Coulson and Soper (1989), in their discussion of protocols for the introduction of biotic agents, list several of the more successful classical biological control projects in the United States. I have selected a few projects as examples of the financial benefits derived from the introduction of natural enemies (Table 1).

The average cost of the basic research leading to the control of each of these pests was about $1 million. The price tag varied depending upon the circumstances surrounding the specific project. Unfortunately, entomologists are, for the most part, poor economists and seldom determine the cost/benefit ratio of a successful project. However, for those few projects that were analyzed financially, the results more than justified the initial investments.

A noted biocontrol scientist recently published a score card for classical biological control that, I believe, is quite impressive (Rosen 1991). According to his calculations over 1,275 biocontrol projects have been conducted worldwide against 416 species of insect pests. Of these, 156 have been completely successful, 164 resulted in considerable reduction of the pest, and 64 were partially success-

TABLE 1. Examples of success in classical biological control*

| COMMON NAME | SCIENTIFIC NAME | PLANT OR HABITAT | EST. ANNUAL SAVINGS |
|---|---|---|---|
| | **INSECTS** | | |
| 1. ALFALFA BLOTCH LEAFMINER | *Agromyza frontella* (Rondani) | Alfalfa | $13,000,000 |
| 2. ALFALFA WEEVIL | *Hypera postica* (Gyllenhal) | Alfalfa | $48,900,000 |
| 3. BLACK SCALE | *Saissetia oleae* (Oliver) | Citrus | $4,000,000 |
| 4. CITROPHILOUS MEALYBUG | *Pseudococcus calceolariae* (Maskell) | Citrus | (3 and 4 combined) |
| 5. RHODEGRASS MEALYBUG | *Antonina graminis* (Maskell) | Turf | $17,000,000 |
| | **WEEDS** | | |
| 1. ALLIGATOR WEED | *Alternanthera philoxeroides* (Mart.) Griseb | Aquatic | $400,000 |
| 2. ST. JOHNSWORT | *Hypericum perforatum* (L.) | Range | $22,000,000 |

* Adapted from Coulson and Soper 1989.

ful. More than 75 species have been controlled on a permanent basis. Another 89 targets were brought under partial to substantial control. He added that the record for the biocontrol of weeds indicated 267 projects were "directed against 125 species of weeds with similar or better results." In examining these figures it must be pointed out that all too often biological control is considered after all else fails. In other words, we get the rejects.

Many projects ultimately classified as failures never should have been undertaken in the first place. The overall effort and support was insufficient to lead to success. Often the projects were poorly conceived, and the natural enemies had no chance from the start. Unmated females too old to mate were released; parasites were literally dumped out into the field in the heat of the day; specimens were weakened by being held too long in the laboratory before release. These kinds of examples are endless.

## WHY IS CLASSICAL BIOLOGICAL CONTROL AN ENDANGERED DISCIPLINE?

Classical biological control has never been a widely known or well-supported science. The number of totally committed or even properly educated practitioners has never been immense. Today, many scientists, known to some of us as "dabblers," are involved in biological control but only on a part-time basis. According to one of my colleagues there are only about a dozen or so *full-time* researchers in the federal service working in classical biocontrol of weeds. About the same number are involved in the classical approach for the control of insects. However, there are more than 200 entomologists who have one or more fingers in the pie.

Rosen (1991) also pointed out that the limitations of success in biocontrol are not based on the lack of technology or the need for elaborate research. "What we need," he stated, "is simply more dedicated biological control practitioners going out and getting the natural enemies. Basic research should be supported, but should not replace actual exploration."

Another reason for the perilous position of classical biocontrol is the lack of younger entomologists entering the field to replace those who are leaving because of retirement, other opportunities, or the lack of funding and support. Colleagues contacted at several universities confirmed the lack of "new blood" for classical biocontrol. Some professors have gone as far as discouraging students from entering biocontrol studies because of the grim future. Positions for new graduates in classical biological control are just not there, regardless of the public outcry for natural controls and the demands for cleaner environments.

Biotechnology and molecular biology are the "buzz words." It is here that available funding is being placed. The laboratory where I spent my waning years certainly was not heavily funded. I remember one year the budget for my laboratory was increased over the previous year by the appropriation-breaking amount of $1.00! This was at a time when there were new funds for biological control, or so we were told.

I would like to examine briefly several items that I believe have an adverse impact on classical biological control.

### (1) Long Budget Cycle

In the federal system the budget cycle takes too much time to allow for immediate or quick action to meet an unexpected situation. We work on budgets several years in advance. By the time the budget is established the promising targets may have changed and the funds should be redirected. Like any research program, biological control requires a long-term commitment. But the commitment has to be flexible enough to pounce on whatever project has the most promise. With many fishing lines in the pond, the one with the nibble is the one to play.

## (2) Instant Gratification

The consumers or producers wish to have their problems solved now. This puts a limitation on classical biological control. There is a considerable time span from the initial identification of the target pest to the introduction and spread of the natural enemy in the field. During this time lag the farmer may have been forced to turn to the increased use of insecticides to meet his needs. However, let us not forget that other areas of research do not produce instantaneous results either, and they may not be as visible as classical biological control while developing a control for a target pest. Industry tells the world it has a control when the product is placed on the shelf, but their product also took a long time to get to the shelf, maybe 13 years at a cost of $40 million or more (Menn and Christy 1992).

Furthermore, classical biological control is not necessarily the best solution for all areas of agriculture. Cosmetic appearance of the product, a low threshold of acceptable damage to the crop, and the transmission of a disease by the target organism even though in low numbers are some of the considerations that may limit the use of classical biological control in a given situation.

## (3) Lip Service

We all have been subject to silver tongues. Biological control has had its share. We hear about the clamor for biological control and the need for nonpolluting methods of control. We are being told that funds are being allocated for biocontrol and so on. Where are these funds? In a recent article relating to problems in biological control, M. J. Tauber (1991) of Cornell University made the comment that "Even existing (classical biological control) programs are grossly underfunded."

F. E. Gilstrap (1991) of Texas A&M stated that one of the major problems is that "State and federal agencies have not consistently or adequately supported classical biological control." At the laboratory where I worked for several years prior to retirement, there had not been a significant increase in funds at the bench level for years. We did not have to worry about space for a new person. No funds were available to hire a new scientist. The purple loosestrife project discussed by Stephen Hight during this symposium (Hight 1993) is now at the stage when the fruit is to be picked from the tree—the beneficials are to be released—but the meager funds are running out. What little money is left will not permit any of the in-depth research that should accompany a project of this scope and importance.

## (4) Sagging Economy

The sag in funding is hitting us all. There is not much to add except that, as the garment industry notes, some of us sag more than others. Classical biological control certainly could use more support.

## (5) Competition for Limited Funds

This is a combat with little quarter given. Unfortunately, classical biological control is not armed with enough of the appropriate "buzz words" or cutting-edge-of-research concepts to fare well in a world of molecular biology and advanced biotechnology. Classical biocontrol lacks "glitz!" It does not require impressive machinery or complicated laboratory techniques. It has the proven potential for successful control using the time-honored methods of basic entomology and applied ecology. When a project is successful, the problem quietly disappears. Usually the field is not littered with the dead bodies of the pest, indicating that the battle is over and won. Results like these apparently do not have the correct appeal for those who are handing out the big bucks. If this is true, then it is up to biocontrol entomologists to beat our own drum and raise the level of awareness of those who control the purse strings.

### (6) Conversion of Positions to Biotechnology

Biotechnology has been a thorn in the side of classical biocontrol scientists since the word was coined. Biotechnology is well-worth supporting—no question about it—but there must be an equitable emphasis between new research approaches with their potential for brilliant achievement, and the old successful research that has "paid the rent" for these many years. An incisive article by Van Driesche and Ferro (1987) entitled "Will the benefits of classical biological control be lost in the biotechnology stampede?" clearly states the problem facing classical biocontrol in relation to biotechnology. The authors are very concerned that classical biological control is "losing more ground to the glamor of biotechnology." They add that "positions are being lost. Classical biocontrol specialists are being reassigned to other duties . . . As universities and the USDA rush to embrace biotechnology, their neglect of biological control seriously threatens to make a poor situation worse." I have seen this occur within ARS and elsewhere. Individuals making many of the decisions regarding classical biological control do not have an appreciation or an understanding of classical biocontrol. The impression given is that biotechnology will solve the problems, so why waste time looking for natural enemies? Again to quote Van Driesche and Farro, "Biotechnology has yet to control its first pest: classical biological control has worked for 100 years, and yet its potential has barely been touched."

Unless we can alleviate the budget crunch and motivate students to enter the field of classical biological control, this field of research will continue its decline. The science will not be able to meet the present and increasing demands made on it for integrated pest management, sustainable agriculture, organic farming or other approaches that are concerned with agricultural production and, protection of the environment.

## WHY IS CLASSICAL BIOLOGICAL CONTROL A DANGEROUS DISCIPLINE?

Why should classical biocontrol be considered dangerous? The theme of this symposium is biological pollution by exotic invaders. Our purpose in the classical approach to biocontrol is to introduce exotic organisms that are beneficial and are safe for the environment. If we do not have adequate fiscal support and well-trained personnel to conduct the essential research to meet these goals, we are courting disaster. Any major error in the process of introducing a natural enemy could lead to a major environmental disaster that could be national or international in scale. Any activity with that potential must be considered dangerous.

What could cause this dangerous situation? An introduced organism knows no political boundaries. To quote P. Harris (1990), a Canadian biocontrol specialist, "Biological control cannot be used on a trial basis to determine if its effects are desirable." An unwanted creature released into the environment because of inadequate, unprofessional research by insufficiently trained or motivated scientists, research conducted in second-rate facilities, or a misidentification by an overloaded or poorly supported taxonomist may be impossible to erradicate or contain. The results of this scientific crime could be a form of biological pollution that would rival the pollution caused by the gypsy moth, the fire ant, kudzu, or purple loosestrife.

To date, we have a very good record on the importation of invertebrate natural enemies to control pests, thanks to the integrity of the scientists involved and to Lady Luck, especially when classical biocontrol was operating without the strict quarantine controls and importation restrictions that exist today. However, if support for classical biocontrol continues to dwindle but the demands placed on it continue to increase, the potential for pollution increases.

The continuing paucity of students entering the field of biological control can, and will, result in more demands on the professionals already in the field. Eventually these professionals will have to be replaced. Under the present situation the ranks will be filled by scientists poorly trained in the field of biocontrol. Even now we have researchers leading some biocontrol laboratories who have had little or no training in classical biocontrol. They may be top-rate specialists in other fields of entomology and may be the best available to fill these important posts, but they are not dyed-in-the-wool biocontrol entomologists.

Furthermore, due to insufficient funding we are letting some of our best biocontrol practitioners go at a time when the call for nonchemical means of control is increasing. Positions made vacant by retirement are not being filled by biocontrol specialists. They are being filled by laboratory-oriented researchers. One of the most effective classical biocontrol entomologists I have ever known summarized the need for field-oriented research by the comment, "I have never found a natural enemy in the laboratory."

## ITEMS FOR CONSIDERATION

It is much easier to point out the problems than it is to suggest possible approaches to the solutions. Nevertheless, I have included the following as food for thought if there is to be an effort to rejuvenate classical biological control and to make it a safer practice.

**Increased Funding.** This universal cry need not be too loud in biocontrol. Using the dollar standard suggested by Koshland (1986), classical biocontrol is not a "big science" in the $1 billion range or even in the "low income" range of $100,000. We are in the poverty range of $10,000 annually for operating funds at the bench level. Consequently, new life blood would not need an overwhelming transfusion. Just the equivalent cost of a couple of electron microscopes or ultracentifuges would put many of us back in business.

**Redirect Existing Funds.** This can be a painful operation unless the personnel involved are also redirected into an area that is viable and where they are competent and comfortable. This would include combining resources and developing closer cooperation between federal and state organizations and between the states. The available funds may be spread too thin over too many projects that have not produced a "nibble."

**Reexamine the Goals of Classical Biocontrol.** We need more scientists than we have now in the field looking for natural enemies. Too much of today's research in the name of classical biocontrol is taking place in the laboratory at the expense of good basic field exploration and studies. The publish or perish syndrome may be a contributing factor to this situation. Foreign travel does not yield the number of papers that a sedentary lifestyle in the laboratory can produce.

**Reexamine the Targets.** Because an insect or weed is a multimillion dollar pest, it still may not be a suitable subject for classical biological control. Only those pests that appear to be susceptible to control by natural enemies should be considered.

**Dialogue with Industry** to select appropriate targets and avoid the appearance of competition. There are many insects and weeds in areas that are financially impossible to control with acceptable pesticides. Pests of grazing lands, wetlands, forests, and natural areas are ideal targets for the biocontrol approach.

**Examine the Need for and Use of Existing Quarantine Facilities.** Increase federal support for those facilities that should be national in scope and function. A combination of use of federal laboratories by trained state and federal personnel may alleviate the overload during peak seasons or peak periods of a project. A proliferation of quarantine stations may increase the danger of pollution above any acceptable level. Competent quarantine officers are a rare breed. Furthermore, it would help eliminate the less-than-ideal quarantine stations along with cavalier attitudes that presently exist in some areas. A quarantine operation program must have first-class facilities and personnel. Second class is not good enough.

**Greater Publicity of Successes.** Each time there is a complete success in classical biological control, the project is closed and it fades away. Unfortunately, funds are rarely available for the proper evaluation of a project before it is terminated. Little is heard of the role biocontrol played in the control of the organism. "No one remembers costs they no longer have to pay" (Van Driesch and Ferro 1987). There should be a distinct effort to publicize the role of biocontrol and its importance. Scientist-to-scientist dialogue does not raise funds.

In summary, we must conduct classical biological control as it should be done. The margin of error is small with the best of precautions and the best-trained scientists. This field of research need not be ashamed of past accomplishments, as they have been impressive, if essentially unknown. However, if support is continued at the present scale, or reduced, and this field of endeavor is not made more attractive to our young dedicated researchers, classical biological control will continue to be endangered; it will not be able to meet the demands of society, and it may even cause biological pollution that could devastate our environment and erase any benefits that an excellent but unsung record has produced.

## ACKNOWLEDGMENTS

My thanks to R. Schroder and S. Hight, USDA, ARS, Beltsville, Md.; M. Tauber, Entomology Department, Cornell University, Ithaca, N.Y., and F. Gilstrap, Department of Entomology, Texas A&M University, College Station, for constructive comments and suggestions. Special thanks are due the numerous researchers at different locations who spoke frankly of the conditions within their organization and their concerns for the future of classical biological control.

## REFERENCES

COULSON, J. R. and R. S. SOPER. 1989. Protocols for the introduction of biological control agents in the U.S. Pages 1–35 *in* Plant protection and quarantine. R. P. Kahn (ed.). Special Topics, vol. III CRC Press, Boca Raton, Fla.

GILSTRAP, F. E. 1991. Comments on Ehlers's "Revitalizing Biological Control." Newsletter, Nearctic Region. Sect., Intern. Org. Bio. Control, Summer, No. 37. p. 16–17.

HARRIS, P. 1990. Environmental impact of introduced biological control agents. Pages 289–300 *in* Critical issues in biological control, M. Mackauer, L. E. Ehler, and J. Roland (eds.). Intercept Ltd., Andover, Hants, U.K.

HIGHT, S. D. 1993. Control of the ornamental purple loosestrife by exotic organisms. Pages 147–48 *in* Biological pollution: the control and impact of invasive exotic species, B.N. McKnight (ed.). Indiana Acad. Sci., Indianapolis.

KOSHLAND, D. E. 1986. "To lift the lamp beside the research door." Science 233 (no. 4767):609.

MENN, J. J. and A. L. CHRISTY. 1992. New directions in pest management. Pages 409–30 *in* Fate of pesticides and chemicals in the environment. J. L. Schnoor (ed.). John Wiley & Sons.

ROSEN, D. 1991. Forum. Newsletter, Nearctic Region. Sect., Intern. Org. Bio. Control, Summer, No. 37, p. 14–15.

TAUBER, M. J. 1991. Forum. Newletter, Nearctic region. Sect., Intern. Org. Bio. Control, Summer, No. 37, p. 15–16.

VAN DRIECHE, R. G. and D. N. FERRO. 1987. Will the benefits of classical biological control be lost in the "biotechnology stampede"? Amer. J. Alternative Agric. II(2):50, 96.

# Invasive Ecological Dominants: Environments Boar-ed to Tears and Living on Burro-ed Time

Bruce E. Coblentz[1]

ABSTRACT: Human-assisted movement of organisms between realms and the resulting global homogenization of biotas since European colonization began are unprecedented in the biotic history of the planet. Of the three basic forms of environmental perturbation: inappropriate resource use, pollution, and exotic organisms, only exotic organisms are permanent in ecological time. Cessation of polluting or inappropriate resource use results in improved conditions; in contrast, most exotic organisms once established in a community are frequently permanent. Exotic herbivorous mammals, particularly in depauperate communities such as oceanic islands, often become ecologically dominant and can result in wholesale extinctions within several trophic levels as well as physical degradation of the environment. In insular scenarios, numerous extinctions are guaranteed from the instant of introduction, even if 200 years might be required for the individual to perish; thus extinction of an endemic could be said to occur on the day of introduction of the exotic. In mainland environments, exotic herbivores are more likely to cause extirpation of sensitive or conservative species in addition to degradation of the physical environment. Feral equids in North America are particularly guilty of causing these effects, yet are protected by Public Law 92–195, the Wild and Free-Roaming Horses and Burros Act of 1971. Regardless of 0.5 million years of separate evolution of Old and New World equids, the United States Congress declared feral equids in the western states to be "an integral part of the natural system of public lands." Thus, by law, Congress decided it could override (1) the evolutionary process and (2) the extinction of North American equids at the close of the Pleistocene.

Wild boar/feral pigs (*Sus scrofa*) have extensive negative effects on native taxa of several trophic levels in a remarkable breadth of environments. Pigs alter native flora through direct consumption, especially of palatable fruits and seeds and sensitive herbaceous flora with highly savory underground parts, and also through their rooting and trampling activities. Pigs also consume soil invertebrates and any small vertebrates they happen to catch and are particularly destructive to ground-nesting vertebrates. Pigs have especially noticeable effects on oceanic islands and were a major factor in the failure of recruitment of giant tortoises (*Geochelone*) and the extinction of land iguana (*Conolophus*) in the Galapagos.

Saving intact biotic communities simply will not be possible when they are negatively affected by exotic herbivorous mammals. In an environment lacking sufficient predation pressure or other checks on an introduced large herbivore, the introduced species can rapidly become ecologically dominant and cause extensive reduction of biodiversity and habitat degradation. When an invasive exotic can be controlled, the only ecologically responsible action for a resource management agency is control, at the least, or preferably eradication. Of equal importance is the need to guard against further introductions and to guard

1. Department of Fisheries and Wildlife, 104 Nash Hall, Oregon State University, Corvallis, OR 97331–3803

against the acceptance of exotics as part of the non-native community in which they might be found. Exotics should never be granted legitimacy.

Several themes of this paper were borrowed from a paper published previously (Coblentz 1990).

## REFERENCE

COBLENTZ, B. E. 1990. Exotic organisms: a dilemma for conservation biology. Conserv. Biol. 4:261–65.

# Exclusion and Eradication of Foreign Weeds from the United States by USDA APHIS

Randy G. Westbrooks[1]

## NEW WEEDS AS BIOLOGICAL POLLUTANTS IN THE UNITED STATES

For several years there has been a growing concern about the introduction of foreign weeds into the United States (Mooney and Drake 1986; Zamora et al. 1989; Eplee and Westbrooks 1990; Schmitz 1990; Westman 1990; Westbrooks 1991, 1992). Such concerns are based on problems that have been caused by exotic species that were introduced in the past. Exotic species, including foreign weeds, have been termed *biological pollutants* because of their unbalancing effects on natural and agricultural ecosystems (Westbrooks 1991). Unlike chemical pollutants that usually begin to degrade soon after their release into the environment, biological pollutants have the potential to persist, multiply, and spread. Freed from competition from co-evolved parasites and predators, alien species often displace the native biota in natural areas. In agricultural settings they tend to complicate existing weed management practices and hinder crop production. The major classes of biological pollutants include:

1. agricultural weeds
2. botanical invaders of natural areas
3. plant and animal diseases (such as chestnut blight and hog cholera)
4. animal invaders
   a. vertebrates (such as fish and wild horses)
   b. invertebrates (such as exotic mussels)
   c. insects (such as exotic fruitflies).

## THE WORLD MOVEMENT OF BIOLOGICAL POLLUTANTS

The potential distribution of plants is generally determined by historical climate changes, modern climate, geographical changes (mountain building, continental drift, etc.), soils, and long-range dispersal mechanisms (Good 1974). Normally, long-distance dispersal of plants is accomplished by adaptations of propagules for transport by wind, water, migratory fowl, and animals (Ridley 1930). However, the rate of interchange between distant regions increased significantly with the advent of agriculture and long-distance travel by man. In particular, oceanic travel and movement of cargo and personal effects by sailing ships provided an efficient new mode for dispersal of hitchhiking plants and animals into previously inaccessible regions of the earth.

Today intercontinental traveling times have been reduced from months to days (cargo ships) and even hours (jet airliners). This has allowed the worldwide

1. Weed Specialist and Station Leader, Whiteville Noxious Weed Station, Whiteville Plant Methods Center, Plant Protection and Quarantine, Animal and Plant Health Inspection Service, United States Department of Agriculture, Whiteville NC 28472

movement of species on an unprecedented scale and will continually homogenize biotic communities worldwide if left unchecked. Ecologist Charles Elton (1958) has described the enormous increases in immigrant populations as indications of the "terrific dislocations in nature being caused by the mingling of thousands of kinds of organisms from different parts of the world." He contended that such invasions could cause the biological world to become simpler and poorer as aggressive immigrant species wipe out or displace native species. According to biologists Harold Mooney and James Drake of Stanford University (1986), it is now clear that many purposeful introductions " . . . have not had the intended effect and further they have been accompanied by many accidental interchanges, some of which have had disasterous economic and environmental impacts."

## AVENUES OF ENTRY FOR BIOLOGICAL POLLUTANTS

Exotic species are being dispersed around the world by man in numerous ways. However, these modes of transport may be characterized as accidental (as a contaminant of a commodity or various means of conveyance) or intentional (as an ornamental or crop, etc.). A few examples of biological pollutants and how they were introduced into the United States include the following species:

**The Asian tiger mosquito** (*Ades albopictus* Skuse), a dangerous disease-carrying mosquito that was probably introduced into the United States in the larval form in water-filled used tires (Livdahl and Willey 1991). In 1989 this mosquito was found in Florida only in Duval County. It is now known in 64 of 67 counties in the state (G. Craig, pers. comm.) and as far west as Texas.

**The Asian gypsy moth** (*Lymantria dispar* L.), which threatens forests in the Northwest, was recently introduced to the West Coast through grain ships from eastern ports in Russia (Gibbons 1992). This strain appears much more dangerous than the European gypsy moth because the female can fly and deposit her eggs many miles from the point of fertilization. Females of the European gypsy moth (same species, different strain; accidentally released in the northeastern United States in the late 1860s) do not fly. Instead, caterpillars of the European strain are spread short-distances by wind currents and long-distances by hitchhiking on motorhomes and other vehicles.

**Johnson grass** (*Sorghum halepense* (L.) Pers.), which is listed as the sixth worst weed in the world (Holm et al. 1977), was repeatedly introduced into the United States as a forage and pasture grass in the early 1800s. By the late 1800s it had become a major weed of row crops throughout the country (McWhorter 1971).

**Zebra mussel** (*Dreissena polymorpha* Pallas) is a small mollusc native to eastern Europe that clogs water intakes and covers almost any other stationary submarine structure. It was discovered in Lake St. Clair and western Lake Erie in 1988, and now occurs in all of the Great Lakes and the major inland waterways (Allegheny, Hudson, Illinois, Mississippi, and Ohio River drainages). It was introduced into the Great Lakes from eastern Europe in the mid-1980s in ballast water from cargo ships (J. McCann, U.S. Fish & Wildlife Service, pers. comm.).

**Paperbark tree** (*Melaleuca quinquenervia* (Cav.) Blake) was introduced from northern Australia into south Florida in the early 1900s first as an ornamental and later to dry up the Everglades for use as farmland. This species now threatens to destroy the Everglades, the "River of Grass" (Westbrooks 1991).

In most cases, incipient populations of such organisms are viewed with curiosity without much concern for their potential as pests until major ecological or economic problems begin to occur (Westbrooks 1991). However, by that time they are usually firmly entrenched and almost always impractical to eradicate.

## EXCLUSION OF FOREIGN AGRICULTURAL PESTS BY USDA APHIS

The Animal and Plant Health Inspection Service (APHIS) is an agency of the United States Department of Agriculture. One of its missions is to prevent the entry and establishment of designated foreign pests into the United States. The types of foreign pests targeted for exclusion by APHIS include certain insects, plant diseases, animal diseases, molluscs, nematodes, and weeds.

Plant Protection and Quarantine (PPQ) is the operational section of APHIS that is responsible for excluding foreign pests at U.S. ports of entry. Strategies utilized by PPQ to achieve this goal include:

**Prevention** (encouraging or requiring that certain imported commodities be certified as pest free when imported into the United States; i.e., meat imported from other countries must be certified as originating in an area that is free of diseases such as hog cholera, foot and mouth disease, and swine vesicular disease).

**Preclearance** (inspection of certain commodities at the port of export, prior to being shipped to the United States).

**Exclusion** (port of entry inspections and treatments, etc., by PPQ officers at all U.S. ports of entry to detect prohibited pests in imported commodities; treatments to eliminate pest risk).

**Detection** (conducting surveys and communicating with scientists and state agencies for early detection of incipient infestations of prohibited foreign species).

**Containment** (establishment of regulatory rules and programs to prevent the spread of prohibited species out of infested areas).

**Eradication** (elimination of incipient infestations of prohibited species by appropriate means).

Management strategies for species that become firmly established include traditional control and biological control. Various agencies and institutions utilize some or most of these strategies in dealing with exotic plants. However, strategies for prevention, preclearance, and exclusion are almost exclusively within the domain of USDA APHIS (see Fig. 1).

USDA APHIS has been charged by the Congress with exclusion of foreign pests that are of agricultural significance. However, *exclusion of foreign species that threaten natural areas* is one potential expansion of this responsibility. Eradication, traditional control and biological control of exotic invasive species in natural areas within the United States may logically become the responsibility of a number of agencies or institutions, including APHIS.

## FOREIGN PEST INTERCEPTIONS BY USDA APHIS

Over the past few decades, USDA inspectors have become quite proficient at detecting prohibited insects, diseases, and other pests that infest or contaminate imported cargo at U.S. ports of entry. This expertise is generally learned on-the-

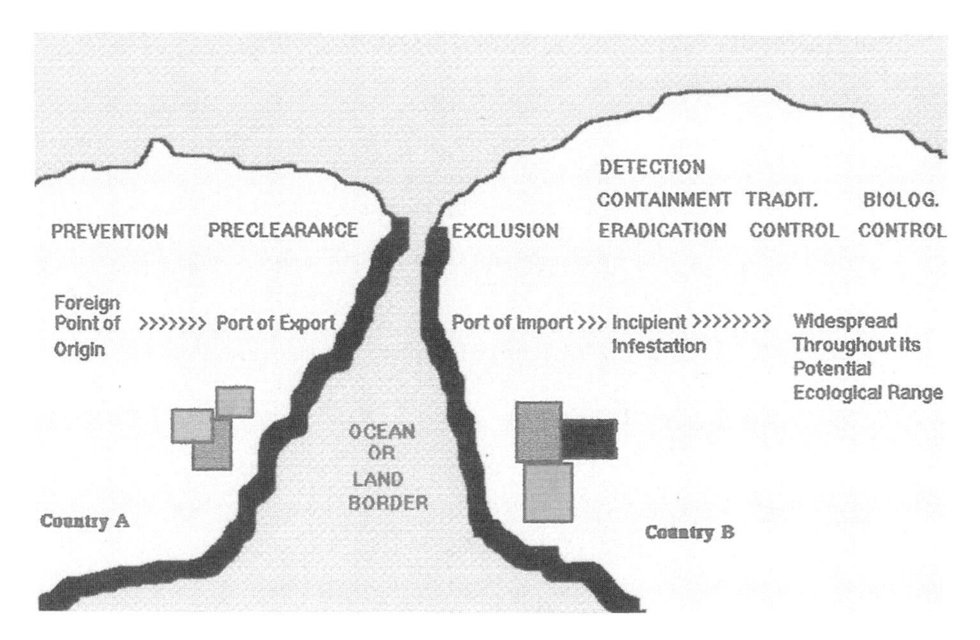


FIGURE 1.—Management strategies for dealing with foreign pests (biological pollutants).

job as inspectors intercept new pests and share the information with each other. For example, if prohibited wood-boring insects are found to infest wood crating around marble from Italy by an inspector in Charleston, South Carolina, a special notice will be sent to other ports using electronic mail. Such interception alerts allow inspectors to concentrate their efforts on cargo from certain countries with a history of pest risk. Over time, such information has accumulated into an extensive interception record that guides inspectors in their daily work. In 1987 and 1988, PPQ officers nationwide made a total of 82,560 interceptions of prohibited insects, 16,400 prohibited plant diseases, and 1,247 prohibited weeds (USDA APHIS interception records).

## AUTHORITY TO REGULATE NOXIOUS WEEDS WITHIN THE UNITED STATES

APHIS was assigned the general responsibility for excluding foreign weeds in 1976 and responsibility for inspecting imported seeds in 1982. Laws that authorize federal agencies to regulate weeds include the following:

**Federal Seed Act (1939).** Regulates purity and truth-in-labeling of imported crop and vegetable seeds; sets tolerance for contamination by nine species of crop and rangeland weeds.

**Organic Act (1944).** Authorizes eradication of incipient infestations of pests.

**Federal Plant Pest Act (1957).** Provides authority to regulate the introduction of economically important parasitic plants.

**Federal Noxious Weed Act (1974).** The FNWA authorizes exclusion of designated foreign weeds that do not occur in the United States or are of limited distribution within it; regulation of interstate movement of designated species under quarantine; cooperation with state and local agencies or individuals in eradicating incipient infestations; control of noxious weeds on federal lands.

## WEED SPECIES CURRENTLY LISTED AS FEDERAL NOXIOUS WEEDS

In 1976, 26 taxa of foreign weeds were designated as Federal Noxious Weeds (FNW) by the U.S. Department of Agriculture. By 1984 the FNW list had been expanded to include 87 species, plus all species of *Aeginetia*, *Alectra*, and *Striga*, plus all species of *Cuscuta* and *Orobanche* that are not native to the United States. The most recent addition to the federal list was paperbark tree [*Melaleuca quinquenervia* (Cav.) Blake], on April 13, 1992. Table 1 lists all taxa now designated as Federal Noxious Weeds.

## FOREIGN WEED EXCLUSION BY USDA APHIS. THE EARLY YEARS

Between 1976 and 1988, APHIS personnel were provided a list of target species, a short list of high-risk commodities, and sampling procedures for inspecting commodities for noxious weeds to guide them in enforcing the FNWA. During this period, greasy (raw) wool, oil contaminated equipment, aquatic plant shipments, and seed shipments were recognized as high-risk vectors for introducing foreign weeds into the United States (Westbrooks and Eplee 1987). However, little information was provided about the target species themselves or which commodities might act as vectors in their dispersal during transport and/or storage. Without photos, seed samples, and pertinent information on each species, early FNW interceptions were usually made by chance rather than design. In April 1979, PPQ officials in Oakland, California, rejected entry of a commercial shipment (909 kg) of kikuyu grass (*Pennisetum clandestinum* Hochstetter ex Chiov.) from the New Hebrides Islands (now called Vanuatu) (APHIS PPQ interception records). This was the first time a noxious weed had been denied entry under the provisions of the FNWA.

TABLE 1. Plant taxa designated as Federal Noxious Weeds by the United States Department of Agriculture.

| Taxon | Common name |
|---|---|
| **AQUATIC SPECIES** | |
| *Azolla pinnata* R. Brown | MOSQUITO FERN |
| *Eichhornia azurea* (Swartz) Kunth | ANCHORED WATER-HYACINTH |
| *Hydrilla verticillata* (L. f.) Royle | HYDRILLA |
| *Hygrophila polysperma* T. Anderson | MIRAMAR WEED |
| *Ipomoea aquatica* Forsskal [ = *I. reptans* (L.) Poir.] | WATER SPINACH, SWAMP MORNING GLORY |
| *Lagarosiphon major* (Ridley) Moss | OXYGEN WEED |
| *Limnophila sessiliflora* (Vahl) Blume | AMBULIA |
| *Monochoria hastata* (L.) Solms-Laubach | GIANT ARROWHEAD |
| *Monochoria vaginalis* (Burman f.) C. Presl | |
| *Sagittaria sagittifolia* L. | |
| *Salvinia auriculata* Aublet | GIANT SALVINIA |
| *Salvinia biloba* Raddi | GIANT SALVINIA |

TABLE 1. (cont.)

| | |
|---|---|
| *Salvinia herzogii* de la Sota | GIANT SALVINIA |
| *Salvinia molesta* D.S. Mitchell | GIANT SALVINIA |
| *Sparganium erectum* L. | EXOTIC BUR REED |
| *Stratiotes aloides* L. | WATER-ALOE, WATER LETTUCE |
| **TERRESTRIAL SPECIES** | |
| *Ageratina adenophora* (Sprengel) King & Rob. [ = *Eupatorium adenophorum* Spreng.] | CROFTON WEED |
| *Alternanthera sessilis* (L.) R. Brown ex DC. | SESSILE JOYWEED |
| *Asphodelus fistulosus* L. | ONIONWEED |
| *Avena sterilis* L. (including *A. ludoviciana* Durieu) | ANIMATED OAT, WILD OAT |
| *Borreria alata* (Aublet) DC. | |
| *Carthamus oxycantha* M. Bieberstein | WILD SAFFLOWER |
| *Chrysopogon aciculatus* (Retz.) Trinius | PILIPILIULA |
| *Commelina benghalensis* L. | BENGHAL DAYFLOWER |
| *Crupina vulgaris* Cassini | COMMON CRUPINA |
| *Digitaria abyssinica* (A. Rich) Stapf. [ = *D. scalarum* (Schweinfurth) Chiov.] | AFRICAN COUCH GRASS, FINGER GRASS |
| *Digitaria velutina* (Forssk.) Palisot de Beauv. | VELVET FINGER GRASS, ANNUAL COUCH GRASS |
| *Drymaria arenarioides* Humboldt & Bonpl. ex Roemer & Schultes | LIGHTNING WEED |
| *Emex australis* Steinhall | THREE-CORNERED JACK |
| *Emex spinosa* (L.) Compdera | DEVIL'S THORN |
| *Euphorbia prunifolia* Jacquin | PAINTED EUPHORBIA |
| *Galega officinalis* L. | GOATSRUE |
| *Heracleum mantegazzianum* Sommier & Levier | GIANT HOGWEED |
| *Imperata brasiliensis* Trinius | BRAZILIAN SATINTAIL |
| *Imperata cylindrica* (L.) Raeuschel | COGON GRASS |
| *Ipomoea triloba* L. | LITTLE BELL, AIEA MORNING GLORY |
| *Ischaemum rugosum* Salisbury | MURAINO GRASS |
| *Leptochloa chinensis* (L.) Nees | ASIAN SPRANGLETOP |
| *Lycium ferocissimum* Miers | AFRICAN BOXTHORN |
| *Melaleuca quinquenervia* (Cav.) Blake | PAPERBARK TREE |
| *Melastoma malabathricum* L. | |
| *Mikania cordata* (Burman f.) Robinson | MILE-A-MINUTE |
| *Mikania micrantha* Humboldt, Bonpland & Kunth | |
| *Mimosa invisa* Martius | GIANT SENSITIVE PLANT |
| *Mimosa pigra* L. var. *pigra* | CATCLAW MIMOSA |
| *Nassella trichotoma* (Nees) Hackel ex Arech. | SERRATED TUSSOCK |
| *Opuntia aurantiaca* Lindley | JOINTED PRICKLY PEAR |
| *Oryza longistaminata* Chev. & Roehrich | RED RICE |
| *Oryza punctata* Kotschy ex Steudel | RED RICE |
| *Oryza rufipogon* Griffith | ASIAN COMMON WILD RICE |
| *Paspalum scrobiculatum* L. | KODOMILLET |
| *Pennisetum clandestinum* Hochstetter ex Chiov. | KIKUYU GRASS |
| *Pennisetum macrourum* Trinius | AFRICAN FEATHER GRASS |
| *Pennisetum pedicellatum* Trinius | KYASUMA GRASS |
| *Pennisetum polystachion* (L.) Schultes | MISSION GRASS, THIN NAPIER GRASS |
| *Prosopis alapataco* R. A. Philippi | ALAPATACO |

TABLE 1. (cont.)

| | |
|---|---|
| *Prosopis argentina* Burkart | ALGAROBILLA |
| *Prosopis articulata* S. Watson | — |
| *Prosopis burkartii* Munoz | — |
| *Prosopis caldenia* Burkart | CALDEN |
| *Prosopis calingastana* Burkart | CUSQUI |
| *Prosopis campestris* Grisebach | ALGAROBILLA |
| *Prosopis castellanosii* Burkart | — |
| *Prosopis denudans* Bentham | ALGARROBO PATAGONICA |
| *Prosopis elata* (Burkart) Burkart | ALGARROBITO |
| *Prosopis farcta* (Solander ex Russell) Macbride | ACATIA |
| *Prosopis ferox* Grisebach | CHURQUI |
| *Prosopis fiebrigii* Harms | — |
| *Prosopis hassleri* Harms | ALGARROBO PARAGUAYO |
| *Prosopis humilis* Gillies ex Hooker & Arnott | BARBE DE TIGRE |
| *Prosopis kuntzei* Harms | JACARANDA |
| *Prosopis pallida* (Humboldt & Bonpland ex Willdenow) Humboldt, Bonpland & Kunth | KIAWE |
| *Prosopis palmeri* S. Watson | PALO DE HIERRO |
| *Prosopis reptans* Bentham var. *reptans* | MASTUERZO |
| *Prosopis rojasiana* Burkart | — |
| *Prosopis ruizlealii* Burkart | — |
| *Prosopis ruscifolia* Grisebach | VINAL |
| *Prosopis sericantha* Gillies ex Hooker & Arnott | ALBARDON |
| *Prosopis strombulifera* (Lamarck) Bentham | ESPINILLA |
| *Prosopis torquata* (Cavanilles ex Lagasca y Segura) DC. | LATA |
| *Rottboellia cochinchinensis* (Lour.) Clayton [ = *R. exaltata* L. f.] | ITCH GRASS, CORN GRASS, RAOUL GRASS |
| *Rubus fruticosus* L. (complex) | WILD BLACKBERRY |
| *Rubus mollucanus* L. | WILD RASPBERRY |
| *Saccharum spontaneum* L. | WILD SUGARCANE |
| *Salsola vermiculata* L. | WORMLEAF SALSOLA, MEDITERRANEAN SALTWORT |
| *Setaria pallide-fusca* (Schumacher) Stapf & Hubbard | CATTAIL GRASS |
| *Solanum torvum* Swartz | TURKEYBERRY |
| *Tridax procumbens* L. | COAT BUTTONS |
| *Urochloa panicoides* Beauvois | LIVERSEED GRASS |
| **PARASITIC SPECIES** | |
| *Aeginetia* spp. (all species) | |
| *Alectra* spp. (all species) | |
| *Cuscuta* spp. (Dodders), other than the following species: | |
| *C. americana* L. | |
| *C. applanta* Engelmann | |
| *C. approximata* Babington | |
| *C. attenuata* Waterfall | |
| *C. boldinghii* Urban | |
| *C. brachycalyx* (Yuncker) Yuncker | |
| *C. californica* Hooker & Arnott | |
| *C. cassytoides* Nees ex Engelmann | |
| *C. ceanothii* Behr | |
| *C. cephalanthii* Engelmann | |
| *C. compacta* Jussieu | |

TABLE 1. (cont.)

C. *corylii* Engelmann
C. *cuspidata* Engelmann
C. *decipiens* Yuncker
C. *dentasquamata* Yuncker
C. *denticulata* Engelmann
C. *epilinum* Weihe
C. *epithymum* (L.) L.
C. *erosa* Yuncker
C. *europaea* L.
C. *exalta* Engelmann
C. *fasciculata* Yuncker
C. *glabrior* (Engelmann) Yuncker
C. *globulosa* Bentham
C. *glomerata* Choisy
C. *gronovii* Willdenow
C. *harperi* Small
C. *howelliana* Rubtzoff
C. *indecora* Choisy
C. *jepsonii* Yuncker
C. *leptantha* Engelmann
C. *mitriformis* Engelmann
C. *nevadensis* I. M. Johnston
C. *obtusifolia* Humboldt, Bonpland & Kunth
C. *occidentalis* Millspaugh ex Mill & Nuttall
C. *odontolepis* Engelmann
C. *pentagona* Engelmann [= C. *campestris* Yuncker]
C. *planiflora* Tenore
C. *plattensis* A. Nelson
C. *polygonorum* Engelmann
C. *rostrata* Shuttleworth ex Engelmann
C. *runyonii* Yuncker
C. *salina* Engelmann
C. *sandwichiana* Choisy
C. *squamata* Engelmann
C. *suaveolens* Seringe
C. *suksdorfii* Yuncker
C. *tuberculata* Brandegee
C. *umbellata* Humboldt, Bonpland & Kunth
C. *umbrosa* Beyrich ex Hooker
C. *vetchii* Brandegee
C. *warneri* Yuncker
*Orobanche* spp. (BROOMRAPES), other than the following species:
O. *bulbosa* (Gray) G. Beck
O. *californica* Schlechtendal & Chamisso
O. *cooperi* (Gray) Heller
O. *corymbosa* (Rydberg) Ferris
O. *dugesii* (S. Watson) Munz
O. *fasciculata* Nuttall
O. *ludoviciana* Nuttall
O. *multicaulis* Brandegee
O. *parishii* (Jepson) Heckard
O. *pinorum* Geyer ex Hooker

TABLE 1. (cont.)

| |
|---|
| *O. uniflora* L. |
| *O. valida* Jepson |
| *O. vallicola* (Jepson) Heckard |
| *Striga* spp. (WITCHWEEDS, all species) |

Since the mid-1980s, PPQ officers at major international airports, maritime ports, and land border crossings in the United States have become proficient at intercepting certain FNWs. Nine common interceptions include: water-spinach (*Ipomoea aquatica* Forsskal), giant hogweed (*Heracleum mantegazzianum* Sommier & Levier), turkeyberry (*Solanum torvum* Swartz), onionweed (*Asphodelus fistulosus* L.), wild oats (*Avena* spp.), *Euphorbia* spp., *Pennisetum* spp., *Setaria* spp., and *Oryza* spp. (only rice spikelets with a red pericarp or bran) (Fowler 1990, 1991). These species are generally intercepted in passenger baggage and with lesser frequency as a contaminant of imported cargo. As such, they represent the core of an expanding APHIS Noxious Weed interception record.

## THE NOXIOUS WEED INSPECTION SYSTEM

Even without some assistance or training, it is likely that APHIS inspectors would eventually become proficient at intercepting most easily identifiable or detectable FNWs. However, during this time additional undesirable foreign weeds may be introduced and become established. To minimize this threat, APHIS PPQ has taken a proactive approach to exclusion and eradication of foreign weeds.

APHIS's ability to exclude foreign weeds has been enhanced by the development of a Noxious Weed Inspection System (NWIS). NWIS was designed by PPQ Methods Development to assist PPQ officers in planning and conducting inspectional activities to detect foreign weeds at U.S. ports of entry. The NWIS makes associations between FNWs and commodities that originate together in agricultural or natural settings. This allows officers to concentrate their efforts on truly high-risk commodities that may harbor foreign weeds. The NWIS is comprised of a Federal Noxious Weed Inspection Guide, a Federal Noxious Weed Identification Guide and a Noxious Weed Seed Collection. Each PPQ work station at all U.S. ports of entry is scheduled to receive at least one set of these resource materials (Eplee and Westbrooks 1991).

### Federal Noxious Weed Inspection Guide

The NWIS Inspection Guide is a comprehensive manual that identifies high-risk commodities from certain countries for inspection. The main tenet of the Inspection Guide is that "certain weeds are likely to be associated with certain commodities from certain countries" (Westbrooks and Eplee 1987). Table 2 provides general associations between different types of weeds and commodities. The Guide was developed to provide answers to numerous questions that might be posed by a PPQ officer at a port of entry. Topics covered by the guide include FNW distributions by weed and country, diagnostic features, basis for being listed as a Federal Noxious Weed, habitat, plant part likely to be intercepted, possible avenues into the United States, notes of interest on each species, and important literature references. It also provides a summary of all FNW interceptions that have been made by PPQ inspectors since 1984.

TABLE 2. Potential associations between different types of weeds and commodities.

| TYPE OF COMMODITY | EXAMPLE | TYPICAL ASSOCIATED WEEDS |
|---|---|---|
| Pasture Product | Raw Wool | Pasture Weeds |
| Pasture Product | Grass Seed | Pasture Weeds |
| Row Crop | Corn | Weeds of Cultivation |
| Aquatic Plants | Water Hyacinth | Aquatic Weeds |
| Wood Products | Bark | Weeds of Forests |
| Machinery | Army Tank | Any Type in Soil Contamination |

### Federal Noxious Weed Identification Guide

The NWIS Identification Guide is comprised of technical monographs on each FNW species. Each monograph was written by a specialist for use by PPQ personnel and other interested persons. Currently, botanical terminology utilized in the original monographs is being revised to make the ID Guide more "user friendly" for non-botanists. Monographs are also being developed for 10 species that were not included in the original contract. Figure 2 (*Orobanche minor* Smith) is an example of line drawings that were prepared for each monograph.

### Noxious Weed Identification Collection

The NWIS Identification Collection is a set of devitalized propagules of representative Federal Noxious Weeds and all Federal Seed Act Weeds. All together, the set is comprised of:

— Propagules of 69 representative Federal Noxious Weeds;
— 35 mm slides of six species of Federal Noxious Weeds that are most likely to enter the United States in the vegetative form (five aquatics; one cactus), and
— Propagules of eight Federal Seed Act Weeds.

Seeds of most FNWs used in the collections were obtained under federal and state permit from various sources overseas. However, a few were obtained from domestic sources within the United States. Most samples used in the collection were devitalized with microwave energy.

Typically, seeds were placed on moist filter paper in a covered petri dish. Microwave energy (700 watts) was then used to coagulate proteins in the seed embryo. Exposure times varied from 3–5 minutes, depending on the type of seed. The object of the operation was to devitalize the embryo without altering surface features of the seed. Quality control measures involved viability tests of a random sample of seeds from each microwaved lot. These included germination tests and staining with tetrazolium red to indicate enzyme activity.

For ease of handling and long-term storage, a few devitalized seeds and a name label are sealed into clear mylar coin pouches (5 cm x 5 cm). The samples are arranged alphabetically by genus, within family, in 35 mm slide holder sheets, and stored in a three-ring binder.

## DETECTION OF INCIPIENT INFESTATIONS WITHIN THE UNITED STATES

Early detection is a critical element of any national strategy for dealing with alien weeds. A few methods that may be employed for detecting incipient infesta-

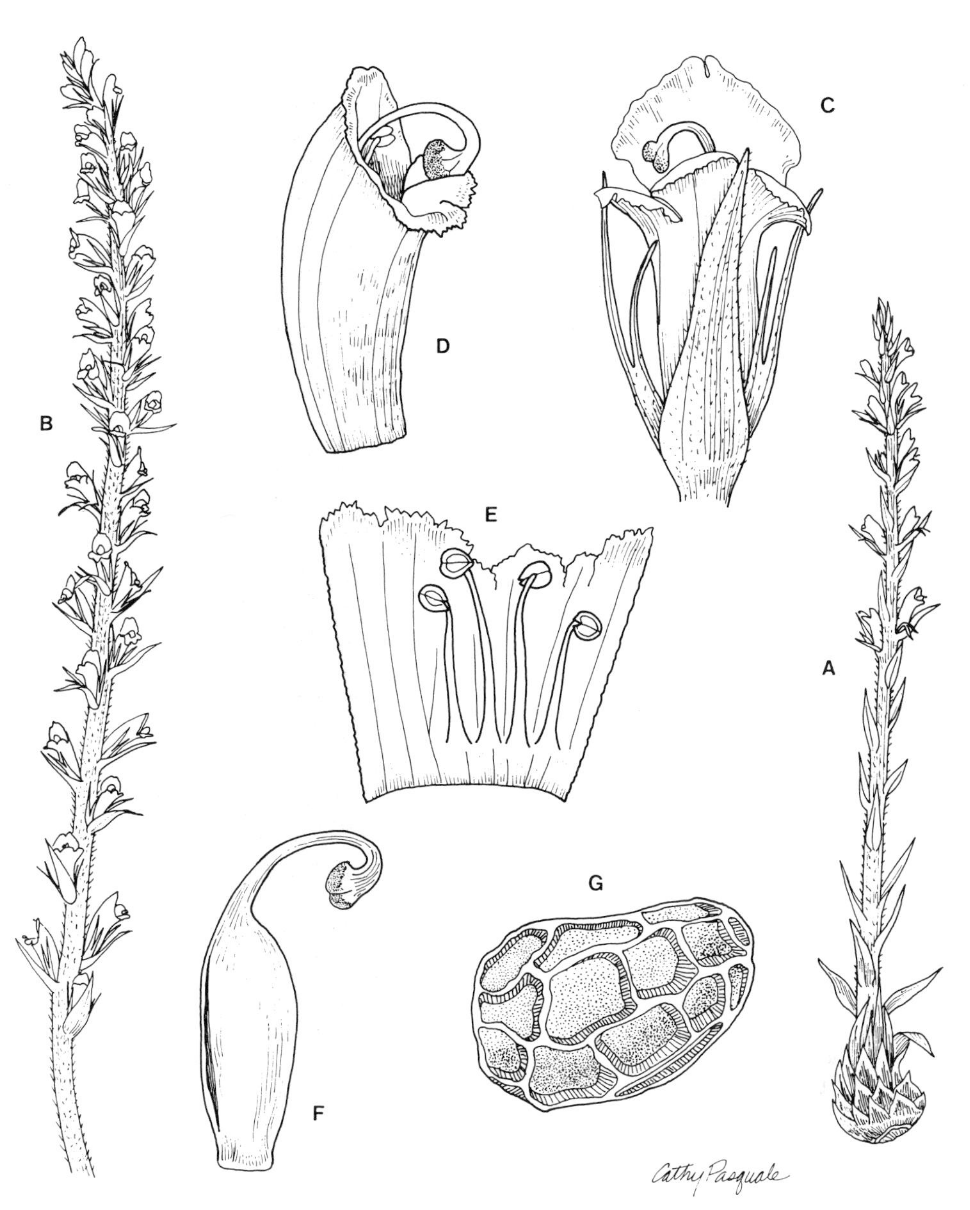


FIGURE 2.—*Orobanche minor* Smith. A. Small plant, showing swollen base and alternate scales (average height = 20 cm); B. Inflorescence of larger plant (26 cm); C. flower, showing bract, 2-parted calyx and corolla (8.3 mm); D,E. corolla, whole (D. 7.8 mm) and dissected (E. 7.8 mm) to show stamens; F. pistil (7.4 mm); G. seed, showing details of surface (0.7 mm).

tions include use of herbarium records to locate sites of collection, detection surveys of high-risk areas, and education of people who work in areas that are susceptible to invasion (e.g., extension agents, botanists, weed scientists, wildlife officers, etc.).

Currently APHIS is conducting a nationwide project to document all FNW specimens on file at 133 major herbaria in the United States. The main purpose of the survey is to identify sites where FNWs have been collected in the United States and to expand our knowledge of their worldwide distribution. Another objective is to encourage association between local PPQ personnel and professional botanists. A few notable specimens that have been detected so far in the survey are shown in Table 3.

TABLE 3. A sampling of Federal Noxious Weed voucher specimens noted in selected U.S. herbaria, 1991–1992.

| FNW SPECIES | HERBARIUM | SITE OF COLLECTION | YEAR COLLECTED |
|---|---|---|---|
| *Cuscuta japonica* Choisy | Clemson University | Pickens Co., S.C. | 1971 |
| *Galega officinalis* L. | University of Kansas | Saunders Co., Neb. | 1972 |
| *Orobanche minor* Smith | Clemson University | Aiken Co., S.C. | 1969 |
| | University of Georgia | Baker Co., Ga. | 1985 |
| | University of Virginia | Washington Co., Va. | 1989 |

Recent or otherwise notable specimens are investigated further if sufficient information is provided on the location of the collection site. Micro-infestations (1 ha or less) of Federal Noxious Weeds that were detected through the national herbarium survey and reconfirmed by on-site visits in recent months include:

**Clover broomrape** (*Orobanche minor* Smith) in five small infestations (1 ha or less) at the following locations:

1. At a private residence, North Augusta, South Carolina
2. On the campus of Erskine College, Due West, South Carolina
3. At a private residence in central, South Carolina
4. Intermittently along Highway 37, Baker Co., Georgia
5. In the Clinch Mountain Wildlife Management Area, along State Road 747, Washington Co., Virginia

**Japanese dodder** (*Cuscuta japonica* Choisy), a 1 ha infestation in the Clemson University Botanical Garden, Clemson, South Carolina.

## FEDERAL/STATE COOPERATIVE NOXIOUS WEED ERADICATION PROJECTS

After an infestation is confirmed, an investigation of the site is made by APHIS scientists and cooperating state personnel to assess its size, extent, agronomic significance, and potential for spread. A report on these findings with recommendations for action (eradication/no action, etc.) is then provided to the APHIS administration and cooperating state agencies for consideration.

If prohibited weeds are detected and determined to be of limited distribution (a single site or identified multiple sites), eradication is considered as one option. Biologically sound methods and procedures for eradicating an infestation are often complex and must be customized to fit the target site(s) and species. However, general strategies employed to achieve eradication include:

1. Detection surveys and delimiting surveys
2. Containment of the target species within infested areas
3. Control activities to prevent further reproduction
4. Elimination of seeds in the soil by:

(a) Stimulating the seeds to germinate (seedlings are usually easier to kill than seeds or mature plants; parasitic seedlings die without a host)
(b) Devitalizing the seeds *in situ* in the soil (fumigation, etc.)

5. Appraisal surveys

Generally speaking, traditional control and biological control are considered after eradication is ruled to be an unacceptable or impractical option.

## Witchweed Eradication in the Carolinas

Witchweed [*Striga asiatica* (L.) O. Kuntze] is a root parasite indigenous to Africa that attacks a number of grass crop plants, including corn (*Zea mays* L.), sorghum (*Sorghum bicolor* L.), rice (*Oryza sativa* L.), and sugarcane (*Saccharum officinalis* L.) (Saunders 1933). This species was first discovered in the Western Hemisphere, in Columbus Co., North Carolina, July 1956, parasitizing corn (see Figs. 3, 4). At that time, American scientists knew little about the plant or how to control it. However, with the aid of A. R. Saunders from South Africa, state and federal researchers soon learned about its devastating potential as a plant pest. Its potential spread into the Midwestern United States was deemed a significant threat to corn and sorghum production. Accordingly, a cooperative federal, state, and farmer quarantine program was established and first funded in 1958 to eradicate the weed and to prevent its spread from the infested area (Eplee 1981). Early surveys determined that the plant existed in four counties in the eastern Carolinas. Since then witchweed has been documented on about 162,000 ha (625 $mi^2$) in 39 counties in eastern North and South Carolina.

In 1959, a USDA research station was setup near Whiteville, North Carolina, then near the center of the infestation, to develop methods and procedures for eliminating witchweed from the United States. The Whiteville Plant Methods Center still conducts applied research on witchweed and provides technical support to the eradication program. However, the mission of the Center has now been expanded to developing methods and procedures for foreign weed exclusion and provides technical support on other APHIS domestic weed eradication projects as well.

As with other pests, the strategies for eradication of witchweed involve *detection and delimiting of incipient infestations* (through walking, equine, and motorized surveys), *prevention of spread* (regulation of soil-contaminated equipment and commodities that might spread the microscopic seeds out of the infested area; for example, soil contact commodities such as peanuts and cucumbers), *denial of reproduction* (by chemical treatment of emerged witchweed plants), and *elimination of seeds in the soil* (through soil injection of ethylene gas to stimulate suicidal germination of the seeds,[2] the use of false host crops[3] and by soil fumigation with methyl bromide or other soil fumigants).

Prevention of spread of witchweed from the quarantine area in the Carolinas has been accomplished by regulation of soil-contaminated equipment and soil contact commodities such as cucumbers, peanuts, and watermelons. Authority to regulate such items is provided under federal and state quarantines (USDA 1976). However, the cooperation of landmanagers and others who work on in-

2. The seeds of witchweed require an after-ripening period of 18 months, a pre-conditioning period (with adequate moisture and soil temperatures of at least 18°C) of three weeks, *plus* the presence of chemical exudates from a host root for germination to occur. The absence of any one of these factors precludes germination.

3. Two well-documented false hosts (plants that stimulate witchweed seeds to germinate but do not permit the haustorium of the parasite to attach to their roots) include cotton (*Gossypium hirsutum* L.) and soybean (*Glycine max* L.).

FIGURE 3.—Witchweed [*Striga asiatica* (L.) O. Kuntze] parasitizing corn. FIGURE 4.—(Inset) Closeup of witchweed showing red flowers [*Striga asiatica* (L.) O. Kuntze].

fested sites, such as utility and highway maintenance personnel, is essential for the regulatory program to succeed.

The need for public support and cooperation was demonstrated in 1990 when it was learned that a farmer had been transporting farm equipment back and forth between an infested property in Bladen Co., North Carolina, and a farm in DeKalb Co., Indiana, during the mid-1980s. Fortunately the parasite has never been detected on the property in Indiana. The track record of the program in preventing the spread of witchweed demonstrates the commitment of affected farmers, as well as state and federal personnel who work with them.

Chemical control strategies include the use of pre- and post-emergent materials for control of the parasite and weed host plants such as crabgrass (*Digitaria sanguinalis* (L.) Scopoli) and Johnson grass (*Sorghum halepense* (L.) Persoon). The growth regulator herbicide, 2,4–D [(2,4–dichlorophenoxy) acetic acid], was developed as a post-emergence treatment early in the program and continues to be important in the elimination of the parasite from the Carolinas. Chemical treatments are applied by private contractors and by program personnel throughout the growing season.

Currently the witchweed infestation has been reduced to about 23,000 ha (88 $mi^2$) in 13 counties in North Carolina and in 4 counties in South Carolina. Hectares that remain infested are expected to be released from quarantine by 1995 if resources continue to be available at their present level.

## Other Federal/State Cooperative Noxious Weed Eradication Projects

Over the past several years, APHIS has been involved in cooperative federal/state efforts to eradicate seven other FNWs from the United States. These include:

**Asian common wild rice** (*Oryza rufipogon* Griff.), a perennial red rice with rhizomes; in a 0.1 ha infestation at Royal Palm Hammock in the Everglades National Park, south Florida; control efforts by the state appear to have eliminated the population; no emerged plants have been detected at the site since the fall of 1991; monthly appraisal surveys will continue for an additional two years.

**Branched broomrape** (*Orobanche ramosa* L.), a root parasite of several broadleaf crops including tomatoes and carrots; first documented along Texas Highway 123 in Karnes Co., Texas, in 1981; maximum infestation: 263 ha (1 $mi^2$); now significantly reduced in density.

**Common crupina** (*Crupina vulgaris* Cassini), an invasive rangeland weed; first noted in south central Idaho in 1968; now occurring on 25,000 + ha (97 $mi^2$) in California, Idaho, Oregon, and Washington.

**Goatsrue** (*Galega officinalis* L.), imported in 1891 by a professor at Utah State University for forage crop research; abandoned when discovered to be toxic; now occurring on 15,789 ha in Cache Co., Utah; infestation significantly reduced by 1991.

**Hydrilla** [*Hydrilla verticillata* L. f. (Royle)], introduced into the United States in the late 1940s as an aquarium plant; originally in about 192 km (119 miles) of canals of the Imperial Irrigation District, Imperial Co., California; now reduced to 100 km (62 miles), but expanding its range elsewhere on the continent.

**Mediterranean saltwort** (*Salsola vermiculata* L.), an alternate host for curly top virus, was imported from Syria in 1969 by a Syrian graduate student at the University of California; research plots were abandoned upon completion of the study; now occurring sparsely in an elliptical band stretching through 1,500 ha (5.8 $mi^2$) of rugged terrain in the Temblor Mountains, west of Bakersfield, in San Luis Obispo Co., California.

**Catclaw mimosa** (*Mimosa pigra* L. var. *pigra*) in five small infestations in south Florida.

Federal/state cooperative eradication projects initiated during 1991–1992 include:

**Japanese dodder** (*Cuscuta japonica* Choisy), in a 1 ha infestation in the Clemson University Botanical Garden, Clemson, South Carolina; primarily parasitizing kudzu (*Pueraria lobata* (Willd.) Ohwi) (Fig. 5).

**Clover broomrape** (*Orobanche minor* Smith), in five small infestations in Georgia, South Carolina, and Virginia.

**Wild sugar cane** (*Saccharum spontaneum* L.), occurring on the Herbert Hoover Dike, along the southeastern shore of Lake Okeechobee; escaped from research plots at the USDA Sugarcane Field Station at Canal Point, Florida.

## SUMMARY

Alien agricultural weeds and botanical invaders of natural areas are biological pollutants that threaten agricultural production and the biodiversity of natural ecosystems in the United States. Efforts are now being made by USDA APHIS to exclude foreign weeds of agricultural significance from the United States. APHIS is also cooperating with affected states to eradicate incipient infestations of 11 FNWs from the United States. A national herbarium survey has also been undertaken to identify sites where FNWs have been collected in the United States in recent years. Sites of notable specimens are being investigated and targeted for eradication whenever practical.

FIGURE 5.—Kudzu covering a field and trees near Athens, Georgia. (photo by James Strawser)

## REFERENCES

ELTON, C. 1958. The ecology of invasions by animals and plants. Chapman and Hall, London.

EPLEE, R. 1981. Striga's status as a plant parasite in the United States. Pl. Dis. 65:951–54.

EPLEE, R. and R. WESTBROOKS. 1990. Federal Noxious Weed initiatives for the future. Proc. Weed Sci. Soc. NC. p. 76–78.

———. 1991. Recent advances in exclusion and eradication of Federal Noxious Weeds. WSSA Abstracts 31:31.

FOWLER, L. 1990. APHIS foreign weed interceptions. WSSA Abstracts 30:34.

———. 1991. APHIS foreign weed interceptions. WSSA Abstracts 31:30.

GIBBONS, A. 1992. Asian gypsy moth jumps ship to United States. Science 255:526.

GOOD, R. 1974. The geography of the flowering plants. Longman Group, Ltd., London.

HOLM, L., D. PLUCKNETT, J. PANCHO ET AL. 1977. The world's worst weeds. Univ. Hawaii Press, Honolulu.

LIVDAHL, T. and M. WILLEY. 1991. Prospects for an invasion: competition between *Aedes alvopictus* and native *Aedes triseriatus*. Science 253:189–91.

MCWHORTER, C. 1971. Introduction and spread of Johnson grass in the United States. Weed Sci. 19:496–500.

MOONEY, H. and J. DRAKE (eds.). 1986. Ecology of biological invasions of North America and Hawaii. Springer-Verlag, New York, N.Y.

RIDLEY, H. 1930. The dispersal of plants throughout the world. L. Reeve & Co., Ltd., Kent, Great Britain.

SAUNDERS, A. 1933. Studies in phanaerogamic parasitism with particular reference to *Striga lutea* Lour. Union South African Dept. Agric. Sci. Bull. 128.

SCHMITZ, D. 1990. The invasion of exotic aquatic and wetland plants in Florida: history and efforts to prevent new introductions. Aquatics 12:6–13, 24.

USDA APHIS PPQ. 1976. Witchweed Quarantine No. 80. U.S. Department of Agriculture, Animal and Plant Health Inspection Service, Plant Protection and Quarantine. Hyattsville, Md.

WESTBROOKS, R. 1991. Plant protection issues. I. A commentary on new weeds in the United States. Weed Technology 5:232–37.

———. 1992. Regulatory exclusion of Federal Noxious Weeds from the United States of America by USDA APHIS. Pages 110–13 *in* Proceedings of the 1st International Weed Control Congress, Feb. 17–21, 1992, Monash University, Melbourne, Victoria, Australia. International Weed Science Society.

WESTBROOKS, R. and R. EPLEE. 1987. Effective exclusion and detection of Federal Noxious Weeds. WSSA Abstracts 27:34.

———. 1991. USDA APHIS Noxious Weed Inspection System. 1991 update. WSSA Abstracts 31:29.

WESTMAN, W. 1990. Park management of exotic plant species: problems and issues. Conserv. Biol. 4:251–60.

ZAMORA, D., D. THILL and R. EPLEE. 1989. An eradication plan for plant invasions. Weed Tech. 3:2–12.

# Legal Avenues for Controlling Exotics

Faith Thompson Campbell[1]

As the papers presented here have shown, there is a long history of concern among scientists about the impacts of invasive exotic species. As a consequence, pieces of a control system are already in place, but many gaps remain, and too often laws are not effective even where they apply. Thanks largely to the zebra mussel, the issue is now reaching some political decision-makers. It is up to us to seize this opportunity to develop a comprehensive system and push for its adoption.

I believe that a program to contain the inroads of alien species must be based on certain principles:

— The approach of conserving the natural diversity and functioning ecosystems through building a network of protected preserves will not be adequate to deal with the threat posed by exotics. A focus on the total landscape is needed.
— "Ecosystem functions" are viewed too narrowly by many people; concern about ecosystem functions, however, does help focus attention on the important roles played by non-charismatic organisms, like invertebrates.
— An ounce of prevention is worth a pound of cure (but I do not advocate giving up if an exotic species is already established).
— The "polluter pays" (meaning establishment of a "clean" list and placing the burden of proof of safety on the would-be importer of any additional exotic organisms).
— Benefit/risk assessments based solely on economic factors will not adequately address the long-term threats posed by exotic species. We must give more weight to long-term harm to often diffuse constituencies, such as the American public, and less weight to short-term benefits which are often for a specific constituency.
— The program must be comprehensive, including mammals such as feral pigs deliberately introduced as targets for hunters and plants deliberately introduced to prevent soil erosion, as well as insects, arachnids, fungi, and viruses that arrive hidden in shipments of fruit, timber, nursery stock, etc.
— We need a national program because the threat and solutions are similar across the nation, and a piecemeal approach will not be successful.

Although the program must be comprehensive, the responsibility for various aspects of a federal program will probably remain divided among several agencies and congressional committees. As a result, it may require several separate bills to codify a comprehensive program.

Experts agree that the preferred approach contains the following elements:

**Prevention:** keeping proven or potentially invasive organisms out of foreign commerce so that they are not spread to new regions of the world

**Exclusion or Quarantine:** stopping proven or potentially invasive organisms at the borders

1. Natural Resources Defense Council, 1350 New York Ave., N.W., Washington, DC 20005

**Control of** those species already here or which will get in regardless of the efforts described above

Containment programs should use a mix of control efforts (including biological controls) that are as species specific and environmentally benign as possible—but that get the job done.

## WEAKNESS OF THE CURRENT SITUATION

Fifteen years ago, a National Academy of Sciences report found that "In the area of pest control research, priorities have often been set under political pressures for immediate answers, with too much regard for short-term problems and too little consideration for broader management objectives. Part of this problem arises from portions of the [law] that define policy, and create a division between the administration and research arms of the [agency]. The Act also tends to promote the 'action' attitude that may inhibit solutions other than short-term, direct chemical control" (National Academy of Science 1975).

The need is for a comprehensive, coordinated, permanent program. However, achieving this goal is hindered by the division of responsibilities among various agencies and congressional committees. For this reason, action to exclude or control invasive exotics will involve several separate bills. Nevertheless, we must remember our goal of a comprehensive program—and work toward it with a missionary fervor.

## NEEDED COMPONENTS OF COMPREHENSIVE PROGRAM

Biological pollution is an international problem. The sources of invasive exotics are usually other countries or international waters. The harm done is international in scope. Moreover, we must recognize the possibility of encountering complaints under the General Agreement on Tariffs and Trade (GATT) that stricter quarantine measures are an unacceptable imposition of one country's environmental standards on others.

For the "prevention" principle mentioned above to be effective, we need to improve implementation of the existing phytosanity convention[2] and adopt new conventions intended to restrict trade in organisms not now included—principally those not perceived as threats to row-crop agriculture and ornamental horticulture. These latter constitute the vast majority of vertebrates, invertebrates, and plants.

Such new international treaties should have the following contents:

- — an explicit statement of the purpose, which is to identify proven or potentially invasive organisms and prevent their dissemination;
- — regulations governing intentional and accidental export or import of organisms so identified;
- — funding (collected in part from fees paid by importers) to pay for needed research; and
- — an enforcement mechanism to penalize both exporting and importing countries which fail to implement the treaty.

---

2. See appendix for a brief list of existing statutes and treaties relevant to curbing biological pollution.

## STRENGTHEN DOMESTIC/FEDERAL LEGISLATION

Negotiating and adopting an international treaty—especially one on so esoteric a topic—will take considerable time. Even when the treaty is adopted by a majority of nations, the U.S. will still need improved implementing legislation.

Such legislation should include measures intended to stop or reduce additional introductions as well as provisions to establish, guide, and maintain a control program for invasive exotics which escape the prevention/exclusion net or which are already established here.

Legislation already exists in the realms of "noxious weeds," "plant pests," and "injurious wildlife," but these laws are interpreted too narrowly, implemented inadequately, and place the burden of proving "injury" on the wrong party. The government must show that importation is detrimental, rather than the would-be importer showing that it is safe.

Again the political reality is that we will probably have to try to strengthen existing legislation rather than start from scratch. I suggest that these statutes should be amended to:

— apply restrictions specifically to exotics which threaten natural biodiversity, not just agriculture;[3]
— create a mechanism to force rapid identification of such proven or potentially invasive organisms; and
— mandate a clean list which puts the burden on the would-be importer to prove an introduced organism will not become invasive.

In addition, the Forest Service, the National Park Service, and the Fish and Wildlife Service should be given a statutory role in instructing the Animal and Plant Health Inspection Service (APHIS) regarding exotic organisms threatening native tree species and natural ecosystems. Furthermore, the Lacey Act's application to vertebrates should be broadened to include domestic animals which pose a threat when they become feral.

Both existing and new laws must also prohibit distribution of invasive organisms once they are in the country. Restrictions of domestic distribution must apply to both invasive species from other countries and those native to one U.S. bioregion but that pose a threat in other bioregions.

I would prefer to mandate complete exclusion from the entire United States if a species posed a potential threat to any area of the nation, but if that is politically impossible, federal law should at a minimum restrict sale or distribution of species designated as invasive by federal or state agencies in ways to backup state laws.

In seeking to strengthen existing statutes and enact new ones, we should leave as little discretion as possible to the implementing agency on the requirement that it act to designate and control invasive exotic species. The criteria for designation must also be clearly defined so as to leave the agencies little

3. The injurious wildlife provisions of the Lacey Act do not mention harm to "wildlife and wildlife resources," including vegetation upon which wildlife resources depend. The 1990 amendments to the Federal Noxious Weed Act require each federal agency to designate an office or person responsible for "management of undesirable plant species" and to develop and fund a program for control of such plants on federal lands under its jurisdiction. But to date, these provisions have not resulted in any agency accepting as its mandate an aggressive program to halt further introductions or to contain exotic species already present in a natural ecosystem. Units of the National Park system do carry out aggressive programs to contain exotics.

room to shirk responsibility. The laws should be written so as to tie both identification of problem species and the development of control techniques to scientific expertise. I advocate such a short leash because history shows that the agencies have not acted in the current absence of strict requirement that they do so.

The laws should also require public comment on designations—in part to educate the public and gain its cooperation, in part because nongovernmental experts may be better informed than their governmental counterparts and are often more free to take a strong stance. For the same reasons, the laws should provide for public petition to designate an invasive alien species. The law should also provide for citizen suits to force the agencies to act if they still are reluctant.

Moreover, the government must provide both technical improvements and greatly expanded staffs and budgets so that the responsible agencies can implement the program effectively.

To control those species already here, we should enact a law establishing a national policy to eradicate (or, where that is impossible, control) invasive exotic species designated by the process described above. Because of the daunting size of the task, the law should also mandate development of a priority list based on both species-related and ecosystem-related factors. I reiterate the importance of prohibiting distribution of invasive organisms within the country.

Finally, the law should create and fund a "SWAT team" to attack newly discovered introductions before they become irreversibly established and while complete eradication is possible with relatively little effort.

Again it is essential to provide adequate funds and staff to the responsible agencies so that they can carry out on-going control programs and the research on which to base them. We must escape the past of "crisis" management.

The control program should emphasize species-specific measures. I support introduction of additional exotic organisms (natural enemies) to control invasives, but their introduction must be preceded by careful screening Drea (1993). points out the dangers inherent in underfunded screening programs. Applications of broader-effect methods, including pesticides, must be carefully controlled, site-specific and in full compliance with applicable laws, including the Federal Insecticide, Fungicide, and Rodenticide Act, National Environmental Policy Act, and the Endangered Species Act.

An effective program must include a broad effort to educate the public about the threat posed by invasive exotic species generally and species specific to their own areas.

If we are to build a comprehensive program, we have many difficulties to overcome. The one mentioned most often is the increasing rate of global trade and travel—which inevitably give organisms more opportunities to reach non-native shores.

Second, we must bring about closer coordination with our neighbors. I think it is a reasonable working assumption that any invasive species that enters Canada or Mexico will get to the United States.

Most important, however, we must attempt to build bridges to those who oppose various exclusion and control strategies—and, if we cannot persuade them, be prepared to confront them. These groups include:

1. hunters and fishermen
2. gardeners, nurserymen, and landscape architects
3. farmers and ranchers (who depend on exotic grasses to feed their livestock, especially in the West)
4. aquaculture
5. owners and sellers of exotic pets

6. environmentalists concerned about use of pesticides and introduction of additional exotics or genetically engineered organisms
7. humane organizations opposed to the killing of or causing suffering to sentient animals

Finally, we must attempt to create and strengthen intrusive, wide impact and expensive government regulatory programs at a time when "no new taxes" is the universal political slogan. Not only does everyone want to shrink the government's role, few politicians are willing to embrace any proposal which does not already enjoy a 97 percent approval rating in the polls.

Despite these very real handicaps, I believe the near future holds some promising opportunities as well. One is the growing awareness of the importance of conserving biological diversity in North America, not just the tropics. Another is the expected release of the report authored by the Office of Technology Assessment (OTA) in spring 1993.

How do we go about persuading the political leadership to adopt a program to curb destruction of our biota by invading exotic species? The intellectual foundation for such a program is a commitment to conserving the biological diversity native to any particular region. A logical outgrowth is to avoid any action likely to reduce that biodiversity. Unfortunately, it is highly unlikely that America will have a national policy on biodiversity soon. But the need to act is sufficiently strong that we must try to tackle the invasive alien problem in the absence of such a policy.

As this conference has demonstrated, biological pollution is a national (actually, global) problem, but it has suffered from a lack of attention of policy-makers to date. I believe the reasons are as follows:

1. ignorance about biology generally
2. failure to recognize the value of conserving indigenous species (native biological diversity)
3. the public's failure to recognize or distinguish between exotic species and indigenous species
4. constituencies which support introduction or maintenance of some invasive exotic species (e.g., European boar, deer, and brown trout; new horticultural introductions or "acceptable substitutes" (alien annual grasses on western lands)
5. all these factors reinforced by some biologists and ecologists who are content as long as "ecosystem functions" are still apparently healthy; or even like the "enhanced biodiversity" that they see resulting from introduction—we do not even have unanimity among our most sympathetic audience
6. exotic species are often portrayed as a "local" problem because many specific exotics are causing problems only in certain parts of the country; I contend that that view is incorrect for the following reasons:
   a. the causes and solutions are the same everywhere, even if specific organism (and susceptibility of ecosystem) does vary; we need a coordinated program to address the whole problem
   b. many invasive exotics have long since expanded beyond a "local" area (e.g., kudzu, purple loosestrife, gypsy moth, white pine blister rust, mongoose)

It is up to us to begin overcoming the obstacles that prevent us from successfully dealing with invasive exotic species. We should take advantage of examples which have alarmed decision-makers—zebra mussel, melaleuca, Africanized bees, and now mosquitoes carrying encephalitis and other diseases. But we need to find ways to generalize the decision-makers' understanding, or we will remain stuck with the current situation, described by the National Academy of Science, in which there is an arbitrary focus on a "crisis" species, demands for short-term solution which usually cannot be met, followed by loss of interest and funding.

To educate decision-makers, we need a coordinated campaign. One possible approach is for a scientific body or committee to prepare a series of brief, easily understood fact sheets on certain high-impact invasives with specific descriptions linking these invasions to some human concern (including, but not limited to, economic self-interest). While the statements made in the fact sheet must be documentable and backed by unimpeachable sources, this detail would not be included in the sheets themselves, only brought forward if questions were raised. Each fact sheet would compare the expense of control of widespread organisms to the expense of an effective prevention/exclusion program backed by a SWAT team. One by one these fact sheets would be sent to government, media, and industry opinion-leaders in the expectation that gradually they would internalize our concern about biological pollution. Then they would enact the program and educate the broader public.

## SUMMARY

The ideas expressed here did not originate with me; they are summarized from many people with much greater knowledge and experience. My intention is to stimulate and focus the discussion begun here and in other fora.

I appreciate the opportunity to share in this conference—one of the most exhilarating and informative in which I have ever participated. I look forward to working with its organizers, my fellow participants and other concerned individuals and institutions in devising solutions to these problems and pointing the way for future action—action which is absolutely vital to solving biological pollution.

## REFERENCES

DREA, J. J. 1993. Classical biological control: An endangered discipline? Pages 215–22 *in* Biological pollution: the control and impact of invasive exotic species, B. N. McKnight (ed.). Indiana Acad. Sci., Indianapolis.

NATIONAL ACADEMY OF SCIENCES. 1975. Forest pest control. Washington, D.C.

## APPENDIX

## SUMMARY OF STATUTES (LAWS AND TREATIES) GOVERNING INTRODUCTIONS OF ALIEN SPECIES

### ANIMALS

Lacey Act [18 USC §42]

authorizes the Secretary of Interior to prohibit the importation of mongooses, fruit bats, zebra mussel, and other birds, mammals, reptiles, amphibians, fish, molluscs, and other crustacea which he declares to be "injurious" to agriculture, horticulture, forestry, and wildlife resources (including aquatic and terrestrial vegetation upon which wildlife depends).

also states that one of its purposes is "to regulate the introduction of American or foreign birds or animals in localities where they have not heretofore existed." [16 USC §701]

## PLANTS

International Plant Protection Convention (IPPC) [Article 14 of the Constitution of the Food and Agriculture Organization of the United Nations]
establishes international system under which inspections and quarantines are implemented to prevent dissemination of pests affecting plants resources.

Federal Plant Pest Act [7 USC §§150aa–jj]
prohibits knowing importation or interstate transportation (except with a permit issued by the Secretary of Agriculture) of any plant "pest"; "pest" is defined as any living stage of invertebrates, bacteria, fungi, parasitic plants, viruses, infectious substances, etc., "which can directly or indirectly injure or cause disease or damage in *any* plants or parts thereof." [emphasis added] *The law does not restrict its coverage to diseases or pests harming U.S. agriculture.*

Organic Act [17 USC §§147a–e]
authorizes the Secretary of Agriculture, alone or in cooperation with the states or local jurisdictions, farmers' associations, governments of Western Hemisphere countries and international organizations, to detect, eradicate, control, or retard the spread of plant "pests." (See definition of "pest" under the Federal Plant Pest Act, above.)

Plant Quarantine Act (1912) [7 USC §§151–64a, 167]
authorizes the Secretary of Agriculture to regulate imports or interstate shipments of nursery stock or other plants and plant parts and propagules when necessary to prevent introduction of injurious plant diseases and insect pests; [emphasis added] *the law does not restrict its coverage to diseases or pests harming U.S. agriculture.*

Federal Noxious Weed Act [7 USC §§2801–13]
authorizes the Secretary of Agriculture to regulate the entry of interstate movement of "noxious weeds," defined as a plant of foreign origin that is either new to or not widely dispersed in the United States and that can injure agriculture (crops, livestock, or irrigation systems), other useful plants, navigation, fish, and wildlife, or public health; seed shipments are exempt

1990 amendment [7 USC §2814(e)(7)]
requires each federal land-managing agency to establish and fund a program to manage "undesirable" plants found on lands under its jurisdiction; "undesirable" is defined as plants "classified as undesirable, noxious, harmful, exotic, injurious, or poisonous, pursuant to State or Federal law."

Agricultural Quarantine Enforcement Act (1989)
prohibits the shipping of any plant, fruit, vegetable or other matter quarantined by the Department of Agriculture via [sic] first-class mail; search warrants required to open packages.

Forest & Rangeland Renewable Resources Research Act [16 USC §1642]
authorizes the Secretary of Agriculture to conduct research and experiments to obtain, analyze, develop, demonstrate, and disseminate scientific infor-

mation about protecting and managing forests for a multitude of purposes; §(a)(3) specifies protecting vegetation, forest and rangeland resources from insects, diseases, noxious plants, animals, air pollutants, and other agents.

§1642(b) requires the Secretary to maintain a current comprehensive survey of the "present and prospective conditions of and requirements for renewable resources of the forests and rangelands . . . and means needed to balance the demand for and supply of these renewable resources, benefits, and uses in meeting the needs of the people of the United States. . . . "

Cooperative Forestry Assistance Act [16 USC §§2101, 2102, 2104]

§2101(a) recognizes that "efforts to prevent and control . . . insects and diseases often require coordinated action by both Federal and non-Federal land managers; . . . "

§2102(b) authorizes the Secretary of Agriculture to provide assistance to state foresters to develop and distribute genetically improved tree seeds and to improve management techniques aimed at increasing production of a variety of forest products, including wildlife habitat and water.

§2104 authorizes the Secretary to protect from insects and diseases trees and wood products in use on national forests or, in cooperation with others, on other lands in the U.S.; such assistance may include surveys and determination and organizations of control methods. Programs on non-federal lands can be instituted only with the consent of and with a contribution of resources from the owner. The Secretary may also prescribe other conditions for such cooperative efforts.

## OTHER PERTINENT LAWS

Nonindigenous Aquatic Nuisance Prevention and Control Act of 1990 [16 USC §§4701, 4702, 4711, 4712, 4721–28, 4741, 4751; 18 USC §42]

defines "nonindigenous species" to include any plant, animal, or "other biological material" in an ecosystem outside its historic range; defines a species to be a "nuisance" if it threatens the diversity or abundance of native species, the ecological stability of infested waters, or commercial, agricultural, aquacultural, or recreational activities on such waters.

Executive Order 11987 (1977)

directs federal agencies to restrict the introduction of exotic species into natural ecosystems under their jurisdiction and to encourage states to do the same; directs the Secretaries of Interior and Agriculture to restrict the introduction into any natural system of animals or plants designated as injurious or noxious under the Lacey Act and Federal Noxious Weed Act.

National Park System Organic Act

no specific reference to exotic species, but establishes the principle purpose of national parks to be to "conserve the scenery and the natural and historic objects and the wildlife therein . . . unimpaired. . . . "

# ADDITIONAL PRESENTATIONS AT THE SYMPOSIUM

(LISTED IN THE ORDER OF PRESENTATION)

**African Bees in the Americas** or **How to Invade Two Continents and Become a Multimedia Villain.** Orley Taylor, Department of Entomology, University of Kansas, Lawrence, KS

**Invasive Exotics in Our National Parks: Policy and Practice.** Gary Johnston (read by Lloyd Loope) Wildlife & Vegetation Division, National Park Service, Washington, D.C.

**What Makes a Community Vulnerable to the Entry of Alien Plants?** Richard N. Mack, Botany Department, Washington State University, Pullman, WA

**Releases of Genetically Engineered Organisms: Ecological Concerns.** Rebecca J. Goldburg, Environmental Defense Fund, New York, NY

**The Importance and Relevance of Native Vegetation in the Landscape.** Gerould Wilhelm, The Morton Arboretum, Lisle, IL

**Introduced Species and the Conservation of Biodiversity.** George Rabb, Brookfield Zoo and Chicago Zoological Society, Brookfield, IL

**Rodeo® Herbicide: an Effective Control for Exotic Plants.** Martin D. Lemon, Monsanto Agricultural Company, St. Louis, MO

**Utilization of Established Exotic Species as Biomanipulation Tools for Lake Management.** Thomas L. Crisman, Department of Environmental Engineering, University of Florida, Gainesville, FL

**Coping with Existing and Emerging Problems. Agency Role in Exclusion/Management of Invading Pests** and **Identifying and Managing Pest Risks of Soviet Timber Imports.** Michael J. Shannon, USDA APHIS, Plant Protection & Quarantine, Beltsville, MD

# Index

*Sales Representative*

❧

Lynn Churchill

*Printing Coordinator*

❧

Judy Jarrett

*Production Coordinators*

❧

Kurt Boschen, Nonie Ratcliff
and Chris Tower

*Typeface*

❧

Times Roman

*Compositor*

❧

Weimer Graphics, Inc.

*Printer*

❧

Shepard Poorman Communications
Corporation